Sharing Cities

Urban and Industrial Environments

Series editor: Robert Gottlieb, Henry R. Luce Professor of Urban and Environmental Policy, Occidental College

For a complete list of books published in this series, please see the back of the book.

Sharing Cities

A Case for Truly Smart and Sustainable Cities

Duncan McLaren and Julian Agyeman

The MIT Press
Cambridge, Massachusetts
London, England

This book was set in ITC Stone Serif by Toppan Best-set Premedia Limited. Printed on recycled paper and bound in the United States of America.

Library of Congress Cataloging-in-Publication Data is available.

ISBN: 978-0-262-02972-8

10 9 8 7 6 5 4 3 2

Contents

Foreword

When Mayor of London Ken Livingstone, Mayor Michael Bloomberg from New York, and others set up the C40 Cities Network a decade ago, they had the vision that cities will be the locations where the world's greatest environmental challenges will be solved. As nations continue to stumble and falter and are seemingly unable to make sufficient progress on issues such as climate change, their vision is becoming shared by many more people.

That you can't fix the planet without fixing our cities is obvious, but less obvious is that cities can fix the planet.

A large majority of the population of the global North live in cities already, and city living will become the norm for most of humanity in coming decades. These are the places where most consumption takes place. The energy consumed in our cities to heat our homes and power our transport is driving climate change. The food we import to our cities, particularly meat and dairy produce, is leading to the destruction of wildlife-rich habitats across the globe. The consumer goods that we take for granted in the global North gobble up resources extracted thousands of miles away, far too often with dreadful environmental impact and working conditions that were outlawed in the US and UK over a hundred of years ago. The waste belching out of exhaust pipes, chimneys, and sewage systems is poisoning the air and water that we and every other species on the planet depend on.

Viewed like this our cities are driving us towards a dystopian hell of environmental collapse and gross social inequalities.

But as this book makes abundantly clear, there is the potential for the world's cities to drive a very different future; a future where cities take their environmental and social responsibilities seriously; a future where cities transform themselves and the rest of the world; a future where cities fix not just themselves but also fix the planet.

Central to this more hopeful vision is sharing.

Sharing is not new. The vast majority of us share our journeys to work or play, for example on the subways of America's great cities, or the London Underground, or the Bus Rapid Transit Systems springing up across Latin America. The green spaces in our cities are shared, and their loss or privatization is fiercely resisted. And it isn't so long ago that libraries were where most of us got the books we wanted to read.

But sharing can and must go much further.

The tantalizing prospect offered in this book by McLaren and Agyeman is that we are just starting to embark on a sharing revolution. A revolution which builds upon the digital world of the twenty-first century; that utilizes the ingenuity and imagination that springs from the cross-fertilization of ideas from the diversity of people living in cities; that builds empathy and understanding between people rather than fear and loathing; that leads to much greater levels of sharing of stuff and much greater resource efficiency; that takes naturally evolved cultural traditions of sharing within families and local communities, and reinvents them to enable sharing between citizens and strangers; and that fundamentally transforms the dominant world view that individualism and material possessions are central to what it is to be human.

The northern cities of the United Kingdom led the Industrial Revolution. The thousands of chimneys belching out smoke were seen as progress. That children born in these cities were condemned to live in slums, live short lives, and suffer from illnesses such as rickets due to lack of sunlight was seen by some as a price worth paying. In these cities the chimney stacks and slums have now gone. But as we all know, they have not disappeared. They now dominate many cities in China, India, and other fast-developing nations. If the Sharing Revolution is to be truly transformational, it must not only complete the transformation of the cities of the global North it must also transform cities across the globe.

And it can. In different ways, cities such as Seoul and Medellín are leading the revolution. And sharing is still part of daily life for many people in many cities across the global South. The Sharing Revolution isn't a revolution to be led by wealthy countries and copied by the rest; it is a shared revolution with cities across the world learning from each other. The C40 Network and the Sharing Cities Network run by Shareable.net are testament to this.

Mayors Bloomberg and Livingstone had a vision. The C40 Network that they gave birth to has already enabled the world's largest cities to learn from each other and learn from the most innovative smaller cities across the globe. As cities across the globe fight for and in many cases get greater fiscal and regulatory autonomy, such sharing is more critical. But in this book McLaren and Agyeman offer something new, something exciting, something earth-shattering—that if cities become Sharing Cities then we will not only fix the planet but will also transform the prospects for social justice. Now that's a message well worth sharing.

Mike Childs
Friends of the Earth, London

Acknowledgments

Having known each other since working on sustainability issues in London in the 1980s—Julian in local government, Duncan in the nonprofit world—this book was born in the inspiration and stimulation of conversations over several years about sustainability, urbanism, equity, and justice. Fortunately our emerging ideas found fertile ground in UK Friends of the Earth's Big Ideas Project.

We both—and Duncan in particular—are indebted to them, and in particular to Mike Childs, the leader of the Big Ideas Project. Without their support—moral and financial—neither our original Sharing Cities paper, which garnered great interest around the world, nor this book would have been possible. And in this endeavor we benefited immensely from the initial encouragement and support of the former acquisitions editor Clay Morgan and the Urban and Industrial Environments Series editor Bob Gottlieb—who responded so quickly and positively to our proposal—and latterly from Miranda Martin, Beth Clevenger, Katie Hope, Marcy Ross, and Margarita Encomienda at MIT Press.

From the outset, the opportunity to work with Harriet Bulkeley, Eurig Scandrett, Roman Krznaric, and Victoria Hurth on their papers for the Big Ideas Project helped shape Duncan's thinking as the ideas in this book developed. For Julian, it was discussions with his students at Tufts and activists and academics in the Boston area that sparked his ideas. We would also like to thank Neal Gorenflo and Shareable for their encouragement, and for the flow of informative and insightful commentary on sharing cities and the sharing economy published on Shareable. Research assistance from former Tufts students Adrianne Schaefer Borrego and Abby Farnham was invaluable, especially in compiling the city case studies—we thank you both. Feedback and suggestions from three unnamed reviewers was instrumental in helping us improve our early drafting. Also deserving a mention are Skype and Dropbox, two characters that enabled regular communication between Boston and Västerås, Sweden. Without these sharing tools, our task would have been far more onerous. We'd also like to thank all those friends and contacts who took time to provide feedback on our title suggestions on Survey Monkey. As important as these joint acknowledgements are our individual ones:

From Duncan: My partner Pernilla Rinsell, and our children Alex and Emelie, deserve deepest thanks for their support and tolerance—and maybe when they see their names here, the kids will appreciate what all those hours in front of a laptop were about, even though their screen-time remains tightly rationed. I'd also like to thank my PhD supervisor Gordon Walker for his understanding, even if he possibly didn't anticipate just how much this book would drag me away from my studies.

From Julian: I'd like to thank my partner Lissette Castillo and her daughter Nairobi for their support and love when I had to work on the book. I'd also like to thank my nephew and niece in the UK, Louis and Connie, whose vacation in Boston in summer 2014, while we were finishing up the manuscript, reminded me to have more fun.

Introduction

Our purpose in writing this book is as bold as it is clear. We believe that the world's cities, where the majority of people now live, could become more socially just, more environmentally sustainable and more innovative through the twenty-first-century reinvention and revival of one of our most basic traits: sharing. We will demonstrate how, with modern technologies, the intersection of urban space and cyberspace provides an unrivaled platform for more just, inclusive, and environmentally efficient economies and societies rooted in a sharing culture.

Yet this opportunity is currently being overlooked. Cities have always been about shared space, human interaction and encounter, and the exchange of goods and services through marketplaces and moneylending. A successful city needs good governance and collective civic structures to facilitate and regulate the interface between the shared public realm and private interests, and to enable effective and fair sharing of resources and opportunities. In their more recent incarnations however, *sharing* and *shareability* are typically too narrowly conceived and perceived as primarily about economic transactions: a so-called sharing economy. The opportunity is so much greater than Airbnb, Spotify, middle-class "swishing" and getting "bums on bikes" through urban bikesharing. We will show how a cultural and political understanding—and implementation—of sharing in all its rich variations can overcome the shortcomings of commercial approaches and transform how we think about sharing and cities.

This introduction highlights the challenges and opportunities of humanity's increasingly urban future, sets out our case for sharing cities as a response to those challenges, and introduces some critical concepts associated with what we call the "sharing paradigm" and the necessary sociocultural and political changes needed to bring it about.

Possible City Futures, Challenges, and Opportunities

Ever since the origins of cities, there has been much talk about city futures. In the past 40 years alone David Harvey has focused on social justice and cities;[1] Manuel Castells on urbanism,[2] networking and information;[3] Saskia Sassen on "the global city";[4] Leonie Sandercock on the city as cosmopolis;[5] Richard Florida on the creative class and cities;[6] Charles Landry on the creative city;[7] Jeb Brugmann on the productive city;[8] Susan Fainstein on the just city;[9] Edward Glaeser on the triumphant city;[10] and Harvey again on rebel cities, to name but a few.[11] What all these different visions of urban futures share is hope and an abiding belief that cities could be the best form of organization our species can achieve. Different conceptions abound, some associated with particular authors, whereas others—such as "sustainable," "social," or "participatory" cities—are more general in nature. Planners, architects, activists, and urban consultants promote and refine such ideas; arguably the most current, powerful, and influential urban zeitgeist is the "smart city." Smart cities invest in high-tech information and communication technologies (ICT) to "wire-up" the city and enhance its efficiency, boost the ICT sector as a motor of growth and property development, and attract skilled talent by delivering a high quality of life. One of our aims in this book is to show how truly smart cities must also be sharing cities.

The challenge and opportunity of the sharing city is one and the same, namely that around 53 percent of the world's population currently lives in cities.[12] This is set to rise to 64.1 percent in the global South and 85.9 percent in the global North by 2050,[13] intersecting an even faster rise in populations with access to cyberspace.[14] This rapid rate of urbanization highlights the interlinked economic, social, and environmental challenges of 1 billion people living in extreme poverty, amid rising income inequality and the lack of affordable housing, in a world slowly facing up to the realities of multiple resource scarcities, biodiversity loss, and climate change. According to leading scientists, we are living outside four of the nine planetary boundaries that constitute a safe operating space for humanity: the climate system, biodiversity loss, land-system change and phosphorus and nitrogen cycling.[15] To these planetary (environmental) ceilings within which we should live, Oxfam's Kate Raworth,[16] building on earlier work in Europe and Latin America,[17] adds a "social foundation." As she notes, this social foundation

> forms an inner boundary, below which are many dimensions of human deprivation. The environmental ceiling forms an outer boundary, beyond which are many dimensions of environmental degradation. Between the two boundaries lies an area—

shaped like a doughnut—which represents an environmentally safe and socially just space for humanity to thrive in. It is also the space in which inclusive and sustainable economic development takes place.[18]

In this formulation Raworth reflects one of us who promoted "environmental space," defined as follows: "a rights-based approach that conceptualizes sustainable development in terms of access for all to a fair share in the limited environmental resources on which healthy quality of life depends."[19]

Far from being the antithesis of sustainability that some would have us believe, well-planned and -governed cities are potentially the form of human organization that could keep us within environmental limits while simultaneously building the social foundations prescribed by human rights, dignity, and a decent quality of life. For this to happen, however, we must not only recognize the place of cities in global environmental, social, and economic systems, but also build on the inherent social, economic, and environmental efficiencies of urban living. This means understanding cities as the political, economic, and cultural drivers of global society, and thus linking the sharing of urban spaces to the sharing of global resources. It also means understanding cities in themselves as shared entities with shared public services (such as healthcare, childcare, education, and libraries); shared infrastructural resources (such as shared streets, mass transit, electricity, water and sewerage, and shared spaces (such as public spaces and green spaces). But we go still further in seeing not only a "right to the city" and to the "urban commons" (common resources, managed and sustained by our collective activities), but also a right to *remake* them,[20] as being fundamental to any form of urban social contract worthy of the title "sharing city."

Sharing Cities as Just Sustainabilities

The concept of sharing cities represents yet another powerful expression of "just sustainabilities"—the idea that there is no universal "green" pathway to sustainability, that sustainability is context-specific but justice is an intrinsic element in any coherent route:

[Just] sustainability cannot be simply a "green" or "environmental" concern, important though "environmental" aspects of sustainability are. A truly sustainable society is one where wider questions of social needs and welfare, and economic opportunity are integrally related to environmental limits imposed by supporting ecosystems.[21]

Sharing cities—as we envision them here—represent the nub of the social justice challenge to sustainability, a topic we discuss more fully in chapter 4. Here we simply want to help the reader understand our motivations with a couple of examples. First, a representative from a wildly successful major-city bikeshare program contacted one of us recently with the question: "How can we get more low income and people of color using our bike program"? On the surface, this may seem like a harmless, even altruistic question. It nevertheless belies a deeper problem common in "green" cities discourses and in many sharing economy programs. The problem is simply that most bikeshare programs (and many other sharing economy programs) were never designed with equity or social justice in mind, nor were low-income people involved in the visioning or design of such programs. A recent study found that only 9 of 21 programs reviewed had even factored equity considerations into their station siting.[22] Social justice is typically an afterthought; it is seen as a "retrofit" once the scheme is up and running "successfully" for the targeted "ordinary" users.[23]

Second, Enrique Peñalosa, former mayor of Bogotá, Colombia, hit the nail on the head when he said an advanced democratic city is not one where even the poor own cars, but one where even the rich ride buses. Peñalosa and his fellow former mayor Jaime Lerner of Curitiba, Brazil, were setting practical foundations for the idea of sharing cities by emphasizing the *equity* and *access* dimensions of their innovative bus rapid transit (BRT) schemes, which allowed access to facilities and services irrespective of car ownership and wealth.

The Case for Sharing Cities

When we talk of "sharing cities" we deliberately embrace the ambiguity of the verb and adjective of "sharing." In this book we set out a case for *understanding* cities as shared spaces, and *acting* to share them fairly. In rough outline that case runs as follows.

Humans are natural sharers. Traditional "sociocultural" sharing happens everywhere, but it has largely broken down in modern cities in the face of commercialization of the public realm, the increasingly rapid pace of economic and technological change, and the destabilization and fragmentation of human identities these trends have engendered.

Nonetheless the future of humanity is urban. Demographic, economic, and cultural forces are bringing us together in larger and larger urban regions, particularly in the global South. This is not a disaster for humanity as the physical nature of urban space facilitates—and in some ways necessitates—sharing: of resources, infrastructures, goods, services, experiences, and capabilities.[24] The effects of population density and highly networked physical space are converging with new digital technologies to drive and enable sharing in cities—particularly in novel "mediated" forms. All three come together to provide critical mass in both demand for, and supply of, shared resources and facilities. New opportunities for collaboration and sharing are arising at the intersection of urban space and cyberspace.

With new opportunities for sharing we have new opportunities to enhance trust and rebuild social capital. But they are also creating new spaces in which commercial interests can casualize labor, privatize public services, and capitalize on growing land values through gentrification. In such ways the emerging sharing economy can deepen inequalities and deliver injustice. City leaders therefore need to support and emphasize *communal* models of sharing that build solidarity and spread trust. In other words sharing systems must be designed around equity and justice.

Like any other practice, sharing with equity and justice at the core can naturally shift cultural values and norms—in this case toward trust and collaboration. This can deliver a further dividend, in that increased trust increases social investment in public goods and the public realm, or urban commons. Such an enhanced public realm can in turn directly facilitate more and more efficient sharing with significant environmental benefits.

It also establishes a precondition and motivation for collective political debate that recognizes the city as a shared system. The same measures that enable sharing online, also—if civil liberties are properly protected—enable collective politics online. Again we see the intersection of urban- and cyberspace enabling transformation—this time in the political domain.

In the anticipation of such transformations we suggest that "sharing the whole city" should become the guiding purpose of the future city. Adopting what we are calling the "sharing paradigm" in this way offers cities the opportunity to lead the transition to just sustainabilities.

This offers a radically different vision compared with a global race to the bottom to attract footloose investment capital. It redefines what "smart cities" of the future might really mean—harnessing smart technology to an agenda of sharing and solidarity, rather than one of competition, enclosure, and division.

A Shared Collective Culture

Fundamentally, therefore, our book highlights the importance of the shared public realm in the history and development, and more recently, in the reimagining of politics. We argue that the neoliberal, hegemonic model of development in the modern world prioritizes private interests at the cost of shared interests. Instead, we suggest that a cultural shift is overdue: one that gives much greater recognition and credit to the shared public realm of our cities (both physical and cyber); one that supports a revival of conventional sociocultural sharing—especially of the city as a whole as shared space—as well as a blossoming of novel mediated forms of sharing; and one that recognizes and affirms the ways in which the opportunities afforded to individuals in cities are founded on the collective efforts and actions of whole communities. We share the view that entrepreneurs do not build businesses alone, nor do parents raise children in isolation from the wider community. Regardless of the national culture, both are always forms of co-production.

One point that we must clarify at the outset is that by "culture" we mean something political and not simply something focused solely on human behavior. We do not intend to fall into the "post-political trap"[25] in any of its several forms. This trap underpins the idea that capitalism is unchallengeable as the organizing principle for society. It encourages a belief that we should address social problems through business-led, "smart," technological innovation, rather than through politics. And it implies that "nudging" behavior change among individuals is the way to change norms and culture, rather than by democratically guided regulation, planning, institution building, and structural interventions.

In our understanding of culture we acknowledge the indelible influence of the British and European cultural studies traditions associated with Raymond Williams, E. P. Thompson, and, in particular, Stuart Hall. Earlier scholars equated culture with the *symbolic*; that which is outside politics, society, or economy; yet Hall focused on power, on the ways dominant groups engineer cultural consent to legitimate their hegemony and the ways in which this functions as a persistent ideology. This new focus on power inevitably included politics, engaging with neoliberalism and postmodernism as much as with feminism, cultural identity, race, and ethnicity. Moreover, as French theorists such as Michel Foucault, Jean-François Lyotard, and Pierre Bourdieu insisted, cultural theory is itself political.

We do not discount the symbolic, intangible, and ideational aspects of culture that underpin beliefs, values, norms, and desires. Nor do we downplay the role of shared patterns of behavior, interaction, cognitive constructs, and understandings—developed through education and socialization—that help shape and define identity in (sub)cultural groups, or the way in which group and societal cultures can become forces of collective evolutionary selection. However, we fundamentally recognize culture as political, the site of contestation between different groups in society who compete to ascribe meaning to events, behaviors and information. This then is the terrain over which a "cultural shift" toward the sharing paradigm and sharing cities is emerging, and indeed *needs* to occur.

Defining Sharing: The Sharing Paradigm

Dictionaries agree that sharing encompasses processes whereby we divide something between multiple users; we allow others to consume a portion of, or take a turn using things that are ours; we obtain access to a portion of, or a turn in using, things that belong to others; or we use, occupy, or enjoy a facility, space or resource jointly with one or more others.

Russell Belk, a professor of marketing in Toronto, has helped shape the academic and public discourse around the sharing economy. (While we prefer to use the term "sharing paradigm" throughout this book, we also use the other terms such as "sharing economy," "solidarity economy," and "collaborative consumption" when we refer to these specific aspects or components of the sharing paradigm, or to the work of others using those or other specific terms.) Belk defines sharing from an economic perspective of owned goods, as including voluntary lending, pooling, allocation of resources, and authorized use of public property, but excluding contractual renting, leasing, or unauthorized use of property.[26] We find this framing unhelpful at both extremes. Many formal sharing programs—for both goods and services—involve contracts in some form (for example through membership of carsharing or film rental services). And, more significantly, while sharing on the margins of legality, such as squatting, may not constitute sharing between the formal owner and user of the property; it can still be a collaborative, shared activity between users motivated by norms of equality and justice.

Here, we outline a broad conception of a *sharing paradigm* that includes multiple dimensions: sharing *things* (such as cars, tools, and books); sharing *services* (such as sites for meetings or sleeping); and sharing *activities* or *experiences* (notably political activity, but also others such as leisure). We also include sharing between private individuals as well as collective or state provisions of resources and services for sharing, such as green space, sanitation, city bikes, or childcare. We recognize that sharing can be *material* in nature, or *virtual*; *tangible* or *intangible*; happening in spaces of *consumption* (such as digital music), or *production* (such as community gardens). Sharing can be *simultaneous* in time, as with public spaces, or *sequential* as with recycling material. It can be *rivalrous*, in which the goods or resources are those where use by one person excludes use by another, at

Table 0.1
The Broad Territory of the Sharing Paradigm

	Things	Services	Activities
Individual	Swapping, bartering, gifting	Ridesharing, couchsurfing	Skill sharing
Collective	Car clubs, tool-banks, fab-labs	Childcare, credit unions, time-banks, crowdfunding	Sports clubs, social media, open-source software
Public	Libraries, freecycling	Health services, public transit	Politics, public space

least at the same time, (such as carsharing), or non-*rivalrous* (such as open-source software). The distribution of shares might be by *sharing in parts* or *sharing in turns*.

Mirroring the flourish in creative ways of sharing that we highlight in this book—of things, services, and experiences at individual, collective, and public levels (table 0.1)—is the wide-ranging and ever-growing terminology surrounding sharing. There are a plethora of terms, but they are rarely directly interchangeable. Below we briefly explain our idea of a sharing paradigm and some of the terms we use to define it, and also how it compares with and encompasses a wide range of other commonly used terms and concepts.

Perhaps the most commonly used of those terms is the "sharing economy," but we understand this as only part of the broader and more inclusive concept of our sharing paradigm. A paradigm is a constellation of ideas and concepts that amounts to a worldview, so our use of the term "sharing paradigm" reflects our belief that sharing is, could, or should be something more fundamental to both human and societal development than is encompassed within the more bounded term "sharing economy." It reflects our belief that what we may be witnessing are the seeds of a potentially post-capitalist society.

The sharing paradigm is based on an understanding of the term "well-being." Well-being can refer to both physical and mental health, and to positive mental attitudes (or happiness). We use it to refer to the suite of functionings that people have reason to value—good physical and mental health among them, but also including material pleasure and our ability to make sacrifices for others (which any parent will recognize as potentially more fulfilling than selfish consumption). For our purposes, therefore, well-being depends on building and developing human capabilities for all. The fundamental resources we have available to do that—from breathable air to education, and from energy resources to healthcare—are better conceived and understood as shared commons than as private goods. We may collectively decide that the best way to manage and allocate certain resources is through market economies, or perhaps through public management, but our starting point is the recognition of their collective, shared nature. The sharing paradigm therefore foregrounds ways of thinking based on sharing resources fairly, rather than by ability to pay; treating resources and the environment as the common property of humankind; nurturing the collective commons of human culture and society; and stimulating human flourishing by establishing and enabling the expression of individual and collective capabilities.

Our concerns with the discourse of the "sharing economy" are not just with the intrusion of commerce and money (which are not always inappropriate), but also with the framing of sharing activity as "economic activity" rather than social, cultural, or political activity. This is much more significant than it might first appear. Privileging the economic dimension in this way perpetuates the myth that human society is *founded on*, and *bounded by* the economy, rather than vice versa, and that the environment is simply a source of economic resources, rather than the foundational space in which humans and our societies and cultures evolved and coexist. Moreover, it primes us to seek solutions to our "problems" in markets, in monetized exchange, and in the production and consumption of goods and services, all of which are constrained by economic frames and drivers, rather than by asking searching questions about our primary needs and the whole range of ways in which we can enhance human well-being in just and sustainable ways. In particular the sharing paradigm helps place our focus more strongly on underpinning environmental resources—land, water, clean air—and reveals the way we can share these "commons" fairly, as an inspiration for sharing in the city, and in the economy.

Mapping the Sharing Paradigm

Nonetheless, the "sharing economy" is part of the sharing paradigm. In such framings, sharing represents an important new form or modality of market exchange in which services become the focus of exchange, rights of access replace ownership, and we collaborate with our peers to better fulfill our needs as consumers. The sharing economy also extends into forms of production with new collaborative models especially enabled by the Internet, variously described as peer production, co-production and Wikiproduction. These models are rarely formal cooperatives, but often organize themselves in similar ways. Peer-to-peer (P2P) models can be found in finance, too, as well as in labor processes and all aspects of production, exchange, and consumption.

The business writers Don Tapscott and Anthony Williams identified an emergent trend of mass collaboration facilitated by the Internet.[27] In the disruption and disintermediation of many established businesses, they saw the possibility of new business models based on transparency, collaboration, and open platforms for sharing, with more widespread application to public services and global challenges. They suggest five underlying principles for what they call "wikinomics": collaboration, openness, sharing, integrity, and interdependence. They focus primarily on the business entity, rather than the individual or community. So, for instance, they see sharing as about consumers being more closely involved in production processes, as "prosumers"; or businesses sharing assets "by placing them in 'the commons' for others to use"[28] (as Tesla did with its electric vehicle patents in 2014), or commercially under license agreements. The emergence of "big data" is adding to the incentives for such openness and sharing, but as companies realize the potential value of massive data sets and analysis, data sharing may become more structured as data aggregators and brokers emerge.[29]

Within the sharing economy, the Internet has enabled much "commercial" sharing, where collaborative consumption and production takes place for payment. But there has also been an explosion of "communal" sharing, where goods, services, and skills are donated, swapped, or traded for free, or against an alternative medium of exchange, such as time-dollars. The falling costs of online collaboration mean not-for-profit and community organizations can more easily use mechanisms that were previously largely reserved for commercial purposes, such as large-scale online platforms. Communal sharing can be seen as part of a "solidarity or social economy" that is

> based on democratic control and social justice, not just cooperation and ecological sustainability. It's about sharing power. Solidarity means recognizing our global interdependence and addressing injustices in our communities by replacing dynamics of unequal power with grassroots, cooperative leadership.[30]

The "solidarity economy," according to Ethan Miller of the Grassroots Economic Organizing Collective in Australia, "is an open process, an invitation."[31] As illustrated in figure 0.1, it encompasses a wide range of entities and approaches including lending circles, community crowdfunding, participatory budgeting, community currencies, credit unions, cooperatives, co-working, community gardens, open source projects, art collectives, community land trusts, co-housing, open public spaces, healthcare collectives,

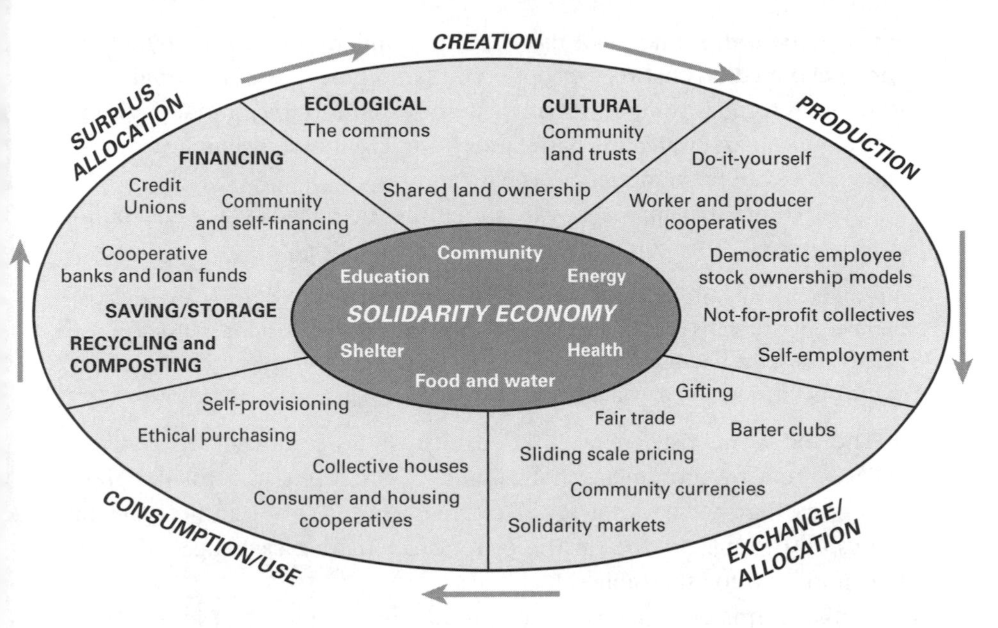

Figure 0.1
The Solidarity Economy. Source: Miller (2010, see note 31, this chapter).

time-banks, community-owned media, libraries, barter markets, Freecycle, free food sharing, and the social and environmental justice organizations that support such approaches, including unions, nonprofits, and progressive businesses.[32] Solidarity economy organizations are not exclusively sharing organizations, but in almost every case—befitting the collective nature of solidarity projects—there is some form of sharing activity.

The solidarity economy also includes growing communal forms of sharing and collaboration in social infrastructure and services such as education and health. These collaborations are sometimes described as "co-production," in this case between citizens and public service providers, while these fundamental collective services can also be called the "core economy." We interpret co-production broadly, as producing and delivering goods and services in a reciprocal relationship between producers and users; recognizing the resources that citizens already have, and delivering spaces, services, and goods *with* rather than *for* users, their families, and their neighbors. Co-production of collective goods extends to the social and cultural milieu of our communities—their physical, social, and cultural environments. These are common resources, managed and sustained by our collective activities forming the "urban commons." In these ways sharing and collaboration are key aspects of the conduct of daily life that underpin social reproduction and social relations between people. The same processes of informal (and sometimes formal) commons management extend to the natural environment—the air and water, the parks and thoroughfares of the public realm—and thus to humankind's relations with nature.

These communal approaches are sometimes described as "the collaborative commons"[33] or "commons-based peer production":

> An emerging and innovative production model in which the creative energy of large numbers of citizens is coordinated, usually through a digital platform, outside of the parameters of the traditionally hierarchical and mercantil (*sic*) organization resulting in the public provision of commons resources.[34]

The social theorist Jeremy Rifkin also includes commercially motived collaboration by prosumers, sharers, and co-producers in a collaborative commons that he sees as a social partner to the high-tech "Internet of things."[35] He suggests that in the commons, "market exchange value" is transmuting into "shareable value."

Similar forms of participatory co-production are emerging in administration and governance, and not just in the form of enabling legal devices such as the "creative commons" licenses for sharing the products of cultural industries. They emerge also in more direct engagement of citizens in the mechanics of government, through participatory budgeting, and in the use of collaborative tools and spaces for political action. In these ways sharing is infusing our institutional, legal and governmental arrangements.

As we shall see, in all these different arenas, models and practices of sharing are part of a contested politics over the reach and nature of commercial markets and relations. Tensions between private interests and the shared public realm are nothing new in cities around the world. In the modern era we see for example, gated communities; guarded shopping malls with dress codes; conflicts between squatters and developers; and competition for road space between private, shared, and non-vehicular transport. In the "sharing economy," commercial models of sharing run the risk of turning people into always-on, sweated commodities whereas communal models promise to return interpersonal relationships to the center of economic activity. In co-produced services, to recognize the importance of public contributions, freely given, means to roll back privatization and marketization, and to resist the enclosure of the natural and cultural commons. In these ways sharing approaches first problematize, then disrupt, and finally reconstruct our mental conceptions of the world and our sociocultural understandings and beliefs, spreading new norms of collaboration and sharing that answer neither to the state nor the market first, but to our fellow humans with whom we share our lives, our communities, our cities, and ultimately our planet. This is the sharing paradigm.

We do not see the sharing paradigm as intrinsically anti-capitalist. Indeed many forms of contemporary sharing are being mainstreamed by conventional capitalist businesses. Yet the dominance of neoliberal capitalism in our lives is problematic, especially where it squeezes out alternative ways of knowing, valuing, and living and disregards and degrades priceless assets (such as our natural environment) or exacerbates social and spatial inequalities and injustices. The urban and political geographer David Harvey suggests that the continued dominance of capital is a product of its ability to constantly shift its development between different arenas of production and reproduction.[36] Harvey is talking of "capital" as an actor or interest group; we use the term throughout this book in the same sense. Harvey identifies seven key arenas—all of which we have already mentioned and will encounter repeatedly: forms of production, exchange and consumption; relations to nature; social relations between people; mental conceptions of the world, embracing cultural understandings and beliefs; labor processes; institutional, legal, and governmental arrangements; and the conduct of daily life that underpins social reproduction.

Contrasting Dimensions in the Sharing Paradigm

To fully understand the scope of the sharing paradigm, we have found it helpful to consider two particular contrasts or tensions within it. These are shown in table 0.2. Although the table divides the territory of the sharing paradigm into four quadrants, in practice these contrasts are not digital binaries but analog gradations that naturally blur into one another. They create what might be best described as four "flavors" of sharing.

Table 0.2
Key Dimensions of the Sharing Paradigm

	(Inter)mediated sharing (learned)	Sociocultural sharing (evolved)
Communal sharing (intrinsically motivated)	"Peer-to-peer" sharing, enabled by not-for-profits, such as Freecycle or Peerby	The "collective commons" including public space and public services
Commercial sharing (extrinsically motivated)	The "sharing economy" of Airbnb, TaskRabbit and Zipcar	The "collective economy" of co-production and open sourcing in business

On one dimension we see a contrast between *sociocultural* or *informal* sharing (typically between family members, friends or neighbors, directly organized by the participants in line with social norms) and *(inter)mediated* sharing, which is mediated through a third party (often using a website or mobile application). Although sociocultural sharing too may be organized online, the distinction we draw is the involvement of the third-party intermediary. Mediated forms of sharing also include centralized models where an organization owns the resources that are shared by multiple users (common in car-sharing companies like Zipcar).

The question of mediation is one of how the sharing process is organized. It can also be seen as a distinction between *learned* behaviors and those that are more an expression of our *evolved* tendencies for cooperation in groups. The other dimension is about why we share, and the *motivations* of the participants. On this second axis we map a contrast between typically extrinsic motivations, notably commercial gain; and intrinsic motivations based in a sense of community, which we label as the commercial–communal axis.

This commercial–communal axis does not simply map the age-old division between market and state, which has structured political debates for decades. Nor does it seek to replace it with a market–community division. Rather it sees sharing as a genuine third way of governance and provision to meet human needs, rooted in collective management of jointly held resources. Sharing behaviors are spreading from the commons into both market and state domains, united by collaborative modes of action in which control or ownership is in some way shared. Neal Gorenflo, the cofounder and publisher of *Shareable* magazine, expresses a helpful distinction between sharing that is transactional and sharing that is transformational.[37] "Transactional" sharing is typically commercial, oriented toward efficiency and asset-sweating: reducing the *prices* users face. "Transformational" sharing however, necessarily involves a shift in power and social relations as well as an increase in *value* for all participants.

In figure 0.2 we illustrate how the diverse terms for sharing applied by different commentators map out across our four flavors of sharing.

As we have argued, the sharing paradigm, with its contestations, challenges, and opportunities, is a broader concept than that of the sharing economy. However, it is still useful to explore some of the terms frequently found in the literature on the sharing economy. Rachel Botsman, the cofounder of CollaborativeConsumption.com, describes the "collaborative economy" as a model "built on distributed networks of connected individuals and communities as opposed to centralized institutions,"[38] transforming production, consumption, finance, and learning. Within this, Botsman

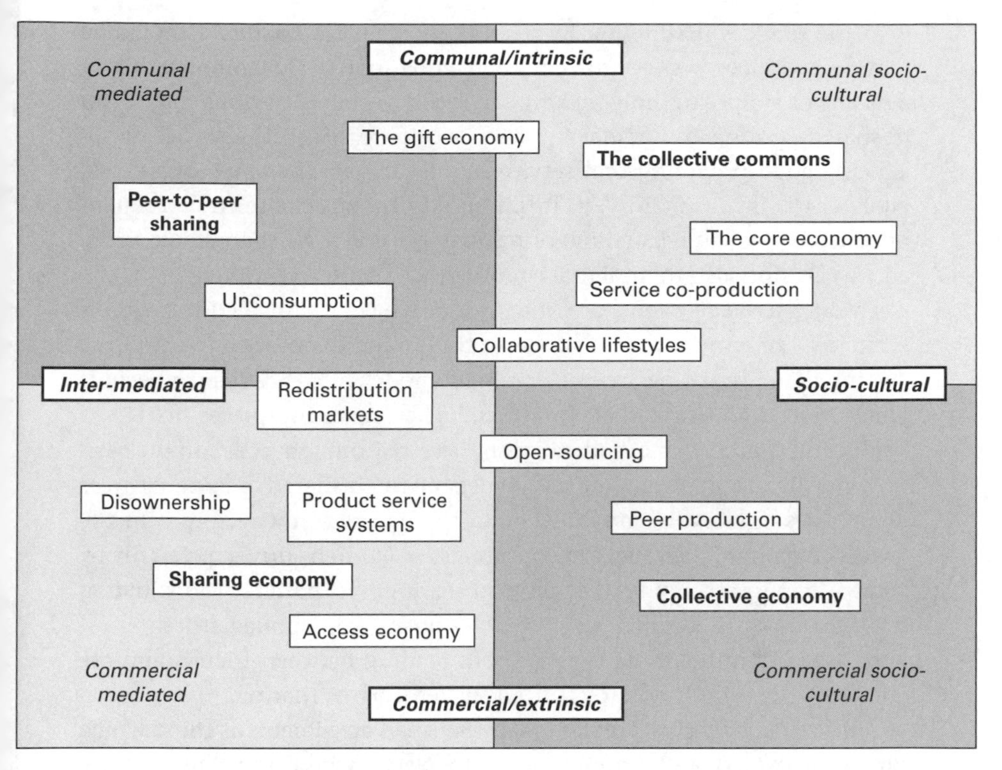

Figure 0.2
The Sharing Paradigm: More than the Sharing Economy

defines "collaborative consumption" as "an economic model based on sharing, swapping trading or renting products and services enabling access over ownership" (within which business-to-consumer (B2C), business-to-business (B2B), and peer-to-peer (P2P) transaction models are all practical). The *sharing economy,* she says, focuses largely on P2P marketplaces to share underutilized assets, including spaces, skills, and stuff for either monetary or non-monetary benefit; on the other hand, the *peer economy* also focuses on such person-to-person marketplaces built on peer trust, but includes those that facilitate direct trade as well as sharing.

Writing with investor Roo Rogers, Botsman focused on collaborative consumption, which they divided into three categories; "product service systems," "redistribution markets," and "collaborative lifestyles."[39]

Product service systems allow companies or organizations to offer the utility of a product as a service without the need for ownership. This is sometimes described as "disownership"[40] and recognizes that the value of a good is in the services it supplies. By treating the car as a means of providing transport services, we recognize that it can be shared. By treating a drill as a means of supplying hole-making services it becomes obvious that it can be shared. As long as the final "product" we desire is seen as a service, virtually any good providing that service can be shared. The value of a video, book, or CD is not (mainly) in its physical form, but in the entertainment provided by viewing, listening, or reading. Naturally we share books, CDs, and DVDs, and also their digital equivalents.

Product service systems are one form of what Rifkin called the "access economy," in which we seek access to things and experiences rather than owning them, rental becomes a dominant model, and experiences become much more significant than products.[41] The marketing professors Fleura Bardhi and Giana Eckhardt distinguish "access" from "sharing" on the basis of ownership: in sharing, they say, ownership too is shared or joint, whereas access does not involve any change in the nature of ownership.[42] In the 'access economy', they suggest, consumers appear to be driven primarily by cost consciousness, not by the values of sharing.[43] Although the distinction helps us understand different modalities of sharing, applied strictly however, it would unhelpfully exclude both lending between friends and the whole of redistribution markets from the territory of sharing.

Redistribution markets are defined by Botsman and Rogers as those which direct pre-owned and unused goods to places where they are needed, expanding the scope of product reuse,[44] or what the journalist Rob Walker calls "unconsumption."[45] In these markets the service available from a product is shared sequentially.

Collaborative lifestyles are defined as those in which people with similar needs or interests band together (physically or virtually) to share and exchange less-tangible assets such as time, space, skills, and money. (These categories overlap in situations where different models of provision may be used to deliver the same end service.) This in turn blurs the boundaries of collaborative consumption into the wider collaborative economy, as collaborations of this ilk are also found production scenarios, where the underlying "productive capital" (of sharing as a means of production) may be as simple and ubiquitous as human muscles and shared knowledge.[46] In contemporary society the falling costs of online collaboration and productive devices such as 3-D printers is extending the convenience and reach of collaborative production. The researcher Kathleen Stokes and her team, writing for the UK innovation charity NESTA include the terms "collaborative education" and "collaborative finance" alongside collaborative consumption in their definition of the scope of the collaborative economy.[47]

Not only do sharing economy approaches often miss the importance of underlying productive capital (and whether it too can be or is shared), but also, like conventional economic analysis, they are typically blind to the collective commons (of physical and virtual public spaces and services) that underlie all phases of production, consumption, and reproduction.

While Botsman and Rogers's categories clearly include communal spaces too, they do not focus attention on such shared commons.[48] Worse, both the media buzz and academic reflection have focused on rapidly growing commercial sharing platforms. And media commentators who have cautioned against the monetization and corporatization of this sphere have often chosen to argue that it is therefore in some way not "sharing." For example, the editor of the Vox Media website, Matthew Yglesias, argues that commercial mediated sharing is not sharing at all, but just constitutes short-term rentals, or secondary markets.[49] And the journalist Sven Eberlein claims, "The whole idea of the 'share economy' seems pretty redundant," and then goes on to ask, "Isn't the sharing of goods, services, property or experience in exchange for money or other agreed upon currency the very definition of 'economy'?"[50]

We disagree. Within the sharing paradigm, however, there is scope both to acknowledge that commercial sharing is a possibility *and* to examine the social and cultural risks and opportunities involved. This illuminates the deficits in the largely monetizing and monetized approaches that currently dominate collaborative consumption. But it also reveals the potential for even commercial sharing to shift both cultural behavior and social norms. Including commercial sharing does not prevent us from also recognizing a wide range of communal sharing approaches, such as "commoning" and "gifting," and the diversity of sociocultural sharing behaviors and domains that occur both within and outside the conventional market economy.

The Harvard professor Yochai Benkler highlights the wide scope of sociocultural or informal sharing that he calls "social sharing."[51] Within social sharing he includes the production of all kinds of goods and services, complementing or substituting for either market or state production systems. But he highlights as well the role social sharing plays in producing norms or rules of reciprocity within communities, and wider standards of human decency and civility across society. Motivations for non-commercial sharing can be equally diverse. Like conventional consumption, sharing and gifting can demonstrate power or status.[52] At another extreme, motivations can be entirely noninstrumental or even mystical or spiritual. Altruistic motivations can be helpfully categorized as directly reciprocal (we give in the expectation of receiving in return) or indirectly (we give—within a community of some form—in the expectation that when we are in need we will receive help from within that community). We use the term "karmic altruism" for the latter form.

Setting the Stage

In this introduction we have set the stage to challenge the reader to think differently about the concept of sharing, the sharing economy in all its forms, and ultimately about a broad sharing paradigm as a foundation for the sharing city. A valuable ally in our evolving thinking has been the thoughtful leadership of shareable.net with its real-time, contextualized, critical, and connected coverage of "the sharing transformation." We necessarily introduce the reader to a wide-ranging and perhaps eclectic set of literatures, some of which we are less familiar with than others. We also introduce the reader to a way of synthesizing those literatures, making connections and linkages where they may not have been made before. Only in this way, we believe, can we not only understand but also, more importantly, act upon the imperative of the sharing city.

The Chapters

Our book consists of five chapters, each preceded by a city case study that introduces some of the themes of the following chapter. A sixth case study precedes our conclusion and addresses the up- and downsides for sharing in rapidly growing cities of the developing world. These city case studies are based on a wide range of secondary sources from academic, professional, and public media. Taken as a group, they provide context, illustrating the diversity and development of sharing in cities around the world today.

In chapter 1 we examine the contemporary revival of sharing as collaborative consumption in mediated and particularly commercial forms in cities such as San Francisco. We report existing surveys of sharing behaviors and note the role of the Millennial generation. We illustrate the booming sharing economy with examples such as food. We focus on key drivers—technical, environmental, economic, and cultural—examining new emerging norms. We explore the economic logics of sharing and examine some of the pros and cons of approaching sharing in this way. The chapter concludes by introducing the risks and benefits of an intrusion of commercial sharing into the social realm.

In chapter 2, following a brief exploration of Seoul, we consider the productive domains of the city. We begin by considering the sociocultural and biological co-evolution of sharing, exploring common features and cultural variations. We then examine key ways in which cities are shared domains of production, reproduction, and exchange, with shared public services, infrastructures, and resources. We outline how services and infrastructures can be co-produced with particular reference to health and education. We also explore co-production in commercial spheres, including peer-to-peer finance and energy, and in cooperatives. We highlight the risks of disowned responsibility and commodification that arise where sharing overlaps public or market provision. We conclude the chapter by discussing how a shared collective commons underpins both commercial and communal city functions.

Chapter 3 follows our case study of Copenhagen, and explores political and cultural dimensions of the sharing city. We discuss the centrality of urban spaces to political movements and how such movements now equally depend on public cyberspace. We explore ways in which sharing can underpin democracy in practice: building social capital, supporting a healthy public realm, and challenging the hold of consumerism on identity. We focus on communal forms of sharing, for example by considering the role of collaborative leisure, and by taking a close look at streetlife and other opportunities for the development of interculturalism. Finally we examine ways in which sharing is emerging in the practices of urban governance, highlighting key challenges to urban democracy in land ownership and taxation.

In chapter 4 we explore the scope for the sharing city to enhance equity and social justice, following an examination of Medellín. The first part of the chapter situates the sharing paradigm in contemporary theories of justice, with particular reference to just sustainabilities, the capabilities approach, and recognition. The second part examines emerging tensions between sharing in practice and justice, highlighting the importance of sociocultural sharing practices. It also illustrates the challenges and opportunities to design sharing for justice and inclusion—with consideration of transport, the commodification of nonmarket aspects of life, stigmatization, casualization of labor, and exclusion of the disadvantaged from sharing practices. We finish with a closer look at urban enclosure, gentrification, and social exclusion.

In chapter 5, following a case study of Amsterdam, we aim to demonstrate how the various domains and flavors of the sharing paradigm could reinforce one another in the sharing city, and to outline the ways in which city administrations could act to deliver such a virtuous cycle. We rebut some common objections to sharing and highlight some genuine obstacles and challenges. We highlight the importance of emergent collective governance and sociocultural norms and explore the opportunity for broader sociocultural transformation through the sharing paradigm. Finally, in the light of this understanding of how sharing might spread, we examine the prospects for implementing and scaling up the sharing paradigm through active policy, planning, and practice at the city scale.

Before concluding we explore the city of Bengaluru (previously known as Bangalore), highlighting the prospects for sharing in rapidly growing cities of the developing world, especially considering the downsides of too enthusiastic a pursuit of "smartness" at the cost of sharing and justice.

Finally in our concluding synthesis, we return to the case we set out above, for the sharing city, and reflect upon some of the ways in which the sharing paradigm offers a powerful alternative to discourses of the smart and competitive global city. In a spirit of collaboration we close with a few suggestions for future transdisciplinary investigation of new themes and possibilities raised by our exploration of the sharing city.

1 Case Study: San Francisco

San Francisco, California, is at the forefront of the modern wave of collaborative consumption with high-tech sharing companies, new sharing startups, and the development of new norms among Millennials. It is home to companies like Twitter (the online social networking and micro-blogging service), Dropbox (the cloud-based storage firm), Airbnb (the online community marketplace for booking accommodations), and Lyft (the ridesharing mobile app), to name but a few. The city's proximity to Silicon Valley's hub of technology innovation has helped power its emerging scene of sharing startup companies. In recent years new high-tech jobs growth in the urban core of San Francisco has outstripped that in the longtime corporate centers of surrounding suburban counties.[1] This shift reflects changing norms among Millennials, the generation providing both the workforce and consumer base of these startups.

San Francisco's urban center attracts young people who are adopting co-working and sharing lifestyles, eschewing car ownership, and reducing consumption. Companies who locate in the city connect better, both with their users and the qualified potential employees choosing to live there. It now appears somewhat unfashionable to start a company in the suburbs.[2]

Starting up in the city is easier with the prevalence of shared workspaces. In 2005, Brad Neuberg and three friends, all of whom were freelancers, rented some common space in San Francisco and set up the first of what soon became known as co-working spaces. Today such spaces can be found in many cities, all "combining the best elements of a coffee shop (social, energetic, creative) and the best elements of a workspace (productive, functional),"[3] catering to a growing market. Freelancers make up somewhat over one-fifth of the US workforce, for example—around 40 million workers.[4]

San Francisco is a leading "smart city," following the advocacy of companies such as IBM and Siemens, and the city government is actively encouraging San Francisco's status as the epicenter of emerging high-tech sharing. However, San Francisco's reliance on the private sector for funding its smart-city goals, although superficially efficient, has resulted in "limited service diversity in terms of social-welfare domains" in comparison with publicly funded efforts in Seoul, and a public-private partnership approach in Amsterdam.[5]

With a growing population—roughly 825,000 in 2012—and a constrained location, San Francisco is becoming a laboratory for how commercial, mediated sharing interacts with public urban challenges. Mayor Ed Lee has attributed the city's rise from the recession in large part to its newfound tech wealth.[6] In 2012 the city launched the Sharing Economy Working Group to examine the economic benefits of emerging sharing enterprises and listen to companies' concerns about policies and regulation.[7] The Working Group was recruited from a diverse array of stakeholders including city departments, community organizations, and sharing companies. The fact that it never formally convened, but operated through "informal discussions with officials"[8] implies that the process was rather less participatory in practice than on paper.

Jay Nath was appointed as the city's first chief innovation officer in January 2012. The position is the first of its kind in the US. Nath's charge is "to introduce new ideas and approaches to make the City government more transparent, efficient, and constituent focused."[9] He is also expected to engage the tech industry in boosting job creation and civic participation. Nath calls cities "the original sharing platform," and referring to the current wave of sharing startups, he claims that the city can be the first to "modernize the regulatory environment" in a way that supports the sharing economy.[10]

The scope of sharing innovation in San Francisco is wide. For example, BayShare member Airbnb created a new tool to allow fee-free accommodation listings in regions affected by a natural disaster in order to quickly deliver emergency housing assistance to displaced residents. Airbnb collaborated with San Francisco's Department of Emergency Management to standardize this tool and ensure that it could be activated in less than 30 minutes.[11] BayShare was subsequently invited to join the San Francisco Disaster Council, working alongside local authorities and emergency service providers on the city's disaster preparedness and resiliency plans.[12]

The Mayor's Office of Civic Innovation boasts another first-of-its-kind initiative, the Entrepreneurship-In-Residence program, which brings selected startup companies together with the city government for 16 weeks to explore ways to make government more efficient and responsive.[13] This collaboration supports startups tackling public challenges, such as energy, education, and delivery of other city services, by providing them with access to co-working space, mentorship from senior public officials, workshops, and training. Six startups made up the 2014 cohort, including Birdi, which makes a smart air device to track carbon monoxide levels and other air quality issues.

Another project of the Office of Civic Innovation is the Living Innovation Zone (LIZ) program, which improves and enlivens public spaces through creative projects and technologies. LIZ strives to build upon the success of San Francisco's "parklets" and other "pop-up" projects (see also "The Crucible of Democracy in chapter 3), which repurpose parts of the streetscape into spaces for people. In these zones, innovators, artists, and designers are provided with real-world opportunities to test the impact of new ideas and technologies. For example, parabolic acoustic amplifiers have been installed at Market Street and Yerba Buena Lane. These have been "adopted by street performers who quietly strum a banjo on one, while hundreds of pedestrians are strolling past on the other side."[14]

More broadly San Francisco provides opportunities at the interface of design and implementation, in which high-tech sharing companies can showcase their innovations. The community participation, enhanced social interaction, and sociocultural development that these projects create provide insight into the potential that high-tech sharing holds for urban spaces. For instance the SF POPOS app helps people discover the city's "privately owned public spaces," providing travel directions and a map that highlights POPOS ranked on various qualities and amenities.[15] Drawing residents and visitors into the nooks and crannies of the city away from commercial tourist areas can also spread economic opportunity across more of the city's neighborhoods.

Skyrocketing San Francisco rents are encouraging some to extend their sharing philosophy into living arrangements. Young professionals in the city and greater Bay Area are taking over leases of grand estates and transforming them into communal living spaces.[16] Jordan Aleja Grader, a co-resident in one 6,825-square-foot mansion, says, "We're seeing a shift in consciousness from hyper-individualistic to more cooperative spaces ... we have a vision to raise our families together."[17] The Open Door Development Group is a real estate investment firm established to buy buildings and convert them into co-living spaces. Its founders argue that they are resisting market forces that threaten the city's diversity through gentrification. By creating curated communities rather than luxury housing, they believe they can build diversity into the plan.[18]

Despite some co-living proponents' desire to maintain and protect social diversity in San Francisco's culture-rich neighborhoods, affordable housing advocates challenge the claim that the sharing economy is alleviating wealth disparities. As landlords realize the exorbitant rent that they can charge this influx of high-tech entrepreneurs, the threat of gentrification for long-time and low-income residents is very real.

In 2014 the city attorney filed suit against two landlords, claiming they illegally converted residential housing to short-term rentals in order to advertise on services like Airbnb. The former residents were evicted using the Ellis Act, a controversial California law that allows landlords to reclaim properties for personal use.[19] Affordable housing advocates are concerned about abuse of this law in the midst of a severe housing shortage. Others point to the predominance of "whole dwellings" and multiple listings on Airbnb as a sign that the platform is not only facilitating a shift of housing away from those in need, but also drifting away from its aim of bringing visitors into shared homes: of San Francisco's 5,000 Airbnb properties, two-thirds are whole dwellings, and around one-third of hosts control more than one listed property. Official analysis confirmed that in 2013 up to 1,960 properties had been removed from the rental market for letting on Airbnb.[20]

In 2014 San Francisco adopted new rules for short-term rentals, broadly seen as enabling the Airbnb model, with some protections.[21] The new law allows only permanent residents to offer short-term rentals, establishes a new city registry for hosts, mandates the collection of hotel tax, limits entire-home rentals to 90 days per year, requires each listing to carry $500,000 in liability insurance, and establishes guidelines for enforcement by the Planning Department. Additional proposals to allow housing nonprofits to collaborate in enforcing the new rules, and quickly sue violators, are under consideration at the time of writing, but these new provisions have yet to overcome concerns over gentrification.

It can also be problematic when startups desire the hip identity that comes with setting up shop in one of the diverse, poorer neighborhoods, and as a result the people who built that unique community and whose struggle has given it its appealing edge are threatened with displacement as living costs rise in response.[22] Even when high-tech workers choose to live in San Francisco and commute to Silicon Valley via "Google buses"—a catchall for private shuttles operated by tech companies in the Bay Area—those shuttles have become their own symbol of economic stratification.[23]

Sharing companies are economically disruptive, but in social terms they may exacerbate existing injustices. San Francisco's ridesharing companies challenge the city's taxi industry, which largely employs lower-income immigrants and people of color. Concerns raised by the incumbent taxi industry have spurred discussions on regulations for car- and ridesharing companies. Taxi drivers argued that carsharing companies were engaged in unfair competition and should be regulated like other taxi drivers.[24] With similar concerns, the California Public Utilities Commission (CPUC) issued cease and desist letters to ride-sharing companies including Lyft, Sidecar, and Tickengo, and subsequently "issued $20,000 fines against Lyft, Sidecar and Uber for 'operating as passenger carriers without evidence of public liability and property damage insurance coverage' and 'engaging employee-drivers without evidence of workers' compensation insurance'"[25]

Later in 2013, after the carsharing industry turned the tables through an intense lobbying campaign, CPUC stamped its seal of approval on ridesharing services. Despite taxi companies' claims of unfair competition, according to Verne Kopytoff of *Fortune* magazine, the sharing firms "convinced regulators to carve out a new category of transportation services for ride sharing."[26] And although 28 basic insurance and safety requirements now apply to ride sharing, "The commission's decision gives the industry a green light across the entire state."

Kopytoff highlights how sharing economy enterprises initially responded to permitting and fee requirements with the attitude, "Your laws are outdated and don't apply to innovative businesses like ours."[27] But more recently these companies have begun to explore how regulation can fit in with their services. "Instead of fighting the system," says Kopytoff, "the companies (with some major exceptions) are beginning to accept that it's better to try to shape the system to their liking as far as possible."[28] The New York University business professor Arun Sundararajan posits that authorities should "delegate more regulatory responsibility to the marketplaces and platforms while preserving some government oversight."[29] Self-regulation, he suggests, is built into the business models and technology of commercial sharing economy businesses.

Others are less sanguine about the extent to which the sharing economy will automatically operate in the wider public interest. With the growth of sharing activities in the Bay Area and legal issues surrounding them, the Oakland-based attorney Janelle Orsi has emerged as a specialist in sharing law, offering legal services for things like shared housing and cooperatives. Orsi, along with the attorney Jenny Kassan, co-founded the Sustainable Economies Law Center to empower local economic exchanges and help people navigate legal barriers within the sharing economy.[30] Orsi sees great potential in the sharing economy to combat income inequality, so long as business structures can be created that return wealth to its users, such as worker-owned cooperatives.[31] Orsi is helping Loconomics, a San Francisco–based business management sharing company, which she describes as "like TaskRabbit if the rabbits owned the company."[32] Josh Danielson, the co-founder of Loconomics, agrees: "A platform helping with self-employment shouldn't be owned by the 1 percent," he says. "We're at a crossroads where technology exists to help the common worker break free from traditional employment models. I felt it was important it be owned by the workers."[33]

San Francisco demonstrates some of the divergent logics of and for sharing—particularly in commercial flavors, but also illustrates an emerging backlash by incumbent companies who see threats to their hegemonic practices. As these regulatory and values-based discussions continue to play out through the sharing project, cities around the world are both following its lead and learning from its challenges.

Sharing Consumption: The City as Platform

At its best, good city-making leads to the highest achievement of human culture.
—Charles Landry

Chapter Introduction and Outline

In this chapter we examine the contemporary revival of sharing as collaborative consumption in mediated (and particularly commercial) forms in cities such as San Francisco, which is actively leveraging its image as a smart, high-tech city. We report existing surveys of sharing behaviors and note especially the current and likely future role of an increasingly researched demographic: the Millennials. We illustrate the boom in collaborative consumption with examples from food sharing in particular.

We then focus on key drivers of this revival: technical, environmental, economic, and particularly cultural drivers, examining the new norms that have emerged in online sharing and are increasingly being exhibited in the real world. We explore the economic logics of sharing and examine some of the pros and cons of approaching sharing in this way; we outline, among other things, how incumbent businesses are responding to the sharing paradigm. The chapter concludes by beginning to examine the risks and benefits of an intrusion of commercial sharing into the social realm.

Sharing, as we discuss more fully in chapter 2, is a sociocultural, evolutionary trait that enabled the development of hunting, agriculture, trade, craft, and manufacturing—and thus cities. Yet with increased marketization, industrialization, and consumerism, cities have become spaces in which this evolutionary form of *sociocultural sharing* has been weakened, especially in more affluent societies, as social capital has been eroded,[1] trust undermined by growing inequality,[2] and the togetherness of cities replaced by private withdrawal.[3] Yet at the same time, a new, distinct, and predominantly urban form of *(inter)mediated sharing* is emerging, exemplified best in "smart" and "wired" cities such as San Francisco (and increasingly Seoul and Bengaluru). In urban societies, but increasingly in others too, with the presence of smartphones and other technologies, we are witnessing a morphing of our evolutionary propensity for *social-cultural sharing* toward *mediated sharing* using the approaches described in the San Francisco case study. In this sense, it could be argued that the emerging sharing economy is simply human nature reasserting itself via these new, wired platforms. As we will see, however, mediated sharing is a contested space, with a range of both *commercial* and *communal* intermediaries seeking to establish themselves.

The Sharing Revival

As of 2014, at least 350,000 people in 34,000 different cities had shared their properties using Airbnb. The San Francisco-based couch-surfing platform earns around $250 million a year from its 11 million users. As a result its putative corporate value now exceeds $10 billion.[4] Airbnb is part of a rapidly growing global sharing economy predicted by Price Waterhouse Coopers to exceed $335 billion annual turnover by 2025.[5] Airbnb is typical in using a brokering model: in exchange for providing the market and services like customer support, payment processing, and providing host insurance, the company takes a 3 percent cut from the host and a 6–12 percent cut from the guest, which varies depending on the property price.[6]

Airbnb is perhaps the leading example of the commercial mediated sharing platforms that are spreading through the urban global North, not only but arguably particularly rapidly, among the younger generation—the Millennials as they are called in the US—signaling a particular evolution in norms.

In recent years, sharing behaviors have spread from cyberspace to urban space around the world. A 2013 survey conducted by the research firm Latitude (in conjunction with *Shareable* magazine) found 75 percent in the US sharing digitally, and 50–65 percent in both the US and UK sharing in other domains (sharing cars, living space, clothes, etc.), albeit often at a small scale.[7] In Germany too, "55 % have experience of alternative forms of ownership and consumption (product-service systems, private business transactions or collaborative consumption)."[8] Moreover, mediated sharing behaviors appear to be multiplying through the establishment of sharing norms and trust building online,[9] the importance of which cannot be underestimated.[10]

Surveys on sharing behavior have been undertaken globally;[11] internationally (comparing Canada, the US, and the UK);[12] nationally in the US,[13]

Germany,[14] and the UK;[15] and in some cities, such as Vancouver, Canada[16]. These surveys generally recognize growing individualism, "with more homes occupied by single people and more cars on the road with single drivers,"[17] and yet they reveal demand for more sharing. For example, "Over half of the UK would love to find ways of being able to share their time and resources within their local community [and] ... one in three people would be willing to share their garden with someone else locally."[18] In the US, more than half of those surveyed have "rented, leased or borrowed the sorts of items people traditionally own in the last two years (52 percent), and more than 8 out of 10 Americans (83 percent) say they would rent, lease or borrow these items, instead of buying them, if they could do so easily."[19] Another survey indicates that "75% of respondents predicted their sharing of physical objects and spaces will increase in the next 5 years."[20]

Online sharing is common globally too. In an online survey of 30,000 Internet users across 60 countries, Nielsen found more than two-thirds were keen to share, with desire to participate significantly higher in Asia (around 80 percent) and Latin America, Africa and Middle East (around 70 percent) than in North America and Europe (around 50 percent).[21]

In the Latitude survey, most participants (78 percent) had also used a local, peer-to-peer web platform such as Craigslist or Freecycle, where online connectivity facilitates offline sharing and social activities. But in Vancouver, Canada, the researcher Chris Diplock found "that less than 10% of respondents reported that they currently lend and/or borrow physical objects or spaces with peers through an online service." However, "70% of this group of respondents agreed that sharing online has helped them share offline."[22] Sharing is expected to grow here too, as "1 out of every 3 people in Vancouver are interested in sharing more with their peers, with individuals 26 to 40 reporting the most interest of all age groups."[23]

Similarly, in the UK, 64 percent report participating in collaborative economy activities, with parents and employed professionals (age 25–54) being more active than pensioners, semi- and un-skilled employees, the unemployed, and those from ethnic minority groups.[24] The patterns are different in the US: "More than any other age group, Americans 55 and older are more likely to engage in this behavior because they don't want to maintain, pay for maintenance, or store the items over time."[25] And:

> Participants aged 40+ were more likely to feel comfortable sharing with anyone at all who joins a sharing community ... however, Millennials were more likely to feel positively about the idea of sharing, more open to trying it, and more optimistic about its promise for the future.[26]

Convenience (46 percent) tops the motivations for older Americans sharing or renting, while cost (45 percent) dominates for younger ones. Younger respondents also cite minimizing waste and conserving the environment as motivations about twice as frequently as older ones.[27] Community building and other intrinsic motivations are widespread. In the Latitude survey "saving money" and being "good for society" were both cited by two-thirds of respondents.[28] And in the UK, reportedly, "75% of us believe that sharing is good for the environment ... [and while] 7 out of 10 people ... say that sharing makes us feel better about ourselves, 8 out of 10 say that sharing makes them happy."[29]

The business analyst Jeremiah Owyang undertook what he describes as "the largest study of the collaborative economy" (90,112 people in the US, Canada, and the UK)[30] and identified three distinct segments: First, *re-sharers*: those who buy and/or sell pre-owned goods online (for example, on Craigslist or eBay), but have not yet ventured into other kinds of sharing. Second, *neo-sharers*: people who use the newer generation of sharing sites and apps like Etsy, TaskRabbit, Uber, Airbnb, and Kickstarter. And third, *non-sharers*: people who have yet to engage in the collaborative economy. But many of these non-sharers report intentions to try sharing services (in particular, re-sharing sites like eBay) in the next 12 months.

Vibrant sharing economies are also emerging in Europe, Australia, South Africa, and some cities in Latin America and the Middle East. In Australia, for example, "new collaborative consumption startups are appearing on a regular basis and participation is growing" according to Chris Riedy of the Institute for Sustainable Futures in Sydney.[31]

The Millennials

The Millennial generation (also known as Gen Y, ages 18–33) forms the largest cohort in history (80 million) in a country, which more than any other, sets global cultural trends.[32] This generation appears particularly active in and receptive to mediated sharing, especially in cities like San Francisco. These younger people hold different values than their parents and grandparents did at their age, including a trend toward minimalism and disownership. Millennials have been coined "the cheapest generation," in that they are not buying cars or homes at the rate their predecessors did. This may mark the beginning of a permanent generational shift in consumption preferences and spending habits.[33] And alongside their comfort with technology and increasing concern for environmental issues, these changing consumer values make Millennials receptive to sharing.[34] Their values extend to sharing public spaces and services. Millennials appear to favor places with high walkability, good schools and parks, and high quality and availability of multiple transit options.[35]

For many Millennials caught in the woes of the job search while bearing heavy student debt, the sharing economy provides an opportunity for flexible work that at least some find fulfilling. Nick Hiebert drives for Lyft, the ride-sharing service based in San Francisco. He is one of many who pass in and out of Lyft's network with daily schedules and obligations that fluctuate.[36] For college graduates like him this flexible revenue stream may appear empowering. Yet as we will see later, even though Lyft "allows its drivers to experience real human interaction with people outside their circles and themselves,"[37] for those with weaker long-term prospects, the flexibility of such work can be a threat rather than an opportunity.

Some researchers interpret the surveys to show that the networked Millennial generation is shifting toward identity defined through "relationships, collaboration, and social interactions and in which 'access is the new ownership'" in which they exhibit "a declining identification with brands in developed economies and in the younger generation."[38] In a paper for UK NGO Friends of the Earth, Victoria Hurth, a marketing specialist at Plymouth University in the UK, and her colleagues argue that in adopting sharing approaches Millennials are beginning to challenge the basic acquisitional way the current economy works, with alternatives that simultaneously reduce resource use and redefine the messages they send to others about their identity.[39]

Dara O'Rourke, a professor at UC Berkeley, chairs the World Economic Forum's Global Agenda Council on Sustainable Consumption. He notes the challenges of emerging Millennial consumers in China, India, and Brazil, as well as in the US and Europe:

> New technologies, and in particular emerging mobile, wearable, ubiquitous, transparent information systems, are radically changing how new millennials see and interact with products, brands, and retailers, what they demand of them, and what they want for their futures. Consumers can know more, and they obviously share much more than ever before. The pace of this change is only accelerating.[40]

O'Rourke sees possibilities that these Millennials will consume like others in the global North, or that they will embrace sharing (rather than owning) and the circular economy. But brands and retailers find that they don't know how to engage with Millennials, even on topics like sustainability and supply chain issues, in part because:

> These new consumers, in fact, bristle at even being called consumers. They think of themselves as makers, users, sharers, and sometimes participants in the production of products, services, content, etc. They have values. They get status from different

things than their parents did. And they want to support products and companies that align with their values.[41]

But this may also reflect the same blind spot in the Millennial psyche that allows the generation to endorse values of inclusion and equality, but reject practical measures to tackle structural discrimination:[42] they are happy to demand the integrated outcome, but don't (yet) understand the vested interests that need to be overcome, and are unready to engage with the messy processes of delivering change. They are, however, embracing collaborative consumption.

Collaborative Consumption

Here we focus on the recent revival—largely Internet mediated—of sharing in contemporary cities. This massively enhances the productivity of shared assets through web-facilitated allocation, and reflects developments in our economic culture with tremendous prospects for the future.

In the last decade conventional second-hand markets for goods have moved online in many countries with the growth of sites like eBay, Gumtree, and Descola aí in Brazil, where consumers can sell, exchange, and buy products and services. Low transaction costs have enabled the establishment of gift-based approaches such as Freecycle, in which usable goods are simply given away online; and more dramatically, the emergence of sharing platforms that allow for the rental of personal goods such as cars, tools, and spare bedrooms, and sale of services such as mealsharing. And this is not just in the West. Xiaozhu.com is a Chinese online service for short-term property rental. In Cairo one can carshare with KarTag, or solicit cleaning or other services on Taskty, while in Seoul (as the case study below details) kozaza.com promotes *hanok*, or traditional Korean houses, in a Korean version of Airbnb, and zipbob.net enables Korean mealsharing.

At the same time the transformation of digital goods markets by enterprises such as Napster, Spotify, LoveFilm, and Netflix has profoundly impacted norms of ownership. No longer do we see it as essential to own a musical recording or a DVD. And similar perspectives are intruding into material goods markets: Why own an expensive ball-gown, if you can borrow a different one for each of the few occasions a year you might wear it? Why have a garage packed with tools or sports gear that is scarcely used, instead of borrowing from a communal hub, or from lists posted online by neighbors? Mike Brown of Boston's GearCommons notes:

We had our "aha" moment to start the company toward the end of a snow shoe marathon, 22-mile race. We were slogging through the end of the marathon, saying "you know we know a bunch of people who wanted to do this race but it didn't make sense to buy shoes just for this race, why doesn't something exist for them to just rent." We also both had someone staying in our Boston apartments through Airbnb while we were at the race, so we were making money while there. So we built GearCommons to provide access to equipment for people who don't want to or need to own it.[43]

This consumer oriented, P2P, sharing economy can be seen as "a bunch of new ways to connect things that aren't being used with people who could use them" and because Internet-based platforms perform this task "radically better than previous systems in achieving higher utilization of the economy's 'idling capacity,'"[44] the sector is attracting serious financial investment. Estimates of the global value of the sector range from more than $25 billion[45] to in excess of $500 billion.[46]

Shared production facilities are booming as well. Shared kitchens, shop fronts, markets, and offices all provide opportunities for small and social enterprises to meet public needs at lower costs. Bill Jacobson, founder and CEO of Workbar, a membership-based workplace founded in Cambridge, Massachusetts, says:

When Workbar started I was running a small tech company, and we had 5 people subletting space and the main tenant just folded overnight, so that's where the idea came about to work around more people with diverse backgrounds, the goal is more about meeting with people in the work space."[47]

Shared working extends beyond knowledge businesses. San Francisco is just one of at least 70 cities across many countries to now have a "fab-lab"—or digital fabrication facility. These combine 3-D printing technologies with access to other tools and equipment, giving entrepreneurs, schools, and communities tools to turn ideas and concepts into reality. With widespread sharing of digital designs under creative commons licensing, the potential for fab-labs to underpin a new wave of co-production and self-provisioning, matched with domestic or community generation of renewable energy, appears amazing.[48] (We discuss this topic in more detail in chapter 2 under the heading "Co-production, Power, and Ownership.")

Breaking Bread

Food sharing illustrates well the breadth and diversity of the boom in collaborative consumption. Sharing—in the process, practice, and product of the kill, the harvest, and the catch—was arguably the fundamental evolutionary source of our collaborative nature. Today food-related sharing activities remain critical: around the world people are sharing seeds, land, produce, recipes, meals and even leftovers and unsold foodstuffs. We discuss shared food production later (see "Growing Together" in chapter 2), but here we focus on the consumption side where mediated sharing—both commercial and communal—is booming alongside traditional sociocultural sharing practices.

Food sharing is still largely sociocultural, at picnics, potluck dinners, religious and ceremonial events, community functions, and street parties echoing the cultural importance of food to ethnic, national, and other identities. Jesse McEntee—the founder of the Food Systems Research Institute—highlights "reciprocal exchanges" of food in parts of the US where the produce from hunting or fishing or one's garden is shared with friends. He calls this form of sociocultural sharing "traditional localism" in that participants obtain food by "non-capitalist, decommodified means that are affordable and accessible."[49] Culturally significant foodstuffs, recipes, and cooking techniques are shared daily in informal interactions in fields and allotments, over garden walls and in kitchens all around the world, as are seeds, growing tips, and assistance. Food is an "intimate commodity," literally "taken inside the body."[50] Moreover:

> Food practices ... are manifestations and symbols of cultural histories and proclivities. As individuals participate in culturally defined proper ways of eating, they perform their own identities and memberships in particular groups. ... Food informs individuals' identities, including their racial identities, in ways that other environmental justice and sustainability issues—energy, water, garbage and so on—do not.[51]

Yet new sharing intermediaries are active in redistributing food in various ways, communally, commercially, and in hybrid forms. Casserole Club is a free food-sharing project, started in two London boroughs.[52] It connects people who like to cook, and are happy to share an extra portion of their home cookery, with elderly neighbors who could really benefit from a hot cooked meal, thus supporting independent living in the community. Casserole provides a mediated sharing platform that relies on volunteers to provide food, and it is free to users. Because the beneficiaries include vulnerable adults, cooks are required to sign up online and undertake a short safeguarding process—including face-to-face identity checks—before they can contact local diners. Casserole is a commercial–communal hybrid, provided on a commercial basis to local authorities by the technology-oriented public services consultancy FutureGov. Local authorities buy into the service for a fixed annual fee and provide it free at the point of use, collaborating with charities working with the elderly to identify potential beneficiaries.

Also in the UK, communal intermediates Foodcycle and Fareshare[53] negotiate with supermarkets and other food businesses to redistribute excess short-life food (fruit and vegetables, baked goods, chilled products) to hundreds of local charities and community projects, whose staff members cook the food they receive on site so they can provide vital and nutritious meals to vulnerable individuals, families, and children. These charities also solicit donations of more durable foodstuffs, such as canned produce, from retailers and the public, helping tackle food poverty. While not without problems such as the stigma associated with giving food past its "sell by" date to those on low incomes, this is a typical example of food sharing as it is currently practiced. Food wastes are also increasingly shared—although much more informally—through freeganism, skipping (dumpster diving), and gleaning. (See "Criminalization: Sharing on the Edge of Legitimacy" in chapter 4, where we explore this topic further.)

In Berlin the largest organic supermarket chain gives surplus food to foodsharing.de, a communal, non-profit, mediated sharing initiative whose web platform enables free foodsharing.[54] Crowdfunded with support from charitable donations, and motivated as a means to reduce waste, it helps people to give away surplus food from events and parties, home cooking and garden produce. The Netherlands' Thuisafgehaald (Shareyourmeal) platform sits between the commercial and not-for-profit space. It provides an online platform on which "home cooks" can sell portions of their meals to interested neighbors. Motivated by the goals of providing an affordable and healthy alternative to takeaway food, and to helping combat food waste and build communities, Shareyourmeal has more than 40,000 users in the Netherlands and Belgium; in 2012, 100,000 takeaway meals were traded on the platform.[55] The not-for-profit company attracted initial funding from foundations and local authorities, but aims to become self-sustaining on a share of the payments made for meals. Revenue is reinvested in the platform, and in supporting volunteers building Shareyourmeal communities in their neighborhoods.

More obviously commercial in motivation are "collaborative dining" platforms such as EatWith and Feastly. EatWith moved its headquarters from Tel Aviv to San Francisco[56] and is now active in 30 countries.[57] Feastly operates in about a dozen US cities including San Francisco. Feastly's founder Noah Karesh calls it a reintroduction of "the original social dining option: the home cooked meal."[58] He says he "wanted to let cooks monetize their passions and provide exciting new food opportunities for eaters." But Feastly, like the charitable sharing outlined above, emphasizes the social and intercultural connections sharing food can form. Feasters (as the site's users are called) apparently "seek authentic food, served around big tables with good people"[59] in the cook-hosts' homes, and are expected to create and post detailed profiles of themselves and their interests on the site (or link to Facebook or other social network profiles). Feastly cooks are individually vetted and commit to follow strict guidelines to make sure every food experience is safe and clean. Cooks set their own prices for the meals, based on the cost of ingredients and compensation for time spent cooking and cleaning, but the company handles payment. Feastly charges a service fee based on the meal cost.

Feastly's developers would appear to have noted the legal controversies surrounding the spread of Airbnb (outlined in the San Francisco case study preceding this chapter). Its terms and conditions[60] place sole responsibility on the cook (host) to comply with all applicable laws, tax requirements, and rules and regulations (including zoning laws, health laws, license laws, and any other applicable laws governing properties and food). Nor does Feastly accept a role as a contracting agent for (or representative of) the host, though it is hard to see how it can take payment to arrange attendance at the dinner venue without taking on some liability. It explicitly encourages hosts to ensure they have appropriate insurance—although it has established a mechanism to recover from diners, on behalf of the host, any validated costs of damage or breakages. The experiences of companies like Lyft and Airbnb suggest that Feastly may run into controversy over insurance as it develops and may find it simpler to take on an insurance or guarantor role.

Despite being commercial-mediated sharing platforms, Feastly and EatWith clearly have potential to help rebuild social capital in consumerist cultures, and it is no surprise to see the development of GrubClub, with a very similar model in London.[61] These dining platforms are all attempting to formalize and commercialize the Cuban *paladare*, or its modern Western incarnation, the underground supper-club.

In 2009 the UK's *Independent* newspaper reported:

> Underground, "pop-up" restaurants in private homes are the latest foodie fad to hit London. From creative cuisine in Kilburn at MsMarmiteLover's The Underground Restaurant to nine-course vegetarian Japanese eating on slouchy sofas in Horton Jupiter's beatnik flat, called The Secret Ingredient, in east London, guerrilla dining is at the cutting edge of eating out, provided you can find the location in the first place.[62]

Collaborative dining opportunities may introduce people to different cuisines, but The League of Kitchens in New York City goes one step further. Described as "an immersive culinary adventure in NYC where immigrants teach intimate cooking workshops in their homes,"[63] it not only covers real culinary skills but also increases encounters and engagement between groups who might not otherwise meet. In turn this can help reduce prejudice and improve intergroup relations.[64] (We explore this topic further in the section "Intercultural Public Space" in chapter 3.)

Underground supper clubs in private spaces thrive on trust. There are no regulations or licenses to protect either hosts or diners; and according to the *Independent*, "most "pop-up" restaurants are free or suggest a nominal donation to cover costs."[65] They rely on word of mouth and social networks, and due to the constraints of space, are likely to place guests together at one or more large tables, adding to the sociability of the event. As with so many aspects of the sharing economy, if commercial mediated platforms like Feastly replace these more sociocultural forms of sharing, social capital will be lost, but if they extend the reach of pop-up dining to new and, especially, intercultural audiences and communities, then they will contribute to the spread of collaborative norms.

As part of Restaurant Day, single-day pop-up restaurants have so far arisen in 42 different countries. Founded in Helsinki, Finland, in 2011, the Restaurant Day movement is intended to promote and celebrate food culture, and stimulate sociability.[66] More generally pop-up restaurants in temporary venues, often at festivals, are seen—like shared kitchens—as a potential entry point for new chefs and new food businesses, and have become more common as crowdfunding has offered a relatively simple way to finance the initial investments needed.[67] As an addition to vibrant, shared public space, whether commercially or socially motivated, they can contribute greatly to a thriving urban commons.

This section has introduced the new wave of mediated sharing sweeping through modern cities, and its most enthusiastic participants—the Millennial generation. Our examples of collaborative consumption in the food sector have illustrated how broad the new sharing movement is, with both communal- and commercial-mediated models crowding into a space long dominated by sociocultural sharing. But why is this happening now?

The Drivers of the Sharing Revival: Botsman's "Big Shift"

Many commentators see the roots of the contemporary boom in sharing in the development of the Internet.[68] The web has both induced and facilitated a significant shift in attitudes toward free sharing, especially among Millennials. This generation has grown up in a "wired, connected world" in which "Real time technologies are re-enabling P2P swapping and trading" and "we are re-wiring our worlds to share."[69] We argue, further, that we are in turn re-wiring ourselves with new sharing norms developed in cyberspace, but increasingly applied in real, particularly urban, space.

The business strategists Joe Pine and Jim Gilmore suggest that ours is "becoming an experience economy that values doing over having" in which technological advances are allowing people to shed the "burdens of ownership" and shift toward shared ownership.[70] Rachel Botsman goes further. She sees sharing not as "a short term trend but a powerful cultural and economic force reinventing not just what we consume but how we consume."[71] She cites four drivers of this transformation, which we both elaborate on and critique below.

1. A Renewed Belief in the Importance of Community

Community remains a contested political space in many countries, but in both real and virtual worlds there is abundant evidence that the tide of individualism—epitomized by Margaret Thatcher's claim in the 1980s that "there is no such thing as society"—is receding. New forms of online communities are booming, while in our cities, contemporary movements for walkability, liveability, shared streets, Complete Streets, and placemaking are (re)building communities by creating places and spaces in common (discussed in greater length in chapter 3). It is difficult to tell whether this is a political backlash, a cyclical swing against the erosion of social capital in our cities, or itself a product of digital communication strategies that have invigorated communities of interest which are in turn reinvigorated in the real world. In all likelihood, all three of these factors are probably significant, but in differing proportions in different locational and cultural contexts.

2. The Torrent of Peer-to-Peer Social Networks and Real-Time Technologies

The web first enabled free sharing both of information (through sites like Wikipedia and Facebook) and of digital media (particularly music and video, through sites like Napster, Kazaa, Flickr, YouTube, and now with Spotify, Grooveshark, and a host of other mainstream legal music sharing services). "Virtual" marketplaces such as eBay, Craigslist, and Etsy have broadened the shared space of functioning exchange markets and have brought producers and consumers closer together as well. Now new technology is also facilitating the sharing and management of material resources. Open access web platforms for offering and booking sharing opportunities combined with real-time smart systems (including RFID, or radio-frequency identification tagging of shared goods) that help match users and allocate shared resources (such as in city bikesharing programs), have transformed the potential for sharing. The tagging of items with cheap RFID chips allows innovations in real-time tracking, stock control, and so forth, enabling the so-called Internet of things.

To technological optimists, the Internet of things "will connect every thing with everyone in an integrated global network,"[72] enabling universal monitoring, analysis, and real-time feedback and control, among other things, and thus creating "smart cities." Rifkin even describes it as the technological "soul-mate" of the collaborative commons: "The very purpose of the new technological platform is to encourage a sharing culture," he suggests.[73] The impacts of such technologies on privacy may make them less effective at encouraging sharing among peers, but it will clearly enable more effective commercial sharing and collaboration between organizations.

Location services like Foursquare are another key digital technology, which alongside mobile payments (PayPal) and social media (Facebook) for confirmation of identity, are transforming collaborative consumption, particularly as smartphone ownership and use rockets.[74] The cellphone company Verizon's new "Auto Share" app offers generic verification and un/locking services. In its initial incarnation it enables users to unlock a rented or shared car with a smartphone. The app locates a vehicle nearby using a global positioning system (GPS) function. The smartphone then functions as a scanner to read the car's quick response (QR) code, which is matched remotely in Verizon cloud servers with a password-protected identification for the driver, automatically unlocking the car and allowing the user to drive.[75] Verizon aims to extend this model to other items that can be fitted with an electronic lock—such as apartments or even laptops or power drills, enabling any sharing startup to "rely on Verizon's infrastructure of ubiquitous connectivity and geolocational tracking to match supply and demand."[76]

The explosion of sharing platforms has been facilitated by an astonishing drop in the costs of launching an Internet venture—which fell by a factor of 100 in the decade from 1997 because of the development of open source tools, cloud computing, and virtual office infrastructure.[77]

Moreover, social networking technology is helping rebuild trust between strangers despite the legitimate concerns over privacy raised by the potential for the same systems to be used for surveillance. The business analysts

Michael Olson and Andrew Connor see Facebook's application programming interface (API) as

instrumental in laying a foundation for trust to grow on the Internet, because it began to remove the anonymity associated with online user identities. Under this new paradigm, peer-to-peer market participants were no longer generic usernames, they were unique individuals with cover photos and a list of friends.[78]

Facebook Connect, which allows users to log in to third-party websites using Facebook made life even simpler for sharing platform users. Moreover, say Olson and Connor, "Sharing economy companies encourage benevolent so-called 'Facebook stalking' because learning about your prospective host or guest (... seeing that they are 'normal' like you) builds a sense of trust among all sharing economy participants."[79]

However, there are downsides in such forced transparency (as we discuss in the section "Civil Liberties" in chapter 4), and in the ubiquity of powerful commercial intermediaries, especially those that seek to banish any need for face-to-face contact in sharing interactions. The digital economy critic Evgeny Morozov argues that intermediaries such as Facebook and Verizon will come to dominate the sharing economy, locking it into a commercial mediated mode.[80]

3. Pressing Unresolved Environmental Concerns

Although the environmental benefits of sharing typically take second place to motivations such as saving money and doing good for the local community,[81] awareness of environmental problems is high in the contemporary world, with Millennials seeing the need for across-the-board rather than the typically reductionist, "focused" approaches. David Weinberger of the Roosevelt Institute's Campus Network (RICN) argues:

Millennials view environmental protection more as a value to be incorporated into all policymaking than as its own, isolated discipline. We are concerned with economic growth, job creation, enhancing public health, bolstering educational achievement, and national security and diplomacy. Young people recognize that each of these concerns is inextricably tied to the environment.[82]

Perhaps partly as a result of this, many of the collaborative and sharing ventures founded by Millennials have integral social purposes. Botsman and Rogers cite Kickstarter, Profounder, Meetup and Wordpress as the products of such young entrepreneurs.[83]

The Millennial generation also takes environmental factors into account as consumers: 69 percent consider social and environmental factors in choosing where to shop, and 83 percent trust companies more that are environmentally or socially responsible.[84] For them as for many people, sharing is understood as a simple positive action, like recycling, which can help tackle climate change and resource depletion. In our cities the immediate pressures of environmental degradation remain visible—particularly in the form of traffic congestion and pollution, making it unsurprising that so many recent sharing initiatives have focused on cars and bikes.

4. A Global Recession That Has Fundamentally Shocked Consumer Behaviors

Olson and Connor suggest that recession helped trigger the takeoff of the sharing economy as consumers delayed "big-ticket" purchases like cars. They argue:

> Although the sharing economy was still in an embryonic stage in 2009, the psychology of the consumer was clearly undergoing a profound transformation. A new paradigm had emerged wherein consumers were owning less stuff and spending less, but still finding creative ways to travel, commute to work, host dinner parties and have memorable experiences.[85]

The financial crash of 2008 and subsequent economic recession brought to a close an era in which politicians could claim "an end to boom and bust." Many economists, such as Tim Jackson—a professor at the University of Surrey, UK—predict that a new era of high growth is inconceivable, as the economic system bumps up against limits to land availability, rising energy prices and climate change.[86] This fundamental uncertainty is reflected in consumer attitudes. The marketing analyst Nigel Piercy and his colleagues suggest we have entered an "'age of thrift' which has radically changed customer purchase behavior, [with] an environment dominated by public skepticism and lack of trust in business and in marketing offers."[87]

In this context, sharing offers an alternative response to the "age of austerity"—figuratively "giving the finger" to the exhortations of politicians to spend and get the consumer treadmill turning again.

Perhaps even more marked again is the behavior of the Millennials:

> A perfect storm of economic and demographic factors—from high gas prices, to reurbanization, to stagnating wages, to new technologies enabling a different kind of consumption—has fundamentally changed the game. ... The largest generation in American history might never spend as lavishly as its parents did, nor on the same things. Since the end of World War II, new cars and suburban houses have powered the world's largest economy and propelled our most impressive recoveries. Millennials may have lost interest in both.[88]

Together these four drivers are heading us to what Botsman calls "the big shift" toward collaborative consumption and sharing[89]—particularly, but not exclusively, in the forms we describe as "mediated." These drivers are not uniquely urban, but the dense populations and physical networks of our cities are critical in enabling sharing as a practical and efficient response to them. It is no coincidence that most sharing services and platforms have launched in specific cities, and are hopping from one city to another as they spread. Under the following headings we further explore some of these drivers and motivations.

Cutting Environmental Impacts

Potential environmental benefits, including resource efficiency and energy savings, are common to many sharing programs. Redistribution markets and product service systems both increase the value society extracts from products before they become waste and enhance the efficient use of inputs into the manufacturing system. Such waste reduction can cut environmental impacts at both ends of the product life, in resource extraction and in waste management. Sharing enhances the potential environmental efficiency benefits of cities while less energy is needed for transportation and production, and less waste is created as everyday products and services are shared among a group.[90] Sharing also might provide the opportunity for a community to make the most of underused land, or maximize use of other existing resources for the enjoyment of the greater public.

There is little detailed research into the environmental implications of sharing. But what does exist suggests the immediate benefits could be significant. The reduced environmental impacts of online sharing of digital music, compared with buying CDs, can be up to 80 percent saving in energy and carbon emissions assuming the purchaser does not burn a physical disc.[91] The benefits are not simply about virtualization. Postal delivery of shared products such as DVDs provided environmental gains over rental stores,[92] and real-world sharing can have substantial results, as shown by studies of redistribution platforms, carsharing, bikesharing and couchsurfing.

Botsman and Rogers report an estimate that 24,000 items are passed on through Freecycle every day (thereby diverting up to 700 tonnes (i.e., metric tons) of material from global waste streams, and averting the extraction or production of perhaps 14,000 to 24,000 tonnes of new raw materials that would be required to replace those products with new ones).[93]

On average, each active Swiss carshare user (with "active" meaning a person who uses a carshare at least once a year) reduces carbon dioxide emissions by 0.29 tonnes each year by carsharing.[94] In the US the figures are (unsurprisingly) even higher. The researchers Elliot Martin and Susan Shaheen surveyed users of a wide range of North American carsharing programs, finding an observed average reduction of over 0.5 tons of carbon dioxide equivalent per household per year and a 44 percent cut in vehicle miles traveled.[95] By including an estimated figure for avoided emissions in situations where joining the program substituted for buying an additional vehicle, the average saving increased to 0.84 tons. About 60 percent of households joining carsharing programs were car-less, and about half of the previously car-owning households became car-less as a result of joining the program. Average carshare vehicles were about 40 percent more fuel-efficient than the vehicles shed by households. On the other hand, about 30 percent of households increased emissions as a result of joining carsharing programs. Although this reduced the *average* climate benefits, it implies a significant improvement in well-being and capability for those households who had no access to a car before joining the program.

Shaheen and her colleagues at the University of California report climate benefits from bikeshare programs, too.[96] Vélib users in Paris cover an estimated 312,000 kilometers per day (equivalent to a saving of 57 tonnes of carbon dioxide from car mileage). Hangzhou Public Bicycle Program users covered an estimated 1.032 million kilometers per day (equivalent to 190 tonnes of carbon dioxide). Only some trips replace private vehicles, with others substituting for walking, public transit, and taxi use, with the proportion varying widely depending on the city. In Denver, B-Cycle claims that 41 percent of bikeshare trips replaced car trips,[97] while a survey of SmartBike (Washington, DC) members found 16 percent of bikesharing trips replacing personal vehicles.[98] Moreover, as the National League of Cities point out, bikesharing also helps reduce congestion and fuel wastage.[99]

Airbnb commissioned the Cleantech Group to assess its environmental impact. They estimated that Airbnb guests use 78 percent less energy than hotel guests and also reduce water consumption.[100] The survey indicated that Airbnb hosts and guests are relatively environmentally aware. Most hosts cut packaging waste by not providing single-use toiletries, and in Europe, 89 percent provide recycling options. Compared to hotel users, Airbnb guests say they are 10 to 15 percent more likely to use public transportation, walk, or bicycle as their primary mode of transport.

Good land sharing offers major environmental benefits as well. Increased density is a land use and environmental aspect of sharing: the greater density, the greater level of sharing. In British Columbia, Canada, research has also shown the criticality of building density for carbon saving. Evaluating data produced by more than 150 of the province's municipalities, the researchers conclude:

By focusing on transportation and increasing density, any community can embark on a low-carbon pathway irrespective of its wealth and population size. ... This finding is particularly important as it suggests that the potentially intransigent factors of income and location need not be barriers to achieving significant GHG [greenhouse gas] reductions.[101]

David Owen, the author of *Green Metropolis*, critiques *Forbes* magazine's 2007 choice of Vermont as the US's "Greenest State" on a similar basis, noting:

Forbes's ranking was unfortunate, because Vermont, in many important ways, sets a poor environmental example. Spreading people thinly across the countryside, Vermont-style, may make them look and feel green, but it actually increases the damage they do to the environment while also making that damage harder to see and to address.[102]

This he contrasts with New York City:

The average city resident consumes only about a quarter as much gasoline as the average Vermonter —and the average Manhattan resident consumes even less, just 90 gallons a year, a rate that the rest of the country hasn't matched since the mid-1920s. New Yorkers also consume far less electricity—about 4,700 kilowatt hours per household per year, compared with roughly 7,100 kilowatt hours in Vermont.[103]

Higher density makes public transit work, too. Metropolitan New York alone "accounts for almost a third of all the public-transit passenger miles traveled in the United States."[104] Density (and shared living)

lowers energy and water use in all categories, ... limits the consumption of all kinds of goods, reduces ownership of wasteful appliances, decreases the generation of solid waste, and forces most residents to live in some of the world's most inherently energy-efficient residential structures: apartment buildings. As a result, New Yorkers have the smallest carbon footprints in the United States ... less than 30 percent of the national average."[105]

Shared production (explored further in chapter 2) also offers environmental benefits. Repair Cafés, now found in 400 locations in 20 countries around the world, are already extending product lifespans, by providing access to equipment and advice to help users repair items they would otherwise have thrown away.[106] Started in the Netherlands in 2009, they are promoted on a nonprofit basis by a charitable foundation. Fab-labs with 3-D printers could also enhance product durability, by allowing the production of a wide range of spare parts without requiring large stocks to be held; 3-D printers can also reduce waste in production, especially where machined parts, conventionally cut out from solid material, can be printed instead. One study of titanium aircraft parts found a 90 percent reduction in direct material use, and reduced the weight of the finished component by 60 percent with consequent energy savings in use.[107]

However, efficiency benefits do not necessarily translate into absolute gains for the environment.[108] Rebound effects—like flying more because Airbnb makes it cheaper to stay in other cities[109]—are part of the normal economic response to efficiency improvements. The biggest environmental benefits will only arrive if sharing changes cultural norms about consumption, encouraging people to reject the rat race.

Reducing consumption of energy and materials is critical to protecting ecosystems and healthy environments for people. Yet for many in the global South, and some in the global North, levels of material consumption remain too low to underpin basic capabilities. Arguably a global convergence of consumption levels is essential, with significant reductions in richer countries but increases among poorer communities. Such a shift demands not only technocratic policy change but also cultural change, since in modern consumerist cultures high consumption is central to many people's identities. Insofar as sharing can shift such identities it will help with this problem. We return to this challenge in the "Putting the 'Citizen' Back in the City" section of chapter 3.

While sharing system participants consistently cite environmental benefits as an important reason for sharing, they are rarely the main driver for behavior change. That is found rather in the intersection of economic benefits and changing cultural norms, which we examine next.

Changing Norms: From Online Sharing to Social Transformation

In seeking to explain the contemporary increase in (mediated) sharing behavior, researchers have identified several factors, often highlighting the ways in which the costs of sharing have been reduced and the ease and potential benefits increased by technology.[110] Yochai Benkler, for instance, suggests that: "The industrial economy shunted sharing to its peripheries - to households in the advanced economies, and to the global economic peripheries. ... The emerging restructuring of capital investment in digital networks ... is at least partly reversing that effect."[111]

We consider some of the economic logics of sharing later in this chapter, but here focus on the sociocultural effects of technological change, and the spread of new norms as a result of sharing online. Surveys typically suggest that sharers are more comfortable sharing in the material world because they have been involved in sharing online. For instance, 78 percent of participants in the Latitude survey "felt their online interactions with people have made them more open to the idea of sharing with strangers, suggesting that the social media revolution has broken down trust barriers."[112]

Online crowdsourcing is seen by Botsman and Rogers as a progenitor of collaborative norms. They suggest that such "online networks bring people together again, making them more willing to leverage ... power in numbers ... [and] spilling off-line, creating change within our cultural, economic, political and consumer worlds."[113] Platforms like Meetup—using the web to stimulate physical get-togethers of people with common interests, and IfWeRanTheWorld—an online platform for volunteering to assist with shared real-world tasks, both reflect and facilitate this process. But the bigger picture is best seen in the significance of online sharing—particularly of digital media, which accounts for over half of all Internet traffic. The vast majority of Internet users have participated in filesharing in some form.[114] Yet for more than a decade filesharing has been highly contested, with copyright owners claiming that much sharing is in breach of copyright and therefore illegal, and most governments making some effort to restrict illegal filesharing. Pioneer enterprises in this space, such as Napster and Kazaa, have been closed down, but the practice—using distributed computer capacity—is still widespread. Usage of one high-profile filesharing site—Pirate Bay—has apparently doubled since governments sought to block access in 2011.[115] As Don Tapscott and Anthony Williams suggest, such trends indicate that young people who have grown up with the Internet are "renegotiating the definitions of copyright and intellectual property."[116]

The persistence of illegal filesharing—especially among Millennials—suggests an irreversible shift in norms, perhaps reflecting growing consumer awareness of the collapsing marginal costs of such goods and services,[117] and a desire for business models that more fairly share the remaining capital and overhead costs of production. In this context the music industry's largely futile legal pursuit of filesharers, even with the assistance of draconian laws such as the UK's Digital Economy Act, say Tapscott and Williams, is stifling creativity and hampering innovation.[118] It also threatens what are increasingly seen as basic rights to access to the Internet and to have a secure online identity.

But online sharing was not and is not just about "things." Nicholas Joh from the Department of Communication and Journalism at the Hebrew University of Jerusalem, documents how the term "sharing" has spread with the development of social networking sites such as Facebook.[119] These sites urge us to "share" our photographs, thoughts, and feelings—potentially publicly, and certainly with what is for many a loose network of friends and acquaintances. This is, as John points out, a "therapeutic" use of the term "sharing," as used widely in self-help groups. Rifkin associates the spread of groups such as Alcoholics Anonymous in the latter half of the last century with a qualitative shift in human empathy.[120] This shift helped the emergence of what he describes as "psychological consciousness" in which individuals in an interconnected but alienating global economy could make empathic connections with one another. Facebook founder Mark Zuckerberg would appear to agree, claiming that "people sharing more—even if just with their close friends or families—creates a more open culture and leads to a better understanding of the lives and perspectives of others."[121]

Such therapeutic sharing online—and the emotional openness it implies—has helped shift the boundary between public and private, facilitating sharing of our cars, homes, and other personal spaces with strangers in the collaborative economy.[122] In turn that shifting boundary potentially signals a cultural shift with respect to the meaning of public space, and the mutuality required for the public domain to function.

Filesharing has stimulated cultural as well as legal contestation. The narratives and discourses used by supporters and opponents of filesharing are telling. The resonance of the term "filesharing" is perhaps:

> Due to its association with the sharing of emotions and related assumptions about how people should relate to one another—with honesty, openness and mutuality. ... [It] is this aspect of sharing that gives the term "filesharing" the rhetorical power that so frightens the entertainment industry and government, and that makes the term such an important one.[123]

"Sharing" has a long history in computing, rooted in turn-based sharing of early processors and subsequently evolving to refer to computing power as a resource used and managed in common—a trend exponentially reinforced with the emergence of "the cloud." In this shift, John argues, "The context has shifted dramatically from one of scarcity of computing resources to one of abundance; ... [and] sharing has shifted from being a structural necessity to an ideology."[124]

Copyright holders talk not of "filesharing" but of "piracy." Yet this discourse, intended to divide filesharers from ordinary law-abiding citizens, may be counterproductive. Piracy is being appropriated as a proud badge of countercultural identity, as seen with the prominent filesharing site The Pirate Bay and the emergence of political Pirate Parties in many countries.[125] The author Kester Brewin argues eloquently that the term "piracy" has always been a signal that "something that should be held in common has been enclosed for private gain," applying as much to music and knowledge as it did to the wealth being seized from the New World by European monarchs in the sixteenth century.[126] He notes that the pirates of the Caribbean in that era were not the only ones "thieving"—they simply "refused to pass on what they stole to the King," and instead shared it fairly among their crew, typically adopting democratic structures on board—elected captains and powerful general assemblies—in direct contrast to the strictly enforced hierarchies of naval and commercial crews. Noting the extraordinary heterogeneity and democratic practices involved, David Graeber—a social anthropologist at Goldsmiths, London—suggests that such "spaces of intercultural improvisation … largely outside the purview of any states," on pirate ships and in American frontier communities of the same period, were a key source of modern democratic values.[127]

The continuing modern-day battle over the term "pirate" echoes the recapture of terms such as "queer" by the LGBT community, as "a means of in-group demarcation to bring members of the targeted group closer together and to remind members of the targeted group that they are, indeed, a targeted group.'"[128] In this sense illegal music downloads are not just an attempt to redefine consumerism but also a way to define self- and group identity for those involved, and indeed a form of countercultural collective resistance.[129] In this it is also a cyber-echo of the role of graffiti in urban youth culture—an open symbol of identity and resistance that helps construct our wider cultural milieu. The sustained cultural popularity of pirates in film and even children's party themes indicates how this counterculture has infected the mainstream.

It is helpful to understand sharing as a cultural discourse as well as a practical phenomenon. As a discourse, sharing is obtaining social power, rooted in the connections being made between sharing on the web; sharing in the sharing economy; sharing of public services, and sharing in emotional and psychological therapy. Sharing is therefore being consistently framed in terms of openness, honesty, empathy, and personal relations. In terms of the cultural impact of mainstreamed sharing practice, it is the revival of these norms that makes both the sharing paradigm so powerful, and the vision of the sharing city so seductive.

This framing is, however, also a risk. For instance, in the digital domain, where information about and access to people is so valuable for advertising, abuse by social networks such as Facebook is a real concern. John argues that "Facebook would appear to be using the word and exploiting our generally positive feelings toward sharing in order to mystify its business agenda," allowing us to be converted into the product Facebook sells to advertisers.[130] The psychotherapist Aaron Balick suggests a pernicious psychological effect is also triggered by social media in that it turns social recognition into a commodity.[131] When we share and post, we are psychologically seeking recognition through "likes" and comments, but this—according to Balick—hollows out our inward-looking aspects, leaving them unfulfilled and unrecognized.

Tapscott and Williams focus on loss of privacy:

> Online or off, our digital footprints are being gathered together, bit by bit ... into personas and profiles and avatars—virtual representations of us. ... This digital shadow is used to provide us with extraordinary new services, new conveniences, new efficiencies. ... But there are also great risks when very little about our lives is truly private.[132]

Our "data exhausts" turn into profiles held both by governments and by corporations. Neither group has adequate safeguards to prevent inappropriate use, snooping, or resale of data for annoying and dubious purposes. Moreover:

> Novel risks and threats are emerging from this digital cornucopia. Identity fraud and theft ... [and] new forms of discrimination. ... Personal information, be it biographical, biological, genealogical, historical, transactional, locational, relational, computational, vocational, or reputational, is the stuff that makes up our modern identity. It must be managed responsibly.[133]

Existing privacy and data protection laws were not designed for a world in which we voluntarily publish digital data on social networks or sharing platforms. Nor for a world with an exponentially growing "Internet of things," recording, monitoring, and sharing data. Nor indeed for a world where data anonymization merely leaves blank spaces that can be filled back in by cross analysis of overlapping data sets.[134] Moreover, the idea of informed consent is complicated by new unexpected uses for data. Some advocates of the sharing economy argue that alongside these technological changes, norms of privacy are shifting toward greater openness and transparency. Rifkin questions whether future generations will even value privacy as it is understood today.[135] We see, rather, an urgent need for new standards of transparency (for organizations) and privacy (for individuals). Tapscott and Williams identify some emerging principles, starting with the basic tenet, "Personal information belongs to the individual."[136] People must be able to find out what data an organization holds about them, to control its use, and to be free to transfer that data to another platform or service provider. (We return to questions about privacy and Internet freedoms in chapter 4.)

More widely, the framing of sharing in terms of prosocial norms potentially makes those norms vulnerable to abuse by so-called sharing companies that are driven instead by the logic of corporate competition and asset sweating. Of course such companies can be expected to win less loyalty and face reputational challenges in the long term—but cost leaders in other sectors have shown that a customer base can be maintained by ruthless cost cutting and exploitation of an effectively captive labor force. Unlike the sharing economy per se, the sharing paradigm challenges both sides of that equation—it can help liberate people from wage-slavery in the labor market, and makes users more aware that behind the product or service there are people like them, with shared values and aspirations. As Cameron Tonkinwise, Director of Design Studies at Carnegie Mellon University, argues, unlike commercial interactions, sharing interactions can be simultaneously social and "awkwardly" economic. We meet "strangers, who … cannot hide behind their roles but are encountered as people."[137]

We hypothesize here, therefore, that with the spread of the sharing paradigm—both within the formal economy and outside it—public tolerance for abuses of human rights, environmental standards (and so forth), in the interest of cheap consumer products, should decline. In other words the norms seeded by online sharing have the potential not only to transform sharing behavior in the material world but also spread new norms into wider economic, sociocultural and political domains.

A Healthier Economy?

We turn now specifically to the economics of sharing. We review first some of the arguments as to why sharing might be seen as economically beneficial and attractive to city leaders, and then delve into the economics of different sharing approaches and resources. By exploring this territory we hope to illuminate the drivers behind sharing, seek insights into whether sharing might decline again in economic recovery, and highlight some of the risks that might make a purely economic approach to sharing counterproductive (as we discuss later in this chapter).

Sharing can deliver economic benefits for cities on at least three levels. First, for individuals and enterprises there can be cost savings or even earnings from shared resources. Surveys consistently show that opportunities to save and make money can motivate sharing.[138] For instance, if someone only needs a car to commute to work, carpooling offers large savings. Airbnb allows householders to make money from a spare room. Credit unions and online funding platforms such as Kickstarter allow individuals to fund new enterprises. At a superficial level it is easy to see why such sharing approaches might flourish in a time of economic insecurity.

Second, at the city scale, sharing can similarly reduce operating or service-provision costs. We discuss this further in chapter 2, but the potential is illustrated by the London borough of Croydon, which cut staff car travel costs by over 40 percent (and cut miles traveled and carbon emissions, too) by replacing fleet vehicles with a Zipcar partnership.[139]

And third, there are also potential system-level benefits for creativity and innovation, as long as resistance from incumbent businesses can be overcome. Open innovation and development processes make firms more effective and competitive,[140] while cluster theory shows how innovation and conventional economic growth can be generated by the interaction and sharing of skilled staff and research resources.[141]

In San Francisco, the success of Silicon Valley is attributed to the clustering of related firms, venture capital, and research institutions in a single locality with effective means for the swift transformation of research into innovative products. This involved the cyclical movement of staff between research and commercial organizations with permeable boundaries and collective learning through multiple connections between funders, researchers, and developers. In effect knowledge, skills, computing power, and even people were being intensively shared in what might be called a "digital commons."[142] More generally researchers suggest that the smartest, most innovative cities are those that attract key "talent" in the knowledge and cultural industries. In turn this is an argument for attractive, living cities that welcome diversity. Studies suggest, for example, that the cities friendliest to gays are also the most innovative, while places in which women are best supported with quality childcare services are naturally those where women can make the greatest contributions to the economy.[143]

The success of key "clusters," however, is not just about the right people but also about how synergies are created from their knowledge and skills by sharing those within a local innovation ecosystem, and building effectively on public sector investments in research and infrastructure.[144] In Seoul, for example, the city has invested heavily in infrastructure and facilities in its Digital Media City, a district in northwest Seoul to attract high-tech business.

In the UK, the rapid growth of science-based industry around Cambridge has made the so-called Silicon Fen, with more than 1,600 businesses, one of the most successful technology business clusters in the world. Many of these businesses have connections with Cambridge University, developing products based in research done in university laboratories. The cluster effect generated by people mobilizing resources and finance happens "through personal connections made at informal levels, rather than being driven or managed by an overarching organization, structures or systems." In this "self-organizing, constructive chaos … who you know is as important as what you know."[145]

It might be thought that growing web connectivity is overtaking the importance of place in such processes. The Internet massively increases the density of possible connections through which innovation can happen and spread.[146] Online collaboration backed by crowdfunding might appear to offer the same chances as real-world clusters to share knowledge and skills to generate innovation. But in practice, it appears that the digital versions are better seen as a complement to conventional face-to-face collaboration. The researchers Roberta Capello and Alessandra Faggian report, for example, that collective learning in the form of cooperation with local suppliers and customers in an urban milieu is a key determinant of firms' innovation.[147] This interdependence between specific urban spaces and the virtual realm underpins development and innovation in other sectors, such as the financial industry, and also in other domains, such as that of activism and protest.[148] (We consider the intersection of urban- and cyberspace in chapter 3.)

Online platforms can open up innovation processes to new ideas and new approaches, but only where a culture of collaborative sharing and learning exists. Open source software works as a model for innovation not only because of the online platforms for collaboration, but because it is not tied down by commercial secrecy and artificial barriers to the sharing of knowledge and resources on those platforms. Cities that create and strengthen a culture of sharing can thus also strengthen their potential for positive economic innovation. Moreover, open sharing of green technology and know-how to minimize duplication of effort and maximize uptake is a critical step in tackling sustainability challenges. Companies need to share innovative practices, technology, and the underlying intellectual property.[149] The Yale professor Dan Kahan similarly sees conventional intellectual property rules as problematic, and sharing as preferable. He says:

> The deadweight losses and administrative costs inevitably associated with intellectual property rights needn't be endured to secure the public benefits of invention. Indeed, university scientists, computer hackers, and other reciprocal producers tend to suspend the free exchange of ideas once they come to suspect that those with whom they are collaborating are intent on appropriating the commercial value of those innovations for themselves.[150]

Conventional intellectual property rules arguably crowd out other incentives, such as social benefit, curiosity, and even scientific prestige, even though the monetary incentives are weak in most circumstances. They also impede diffusion of knowledge, despite the benefits of the Internet.[151]

This section has outlined how sharing today reflects not just technological change, but also an evolution of norms, originating in cyberspace—norms that increase the strength and diversity of the "weak ties" that link modern society. We saw that changing norms about sharing have potential to bring environmental and social as well as economic benefits. But realizing these benefits is not simple. It requires attention to the particularities of place, and a focus that goes beyond the economic.

Economics Isn't Everything

Sharing clearly offers wide-ranging economic benefits, with potentially important consequences for society and the environment.[152] But the current framing of sharing as synonymous with a "sharing economy" has also unhelpfully focused analysis on economic explanations and opportunities of sharing behaviors, and risks succumbing to a form of technological determinism. Such explanations tend to focus on the characteristics of the goods concerned or the platform technologies, rather than the values of the people involved, or the broader sociocultural potential of sharing.

Nonetheless economic factors can help us understand trends in different flavors of sharing. So here we briefly consider issues of spare capacity and relative cost ("lumpiness"), how provider and user can be brought together ("coordination"), the extent to which use of the good depletes its quality ("rivalrousness"), and the extent to which others can be excluded from the benefits of the good ("exclusivity"). These can help us understand the emergence and modalities of sharing, but do tend to sideline any values-led motivations.

Lumpiness

Coffee is not a lumpy good. We can buy it by the ton, or the gram, or even in single servings. But power drills are lumpy—even if we just need one hole, we have to get a whole drill. Bikes and computers too fit into this category. This means they typically have lots of useful spare or "idle" capacity. Cars not only have spare capacity when standing in the driveway, but typically also when in use (spare seats or luggage space). It is said that 80 percent of products in the UK and US are used less frequently than once a month.[153] It is not only moveable products that can have spare capacity, but also land and buildings, and, arguably, intangibles like time and skills. As we shall see a range of initiatives from garden sharing to skill sharing take advantage of this factor.

Utilizing such spare capacity by sharing (in whatever form) increases efficiency. As Benkler explains, even a simple book

> has much more capacity ...—[for] communicating its content—than a single non-obsessive individual can consume. This overcapacity is the source of the secondhand book market (market provisioning), the public library (state provisioning), and the widespread practice of lending books to friends (social provisioning).[154]

Lumpy goods like cars or buildings may also be relatively expensive, and thus people may choose to collectively share the cost of owning a particular item or resource (a vacation home for example).[155] This implies that increased wealth per se might reduce sharing,[156] but it also suggests that sharing will increase if the costs of coordinating the sharing process fall.[157]

Where goods are expensive and lumpy, sharing can increase the choice available to the end consumer and offer greater flexibility at lower cost and greater convenience. The vast majority of the hundreds of billions of dollars spent on marketing have encouraged a view that ownership is preferable to sharing or rental. But the example of car sharing demonstrates otherwise. While a car owner, having paid all the overheads on the vehicle, is effectively limited to using that one vehicle, a carsharer might use a micro-mini (compact) around town one day, and a family estate (wagon) for a trip to the country the next. And as car sharing has become more convenient online, the inconveniences of car ownership have remained the same. An owner still has to deal with registration, maintenance, parking, insurance, and repair. A sharer simply has to worry whether the previous user returned the car on time and in working order.

Sharing products like films and music, cars, or fashionable clothes and accessories widens choice and allows us to more rapidly change our image and status compared with ownership. Bruce Jeffreys of GoGet cars, an Australian carsharing company, lightheartedly contrasts the opportunity of "consumer philandering" with the monogamy of ownership, and the journalist Paul Boutin suggests that "carsharing turns members into automotive swingers, free from having to commit to one model."[158] The particular imagery of these commentators may not motivate everyone, but the central concept is clear. In this respect sharing may be an ideal postmodern model,[159] allowing people to change their image and identity swiftly to keep up with the rate of change in contemporary "liquid life."[160] Elsewhere we highlight the psychological damage done by the acceleration and uncertainty of liquid life, yet if sharing allows people to take more control over their "identity projects" in such circumstances, then it is a very real benefit.

Coordination and Transaction Costs

Problems of coordination can be a wider deterrent to sharing. People choose ownership as a means of retaining control and ensuring access, fearing that a shared good might not be available when they want it.[161] More generally the matching of spare capacity with potential users—especially in real time, for goods such as shared rides—is no simple matter.

Benkler suggests that it was the high transaction costs of coordinating sharing—along with serious information asymmetries, especially about the reputation of potential borrowers—that historically made social sharing among friends and family preferable to secondary rental markets as conventional means of redistributing excess capacity.[162] However, today both transaction costs and information asymmetries are being dramatically reduced by the Internet and the mediated sharing platforms utilizing it. Web-based sharing also grows the pool of potential sharers, making it much more likely that users will be able to find what they want, when they want it. By actively recruiting a diversity of users with complementary demands, rather than only similar users with potentially competing demands, managers or marketers of sharing systems can further facilitate effective coordination.

Before the Internet, the huge transaction costs of communication and coordination also deterred other forms of collaboration. As the technology journalist Clive Thompson notes, "For centuries, people collaborated massively only on tasks that would make enough money to afford those costs."[163] Even businesses only became genuinely multinational in the Internet age. But today, collaboration across space and boundaries is much easier. This is disrupting conventional capitalism, diminishing some of the very reasons that firms exist. (The "transaction costs" theory of the firm, suggests that firms exist because the costs of coordinating independent agents in a production process are so much higher than the costs of managing and directing a group of employees within an organization.) Internet enabled collaboration enables easier coordination, and generates a collective intelligence, which is being exploited by civil society as well as big businesses.

The scale of the system affects coordination costs, too. "Critical mass" creates wide choice in sharing, which allows sharing platforms to aggregate risk across multiple interactions,[164] and also serves as a source of "social proof."[165] Social proofing is the norm-building process in which people see a new behavior demonstrated by their peers and thus feel more confidence to try it. Behavior copying is how cultural habits change.[166] The Internet has reduced the transaction costs of mediated sharing dramatically, enabling critical mass, and particularly enabling selective sharing with strangers (in both nonprofit communal as well as commercial systems), through the establishment and validation of "online reputation." As a result, sharing platforms face lower costs than conventional market alternatives, and "value can be redistributed across the supply chain ... because producers' costs are lower."[167] These benefits are being exploited by both peer-to-peer (P2P) and business-to-peer (B2P) models. These differ primarily in terms of who owns the shared resource: in P2P models users themselves own the shared items and the intermediary platform is merely a broker; in B2P the central intermediary also owns the items. Juliet Schor, a professor of sociology at Boston College, highlights the significance of this for the division of profits or benefits: "With a P2P structure, as long as there is competition, the 'peers' (both providers and consumers) should be able to capture a higher fraction of value. Of course, when there is little competition, the platform can extract rents, or excess profits, regardless."[168]

The same factors have also reduced the costs of sociocultural sharing in comparison with exclusive ownership. Where sharing is sociocultural, practical coordination challenges—such as matching schedules—need to be able to be overcome without disproportionate effort such as holding meetings, and also without surrendering autonomy to some kind of bureaucracy or leadership. Here too, digital coordination using simple and increasingly flexible web and mobile applications typically fit the bill.

Rivalrousness, Exclusivity, and Selectivity

The marketing researchers Cait Lamberton and Randall Rose offer a typology of sharing (table 1.1) which distinguishes the "goods" being shared by the extent to which they are rivalrous (where one person's consumption prevents another from consuming the same good), and/or exclusive (where use is limited to a particular group by some other mechanism).[169]

This helps us recognize different sharing models for different types of good or resource. Most of the sharing examples in this book are of lower exclusivity. In other words, anyone can access them, either freely or by joining a relevant club or organization.

In the lower left hand quadrant of table 1.1 we see what Lamberton and Rose call "open commercial goods"[170] and Botsman and Rogers describe as product service systems and redistribution markets.[171] These encompass a

Table 1.1
Shared Goods Typology

	Lower exclusivity	Higher exclusivity
Lower rivalry	Public goods e.g., public parks, open source software	Club access e.g., country clubs, gated communities
Higher rivalry	Open "commercial"[a] e.g., tool-banks, Freecycle, carsharing	Closed "commercial" e.g., HMOs, frequent-flyer mile sharing programs

Source: Based on Lamberton and Rose, (2012, see note 161, this chapter).
a. Lamberton and Rose call these "open commercial goods," but the category encompasses much not-for-profit sharing.

wide range of models for sharing different resources "in turns," both commercial and communal. Products with high idling capacity, limited use because of fashion or only a temporary need (baby equipment), or diminished value or appeal after usage (a film or book), as well as those that face high startup costs (solar panels), require frequent updating (software), or are expensive to repair or maintain, are all suitable for product service systems.[172] Public libraries, tool banks, and car clubs all fit here.

In the higher left quadrant, we see public goods. Pure public goods—such as a malaria-free environment—are not diminished by use, and it is impossible to exclude others from them. Many goods share public-good characteristics to some degree, and they are not necessarily managed in the public sector. Open source software is as much a partial public good as public parks. More generally, the simple possibility of sharing something tends to indicate: first, that it shares some of the characteristics of public goods; second, that markets are failing to provide it effectively; and third, that we can possibly deliver positive side effects, such as stronger social capital or reduced environmental pollution, by enabling sharing.

This typology is helpful in another way too. It reminds us that the same goods can be managed in different ways depending on cultural, regulatory, and market design factors. For example health services may be shared as public services, or through private insurance, not-for-profit clubs, or exclusive commercial health management organizations. In other words it is less the nature of the good or service that determines the extent and nature of sharing, but the ideology, culture, and policy measures that surround it.

A key factor in assessing the social value of sharing economy approaches is the issue of what the shared good replaces. Things like "a space in a carpool" only become valuable economic goods at all when, *within* the sharing paradigm, they embody the shift to a service rather than a product orientation. In other words, sharing creates value and well-being in spaces previously unreachable by markets. However, as the sharing paradigm has spread and the technologies underpinning sharing platforms have developed, it has become clear that such services can be shared commercially as well as for nonfinancial motives. Ridesharing services now commodify spare seats on car journeys, for instance. In a similar way, the Hong Kong–based startup Slicify is adapting the model of SETI@home—in which PC processing capacity is donated to facilitate research goals—and commodifying that idle or unused computing power to host cloud-based applications for individuals or businesses.[173]

In these respects, contemporary commercially mediated sharing platforms are intruding into what was previously primarily a gift economy. The net social benefit will depend not only on whether they are increasing the general supply of shared goods, but on what they are competing with, and to what extent they are eroding or extending the values of honesty, openness, and mutuality the sharing paradigm seeks to embody. If Lyft and Uber genuinely compete with car ownership and Slicify with commercial cloud computing, then they potentially spread sharing values into new groups and spaces. If instead they simply compete with preexisting gift economies, removing car seats from informal car-pooling, and computer capacity from semicharitable goals, then we should be concerned. So far the evidence seems to be that sharing models are growing rapidly and at least in part eroding commercial competitive markets—as evidenced by the cries of opposition and special pleading from incumbents all over the world.

Understanding Economics Is Not Enough

Such analysis of sharing offers some useful insights, yet underlying it is an economic calculus, an assumption of *homo economicus* that can be unhelpful in considering sharing. Benkler even claims that "sharing of material shareable goods and peer production of software, information, and cultural goods more generally ... resemble an ideal market in their social characteristics, but with social cues and motivations replacing prices as a means to generate information and motivate action."[174]

But trying to promote sharing as an "ideal market" economy is more likely to be counterproductive, encouraging people to engage in it in a purely self-serving fashion. Moreover, to seek to include psychological and cultural motivations in such a calculative approach does not reflect the way those motivations affect us in real life, nor how our evolved nature as reciprocal beings expresses itself in our interactions with fellow humans. As

Kahan notes, "In collective action settings, individuals adopt not a materially calculating posture but rather ... a richer, more emotionally nuanced reciprocal one. When they perceive that others are behaving cooperatively, individuals are moved by honor, altruism, and like dispositions."[175]

Similarly, Graeber highlights that gift economies work on the basis of social and cultural values, not economic ones. In pre-money exchange systems, even "on those rare occasions when strangers did meet explicitly in order to exchange goods, they [were] rarely thinking exclusively about the value of the goods," as such meetings also typically involved cultural exchanges, extending even to the exchange of sexual favors.[176] To Graeber then, *homo economicus* is "an almost impossibly boring person—basically, a monomaniacal sociopath who can wander through an orgy thinking only about marginal rates of return."[177]

Economic analyses may help us understand how sociocultural sharing behaviors evolved, but they overlook the power of evolved behaviors and values—unconsciously and consciously—to shape habits and group norms today. Moreover, if we were to focus primarily on such questions in the design of policy for sharing, we would then be left needing to "retrofit" justice and inclusion. In part, that's because citizens' opportunities to "profit" from the sharing economy are severely impaired by existing inequalities in conventional wealth: you can't rent out a spare room if you are already sleeping four to a room (or don't own your own home), you can't rent out your car if all you have is a beat up jalopy, and you can't even join in on Amazon Turk or Crowdflower for subminimum wages without good web access.

Incumbents Under Threat

Economic analysis can also help us understand how incumbent businesses might react to resist or co-opt sharing. Internet-enabled sharing approaches could displace conventional commercial activities in many markets, in a form of disruptive innovation. Disruptive innovation—by definition—changes the nature of the market concerned. It also often does so starting from what might appear objectively to be a lower quality base. Digital photography, for example, took many years to exceed the print quality of film, but it redefined the product as sharable pictures, viewed on screens, and only rarely printed. Digital cameras allowed users to take many more shots before choosing which to keep or print. These new features came to define the market. Compared to hotels, couchsurfing might also appear lower quality, in terms of comfort and service for example. But it too redefines the product features, making personal contact and variety key differentiating factors.

Couchsurfing may not come to disrupt the hospitality industry as sweepingly as digital photography has replaced film, but the reactions of incumbents to previous disruptive innovations can help us understand what is happening in markets where sharing approaches are growing.

Stephen Sinofsky of Harvard Business School explains that early responses are typically to ignore or deny the threat, especially if the new offering appears inferior.[178] Even as new entrants invest more money and effort developing the new offering in ways that enhance its unique features, encouraging early adopters to buy in, incumbents can still easily miss the threat.

Then comes a phase in which the new entrants start to surpass the incumbents, not only on new features but also on conventional features, and incumbents are forced to respond, through acquisition, strategic partnering, or innovation, which leads to convergence in the products and services offered. In the carsharing sector, car rental companies were well placed to respond, and we have already seen Avis buy Zipcar, Enterprise bag a series of mainly nonprofit carshare operations in Chicago, Philadelphia, New York, Boston, and San Francisco, while Hertz started its own carsharing service in 2008 and then bought Eileo, a carsharing service based in Paris.[179] Car manufacturers have also got in on the act. Peugeot now provides its Mu "mobility solution" which provides rental access to a selection of vehicles, scooters, and bicycles from Peugeot dealers in 14 European cities.[180]

Online travel agencies have also been relatively active in responding to couchsurfing platforms. For instance online travel agency TripAdvisor acquired FlipKey, a leading online platform for the house-swapping / vacation rental marketplace in 2008, and Expedia has developed a partnership with HomeAway which has begun listing HomeAway properties on Expedia's sites.[181]

In finance, responses have been more muted. Some financial institutions sought to rehabilitate their brands by association with popular sharing programs, such as Barclays Bank sponsoring London's "Boris Bikes" until 2015, when Santander took over the deal. Santander has also initiated a partnership with the peer-lending platform Funding Circle. The bank will refer small-business customers looking for a loan to Funding Circle, in return for promotion of its banking services on Funding Circle's website.[182] Such approaches do little to stimulate learning or share skills. On the other hand, some Internet giants have made highly strategic acquisitions to extend their competencies, particularly in the online infrastructure of the sharing economy. For instance, eBay bought Braintree, a mobile payments company, in 2013. Google has also targeted opportunities to profit from its development of self-driving cars—it is heavily invested in Uber through Google Ventures, and in 2013 acquired Waze, a community-based traffic and navigation app on which drivers share real-time traffic and road information.[183] In other sectors, acquisitions and strategic investments appear to have been scarcer—perhaps reflecting a wider gap between corporate cultures and strategies in the conventional economy and the emerging sharing alternatives. The same gap probably explains why we have not seen incumbents launching competing sharing platforms, which Owyang suggests would be a more sophisticated response to disruption.[184]

Such conventional, market-oriented analyses tend however, to miss the potential for deliberate, negative responses: forms of resistance intentionally designed to hamper the emergence of new competitors. By their nature sharing approaches may be less vulnerable to the exercise of raw commercial power in which established companies can undercut competitors' prices, squeeze them out of prime advertising channels, or push their products off supermarket shelves, but they are perhaps more vulnerable than more conventional business competitors to resistance through legal or regulatory channels.

So it is no surprise to see incumbent businesses or key stakeholders in incumbent models, such as organized labor, challenging sharing businesses in the courts, especially where those businesses are most directly threatening established markets—such as taxi operators and hoteliers. And the political influence of incumbents is reflected in the grudging response of many authorities to legal and regulatory obstacles to sharing. Too often cities are proving slow to respond, or are even actively supporting such obstacles in the interests of incumbent commercial operators whose business models are under threat from the growth of sharing.

For instance, in Spain, expansion of sharing in a time of austerity has heightened concerns about losses in tax payments in a growing "black" economy. The Federation of National Transport Bus Companies (Fenebús) has sought to challenge BlaBlaCar, alleging it is unlicensed and fails to pay tax. The Spanish transport authority Dirección General de Tráfico has claimed jurisdiction.[185] The Spanish employers' federation reportedly plans a claim for unfair competition before the National Commission on Financial Markets and Competition. The Hoteliers Association of Madrid is lobbying for regulation of Airbnb with licensing, taxation, and minimum stay requirements of seven days. And the state has already restricted crowdfunding—limiting both the maximum to be raised per project and the maximum permitted for individual investors.[186] Spanish supporters of sharing have described this as a "total war on collaborative consumption."[187]

As in San Francisco, car- and liftsharing operations have been challenged over insurance and liability in several US cities. Similarly, Airbnb's model was declared illegal by a judge in New York—as a breach of the city's regulations on hotels. The judge sought to make an example of one Airbnb "host," fining him $2,400.[188]

These issues are not merely a smokescreen put up by vested interests; there are legitimate concerns that some sharing businesses offer a way for investment interests to circumvent important social protections, such as labor standards or zoning laws designed to keep rental housing affordable. David Golumbia, author of *The Cultural Logic of Computation*, argues:

> The difference between renting one's apartment on Craigslist and Airbnb might seem small, but it's huge: the role of the intermediary converts the effort from an individual one to a corporate one that is all about extracting profit from resources that are not, currently, monetized enough, in the opinion of some venture capitalists.[189]

The cultural theorist Slavoj Žižek raises a parallel concern in his analysis of consumer trends. He argues:

> What we are witnessing today is the direct commodification of our experiences themselves: what we are buying on the market is fewer and fewer products (material objects) that we want to own, and more and more life experiences—experiences of sex, eating, communicating, cultural consumption, participating in a lifestyle.[190]

One commercial sector, venture fund investment, has responded much more positively than any other to the sharing economy. But the interest of venture capitalists is not all good news for the sharing sector. Venture capital investors have flocked to fund emerging high-tech sharing platforms. Google Ventures, Sequoia Capital, Floodgate Fund, and Greylock Partners were all early investors in sharing ventures.[191] Cash from venture funds such as Union Square Ventures, Marc Andreessen, Kleiner Perkins, and Shasta Ventures still substantially outweighs other sources of capital, probably because collaborative consumption models are still seen as risky as well as disruptive, although corporate investments are becoming more significant.[192] This has created tensions between the business models used by sharing platforms and the broader social aspirations of sharing economy participants. The hard-nosed culture of venture capital pushes sharing entrepreneurs to commercialize their business models, moving their social purposes into the background.

In recent analysis, Jeremiah Owyang identifies around 230 collaborative economy startups, funded to an average of $30 million each (excluding Airbnb, Lyft and Uber which together account for $4.3 billion or 37 percent of all funding raised by the sector. The venture funds and other investors providing this funding typically expect a 500–1000 percent return on investment over five to ten years. Even at the low end of the range this implies that, on aggregate, the funded start-ups are expected to generate $58 trillion in returns. Such high expectations create severe pressure on collaborative economy businesses to develop monetized sharing models, which can justify typical commercial "exit strategies" for venture funds, such as equity sales or initial public offerings (IPOs) on the stock markets.[193]

Janelle Orsi suggests that with such commercial business models, sharing platforms are given strong incentives to exploit their providers or users.[194] Even noncommercial sharing models online can be reliant on taking advertising to fund their activities and face temptations to sell the data they gather about users or use it for for targeted advertising. Golumbia argues in much stronger terms, that the sharing platforms are merely a vehicle for such exploitation, as capital commodifies and extends hitherto unmonetized areas of exchange.[195]

Some other commentators, however, think incumbents are simply doomed. In *The Zero Marginal Cost Society*, Rifkin highlights the economic disruption that web-enabled collaboration and sharing are causing by cutting the marginal costs of providing services and well-being.[196] He predicts this trend will continue to near zero, at which point conventional capitalist economic models could no longer function as the organizing approach of the economy. Although capital costs remain, the marginal costs of generating an extra unit of electricity from rooftop solar panels, educating another student on a massive open online course (MOOC), or even printing another widget from a 3-D printer are falling fast. In this situation, standard business models can only continue where oligopolies or monopolies—or other barriers to entry, such as restrictive intellectual property rules—can be maintained, allowing incumbents to charge above-market prices.

Rifkin thinks such resistance will be inevitably swept away by economic forces. But in real, diverse places, cultural and political considerations can interact—either to support or resist such change. Attention to the political and urban dimensions of the sharing paradigm suggests that the transition Rifkin foresees is less certain, and certainly more bumpy. But by considering the role of cities at the intersection of urban- and cyberspace—and the politics that arise there—we avoid falling into the post-political trap of economic determinism.

We believe neither extreme is likely: sharing is genuinely disruptive, but it will not sweep away incumbents and the existing system without major resistance. With falling marginal costs, the significance of regulatory frameworks for prices and profits also grows proportionally. National and international regulatory institutions such as the World Trade Organization are currently dominated by neoliberal approaches. To transform the regulatory and institutional framework is a task for collective politics. As we argue in chapter 3, that becomes possible where cities enable sharing and collaboration in the urban arena. Otherwise—we fear—the economic benefits of falling marginal costs and of sharing will continue to be captured by global and national elites.

This section has explored economic motivations and drivers of sharing. This survey of the issues arising implies that cities which invest in the sharing economy and a supportive public realm can reduce barriers to sharing and obtain positive benefits for the city economy, attracting innovation and entrepreneurship and making more efficient use of the city's resources and those of its citizens. We can begin to see how the intersection of urban- and cyberspace offers a natural habitat for sharing approaches. But our tour of the economics has also revealed how sharing approaches are being resisted by businesses vulnerable to disruption; and co-opted by others, particularly venture funds. If cities are to realize the benefits of the sharing paradigm, we must recognize and understand the social implications of these interactions, and seek to understand the ways in which the emerging sharing economy is part of a cultural or ideological battleground between community and commerce.

Intrusion of Commercial Markets into the Sharing Culture

The sharing economy has sparked a forest fire of excitement in terms of its potential to variously change the way we do business, empower previously powerless people, save resources, and increase our social closeness or civicness. But this conflagration has also exposed the sector to commercial incentives and drivers. Tapscott and Williams celebrate the "professionalization" of eBay and YouTube, with sellers and video makers "making a living" on these platforms.[197] There are however, downsides to this process. From the perspective of the sharing paradigm, eBay's value is as an "online flea-market" enabling the reuse and sharing of pre-loved goods, and YouTube is more important as a space for shared entertainment and campaigning, than as a stepping stone for aspiring Stephen Spielbergs.

The exposure of sharing-company ideologies and cultures to commercial "realities" or opportunities has had widespread effects. For example, Leah Busque initially founded TaskRabbit in 2008 with the vision to "help neighbors connect with other neighbors in real time to share certain skills when they need help."[198] As the company has grown and raised capital investment over the years, however, Busque's vision for TaskRabbit has apparently evolved to "revolutionize the world's labor force,"[199] while the company's user base, which began as regular people outsourcing their odd jobs, seems to be increasingly shifting toward businesses in need of short-term workers.[200]

Such a shift is an increasingly common refrain for sharing businesses, especially those with major stakes held by venture capital. Original community-focused visions, such as using empty seats in cars already on the road (Lyft), connecting neighbors to help each other out (TaskRabbit), and providing spare-room accommodation in crowded cities (Airbnb), have each morphed into something far more commercial in which new capacity—and new needs—are created for the purpose of sharing, rather than existing capacity being shared to better fulfill existing needs.

This shift and continuing rapid growth of commercially inspired sharing platforms has also stimulated controversy and debate, and unleashed a storm of warnings in the media and blogosphere. Critics highlight potential negative social side effects such as casualization of labor or erosion of safeguards in regulation and insurance. They warn against the domination of sector lobbies by for-profit institutions; they fear price-gouging by new "sharing economy" monopolies (such as Uber or Airbnb), worry about tax avoidance and depletion of the tax base, and even dispute whether commercial models constitute "sharing" at all.[201]

Matthew Yglesias is typical:

> Thanks to digital technology, it's now feasible to do what Zipcar does and disperse the cars throughout the city. Since the cars are dispersed, they're more convenient. But none of this is sharing. My neighbor and I share a snow shovel because we share some stairs that need to be shoveled when it snows and we share responsibility for doing the work.[202]

Sascha Lobo also rejects the term "sharing economy," arguing rather that we are entering an era of "platform capitalism."[203] From this perspective, Uber and Airbnb are following in the footsteps of Amazon, Google, eBay, and Facebook, finding new aspects of life to commodify and turn into corporate profit, while building monopoly positions as all-powerful intermediaries.[204]

One common theme running through these critiques is suspicion of commercial motives in spaces of exchange purportedly structured by interpersonal relations. Tonkinwise highlights three far-reaching risks potentially resulting from commercialization of sharing: disruption of existing regulated industries in which employees enjoyed relative security and benefits; further gentrification effectively financed by tenants working in the "sharing economy" as well as in the formal economy; and the further entrenchment of the post-political ideology of "technology as liberator."[205]

In an analogy to "greenwashing," Anthony Kalamar sees a process of "sharewashing" going on, in which the positive image of sharing is deliberately mobilized to conceal the reality of corporate exploitation.[206] Anya Kamenetz sees Peers—the apparently member-driven organization supporting and lobbying for the sharing economy, which claims 240,000 members—as a prime example of fake grassroots community support (often called "astroturfing"). Kamenetz notes:

Peers' director, Natalie Foster, has a world-class progressive resume, having worked with both the Obama administration and the Sierra Club. [But] Airbnb paid a for-profit consultancy to start the organization. The cofounder of Peers, actively involved in the day-to-day operations, is Douglas Atkin, currently also the community manager of Airbnb. Most of Peers' 73 *listed partners* are for-profit sharing economy companies like Airbnb, Lyft, Sidecar, TaskRabbit and RelayRides—only three are nonprofits.[207]

So does the sharing economy benefit ordinary people by "giving them a leg up over corporate actors?"[208] Neal Gorenflo—in discussion with David Golumbia and a representative of the SolidarityNYC collective, promoters of cooperatives in New York City—argued that "as the cost to create, market, and sell an increasingly wide variety of products and services plummets, people have a new system to go to: the sharing economy. Much of what was only possible for large corporations just a few years ago is accessible to ordinary individuals now."[209]

Access to the commercial sharing economy can indeed be beneficial for people, increasing their freedoms and opportunities, which are essential for justice.[210] Yet at the same time, its emergence can be seen as part of a systematic change that undermines and constrains freedoms. Golumbia responded:

Siding with upstart venture capital is not my idea of giving ordinary people "a leg up." The "sharing economy" doesn't have much to do with individuals. Instead, it represents corporate capital doing what it typically does: Monetizing parts of the social world that have previously avoided it.[211]

Similarly, SolidarityNYC representatives were skeptical:

There's a spectrum of sharing economy groups, from cooperatives to private companies like Airbnb. Airbnb is portrayed as helping cash-strapped individuals, which may be true in some cases. But in the long-term, it will likely exacerbate New York City's housing crisis, by allowing landlords to charge more in rent because their tenants can turn to this secondary market to make up the difference.[212]

These more critical views illustrate growing unease at the often-seamless intrusion of venture capital drenched commercial markets into the sharing culture. Golumbia warned:

> Newer "sharing economy" initiatives should be looked at very skeptically. ... While voluntarily sharing some extra space in one's apartment may well be appealing, the prospect of being essentially *compelled* to "share" one's living space *in order to afford it* is much less so.[213]

SolidarityNYC representatives argue: "We should be organizing around economic activities that contribute to community wealth and that include all people. ... Progressives need to ensure that the idea of the "sharing economy" is translated into real policies for economic democracy."[214] This spirit is echoed by progressive commentators like Janelle Orsi, whose organization—the Sustainable Economies Law Center—is one of the three Peers nonprofits, and Sara Horowitz, the founder of the Freelancers Union, which "promotes the interests of independent workers through advocacy, education, and services."[215] Adam Parsons, of Share the World's Resources, argues even more strongly: "A truly sharing society ... is underpinned by systems of universal social protection and requires a strong interventionist role for governments and strictly regulated markets, which would then necessitate ... the removal of profit-maximizing companies out of certain sectors."[216]

Such values-based unease at a perceived expansion of commerce is part of an important emerging debate about the moral limits of the market. As the moral philosopher and Harvard professor of government Michael Sandel notes, this is also about the social norms we might wish to protect from market intrusion.[217] But even if one rejects a moral case against commercialization here, there are also practical concerns raised by the rapid growth of commercially mediated sharing. In the short term, it inflates the danger of an inevitably bursting "sharing stock market bubble" and in the longer term, threatens the "crowding out" of valuable voluntary and community activity.

That threat is brought into focus if we recognize that some variants of the sharing economy may simply be one (albeit more cuddly looking) head of the hydra that is neoliberalism. Neoliberalism as Jesse McEntee and Elena Naumova describe it, is a "political philosophy that promotes market-based rather than state-based solutions to social problems."[218] However, although neoliberal processes, philosophies, and projects are, broadly speaking, driven by market rule and commodification, they typically also involve the co-option or collusion of the state. Neoliberalism is also not an end state, but an ongoing process[219] that is "variegated" or uneven in nature.[220] In this sense, some aspects of the sharing economy may represent part of the variegation, part of the process of neoliberalization, whereas other aspects of the sharing paradigm suggest an alternative to it based in shared resources and commons, collectively governed.

Accusations of destroying labor laws,[221] of getting paid less than minimum wage,[222] and of causing "backdoor gentrification,"[223] and other similar charges are evidence to us that "democracy and social justice will need to be included and protected within sharing economy initiatives from the start if it's going to be more than a [neoliberal] tool for extraction and exploitation."[224] Yet we also recognize substantial opportunities for both the commercial sharing economy and the broader sharing paradigm to stimulate social justice and democracy that make them worth supporting.

In our view—combining as it does the anti-corporate spirit of filesharing and free open source software with the social purpose and altruistic foundations of communal sharing—the cutting edge of the sharing economy is only rarely commercial. Many sharing entrepreneurs are social entrepreneurs first and foremost. Moreover, sociocultural sharing has long informally facilitated the unpaid care, support, and nurturing we provide for one another; and, like the blurred margins between sharing and the commercial economy, the blurred margins between the public sector and sharing are a place where innovative approaches of co-production are multiplying. Mediated sharing platforms, supported by new technologies, could help sustain and expand this segment by extending and strengthening social networks, but they also bring new risks, as we explore in chapter 2.

Summary

In this chapter we have examined the current revival of sharing as collaborative consumption in mediated and, particularly, in commercial forms. We have seen how under the right conditions the city can form a generative growing medium for such enterprises. Cities such as San Francisco are promoting this growth as part of a smart, hi-tech image, despite contestation by incumbent market interests. Many of the household names of the commercial mediated sharing economy, such as Twitter, Dropbox, Airbnb, and Lyft, call San Francisco home. These companies, created mostly by young entrepreneurs, and the consumers that are converging on San Francisco to take part in its emerging sharing economy, are not only changing the city's demographics, they are reinforcing new Millennial generation norms. This generation is the largest in US history and is likely to set trends that have global reach and consequences.

We saw how sharing is being encouraged by environmental concerns, enabled by new technologies and social norms, and is delivering economic benefits that seem likely to be sustained beyond a period of recession. We saw also, however, how a focus on economic aspects of our sharing behaviors—on the characteristics of the goods being shared, rather than on the values of the people involved, or the broader sociocultural benefits of sharing—obscures underlying social tensions and conflicts in the commercial models of sharing. These emerging conflicts raise themes and questions to which we shall return repeatedly in the course of this book: first as we explore the role of commercial sharing within the sharing paradigm, and then as we examine the opportunities for the sharing paradigm to redefine and strengthen the public realm through revived social or cultural sharing, and through greater recognition and enhancement of the urban commons.

Cities recognizing their pivotal role, as San Francisco is doing, will be critical in fully realizing the sharing paradigm, as part of Landry's idea of city-making. "City governments" suggest Orsi and her colleagues, "can increasingly step into the role of facilitators of the sharing economy by designing infrastructure, services, incentives, and regulations that factor in the social exchanges of this game changing movement."[225] And this framework is essential. But with political will, commitment, and public support, sharing cities can do much more (as shown in other case studies), going beyond facilitating commercial sharing. We return to governance in chapter 3 and again in chapter 5, where we examine the enabling roles of city authorities in policy, regulation, and practice to support the sharing paradigm, and the sharing city. First we need to continue our journey through sharing, turning in the next chapter from shared consumption to a focus on shared production.

2 Case Study: Seoul

Seoul, South Korea, is one of the first global cities to officially endorse the sharing economy. The city is home to more than 10 million people living within 234 square miles.[1] With a population density almost five times that of New York City, public issues such as traffic congestion, parking, and housing shortages are magnified.[2] Housing scarcity drives up costs, with the increasingly high deposits demanded in South Korea's *jeonse* rental system dramatically raising household debt.[3] Seoul also has a highly developed technology infrastructure, as capital of a country with the world's highest broadband penetration—97.5 percent of South Koreans have broadband connections, and 60 percent own a smart phone.[4] With this foundation, Seoul is positioning itself as both a leading smart city and a model city for tech-enabled sharing. Yet its approaches to sharing are culturally very different to those of San Francisco, and much less commercially motivated. Sharing in Seoul has a strong sociocultural basis. Some say South Korea has "a 'sharing' culture. It is a special concept … which Koreans call 'jeong' and it is a special kind of love between the people and society. If you don't share you will be seen as a little greedy and have little or no 'jeong.'"[5]

Koreans believe that *jeong* motivates "random acts of kindness between people who barely know each other or total strangers."[6] In this it echoes our concept of karmic altruism:

> *Jeong* is especially used to describe the action of giving [a] small, gratuitous gift—such action is full of *jeong*. A particularly close neighborhood is described as full of *jeong*, in which the neighbors act in a way that displays *jeong*—i.e., helping out and being nice to each other.[7]

Seoul is actively working to cultivate its sharing culture and build the public's trust in sharing enterprises and activities. Seoul's mayor, Park Wonsoon, a political independent and longstanding human rights activist, is a strong driving force behind this official embrace of sharing. In September

2012, the Seoul Metropolitan Government announced its plan to promote a sharing vision through Sharing City, Seoul.[8] This city-funded project aims to make sharing activities accessible to all citizens by expanding physical and digital sharing infrastructure, incubating and supporting sharing economy startups, and putting idle public resources to better use. According to Kim Tae Kyoon, director of Seoul's Social Innovation Division, Sharing City could help citizens regain some of the community that rapid urbanization and industrialization have lost. He told *Shareable* that the ultimate goal of the project is to "share lives among dispersed people, recover trust and relationships, and shape a warm city in terms of people's heart."[9]

Sixty percent of Seoul's inhabitants live in apartment buildings. The director-general of the Seoul Innovation Department, In Dong Cho, sees these densely populated spaces as critical in rebuilding a sense of community.

> In order to regenerate communities in apartment complexes ... we recommend people establish share bookshelves, share libraries, share gardens and common tool warehouses, and to organize community activities through subsidies or grants. ... These movements toward sharing will restore dissolved communities and revive sharing culture in citizens' daily lives.[10]

For instance, the city is facilitating the formation of lending libraries in apartment buildings. In one year 32 such lending libraries were established offering books, tool rental, and repair (plus woodworking programs). The city is also implementing sharing services of its own, notably by opening select city parking lots and buildings to the public during off hours so that the idle space can be used.[11] In the first year 779 public buildings were utilized during idle hours over 22,000 times by Seoul citizens for events, meetings, and more.[12] Seed funding is also being provided, not just "seed resources." The Sharing City project is incubating some 20 sharing startups with office space, technical assistance, and expense subsidies, and thus supporting innovative ideas and encouraging pioneer companies to lay the groundwork for the sharing economy in Seoul.[13] In an echo of San Francisco's Entrepreneurship-In-Residence program, some more-established sharing enterprises will be provided funding to scale up their platform and enhance their services.

Among the startups selected for support through Sharing City are: Kozaza and Lobo Korea, home-stay platforms; Woozoo, a company that transforms older homes into shared housing; Wonderland and Billi, platforms that allow for sharing of underutilized goods; SOCAR, a carsharing service; Kiple, a clothing exchange for children; The Open Closet, a suit distribution service for young job seekers; Living and Art Creative Center, an art and writing education space; and Zipbob, a mealsharing platform.[14]

As the "Korean version of Airbnb," Kozaza has been attracting media coverage with its catchy name, roughly translating as "nighty night."[15] The site offers lodging services and primarily targets foreign tourists who wish to stay at *hanok*, or traditional Korean houses. Kozaza founder Sanku Jo, a 48-year-old senior corporate executive—who incidentally spent over a decade in California's Silicon Valley—says his passion for traditional style *hanok* houses inspired him to start the service.[16] Kozaza not only provides host families with supplemental income and facilitates tourism expansion, but also promotes conservation of *hanok* heritage, a slowly disappearing cultural resource in Korea. But Seoul's approaches to sharing housing are not only targeted at tourists or relieving housing scarcity, they also aim to "reduce the social isolation of seniors" with a new program created "to match young people with idle rooms in seniors' houses."[17]

The word *zipbob* means "homemade meal" in Korean. The Zipbob platform provides a way for people who are interested in eating home-cooked food to connect and share the experience together. Anyone is able to organize a meal gathering online, and if at least seven people sign up to participate, the group meets. Zipbob's founder, the 26-year-old Lynn Park, explains that she works in Seoul's business and investment banking district from 7 to 11 and had become tired of *hoesik*, [18] a term that means "dinner with co-workers," and which is an after-work, at times pressured, socializing activity deeply engrained in South Korean business culture. Park sees the "secret sauce" behind Zipbob's success as the thirst for connection outside of one's circle of co-workers, combined with an appreciation for home-cooked meals, which makes it "more about healing and dining."[19]

Cat Johnson reports on *Shareable* that in the first year, "461 million *won* ($450,000) has been invested in 27 sharing organizations or businesses."[20] And now:

> The Sharing City Seoul project is being rolled out through its 25 boroughs known as *kus*. ... Each *ku* has its own mayor and local government. Because government-endorsed businesses are trusted by citizens, [the city] introduced the sharing economy to two *kus* by endorsing Kiple, the children's clothing exchange. The experiment proved to be successful—Kiple doubled sales in one year.[21]

Seoul is playing a strong role in guiding the development of its sharing economy, both directly and indirectly. As well as actively supporting some, the city has not been afraid to ban what it sees as inappropriate. For instance, as Neal Gorenflo, a member of the city's Sharing Promotion Committee reports, Seoul has banned Uber, instead working to develop its own open source cab-hailing app.[22] Indirectly the city's past investment in an advanced and extensive public transport system helps explain both this decision and why there is little demand for car- or ridesharing services.[23] There are only "564 car sharing locations in Seoul—with over 1,000 cars that have been shared 282,000 times through companies such as SOCAR and Greencar."[24] But the success of mealsharing companies like Zipbob, and the niche they fill in providing a platform for meet-ups and experience sharing, demonstrates the social interaction that people are hungry for.

Sharing approaches like this show that a cultural shift toward access and unique experiences is not limited to the West. Moreover, organizations like these are helping transform the culture of Seoul toward inclusion. We must acknowledge that the city has not had a glorious record in this respect historically. David Satterthwaite, a Senior Fellow at the International Institute for Environment and Development (IIED) describes a review of 40 clearance-and-eviction cases in cities between 1980 and 1993.[25] Eight involved more than 100,000 persons; the largest was the 720,000 people evicted in Seoul in preparation for the Olympic Games. Nor was this a one-time event; from 1960 to 1990, 5 million people were evicted from their homes in Seoul, many several times, often from sites provided after previous evictions.

In 2002, Seoul began construction of the Digital Media City, a new high-tech business district on a reclaimed landfill site in the city's northwest, aiming to stimulate clustering and innovation in digital businesses. Today however, reconstructions in Seoul are as likely to be found in cyberspace as in urban space. Seoul was an "early investor in smart technology, "and one of the first cities to grasp the opportunities to use data harvested from citizens' smartphones and other GPS enabled devices rather than building a new monitoring and sensing infrastructure.[26] Those same devices are the platforms for mobile and web sharing applications.

In contrast with San Francisco, the human aspects of smart cities are at the forefront in Seoul, and the city authorities are much more interventionist to that end. Mayor Park Won-soon's smart-city vision is of one involving "communication between person and person, people and agencies, and citizens and municipal spaces, with human beings always taking the central position in everything. ... [That vision is] characterized by its unprecedented level of sharing."[27] Work to deliver the city's Smart Seoul 2015 plan involves the introduction of electronic currency and contactless smartcards for public transport access and payment, as well as the development of a proposed U-Health Care program to enhance medical access by the disabled and elderly with the use of smart technology and high-tech equipment.[28] In partnership with Creative Commons Korea, the city created an online portal, ShareHub, which provides citizens with information about the Sharing

City project, shares news regarding sharing economy initiatives, and acts as a directory of sharing services.

Technology can help reduce barriers to sharing, not just in collaborative consumption as we already saw in San Francisco, but also in co-production and even co-governance. Here Seoul has a rich heritage on which innovation can build. In the 1990s and early 2000s, various forms of co-management, co-production, and co-governance between the state and the third sector were developed in South Korea. By 2007 they were playing "an increasingly important role in producing and delivering public services such as childcare, healthcare, care for the aged, library service, waste management, education, and community development."[29] Co-production in South Korea is distinctive from European experience in two ways: First, in that it effectively established a state welfare system in circumstances in which welfare had previously been the province of "individuals, families, firms, voluntary organizations and foreign groups."[30] Second it is "being demanded by the public and used by the government to increase the legitimacy of and popular support for ... nascent democracy."[31]

The researcher Jung-Hoon Kim also emphasizes the importance of co-production as an expression of democratic renewal in his study of co-production in waste management in Seoul.[32] In the 1990s, he says, "The coproduction concept was very meaningful. ... Most Korean citizens were focused on real citizenship or citizen consciousness in the process of democratic political developments."[33] In waste management, co-production took the form of voluntary waste reduction and waste sorting for recycling. The city introduced financial incentives, in the form of charges on residual waste, but also established free collection of recyclables. The net effect was to reduce average household expenditure on waste management. Kim highlights the cultural readiness of ordinary Korean citizens—primed by *jeong*—to cooperate. He says, "The goodwill, obligation, and consciousness" of citizens was essential to the success of co-production in waste management.[34]

Co-governance continues today. The city is directing some of its budget in line with citizen input: its Residents' Participatory Budgeting System provided Seoul citizens with the opportunity to direct spending of 50 billion *won* (approximately $47 million) in 2013 to fund projects that were democratically decided upon.[35] In the same year, the Social Innovation Camp in Seoul brought together 50 digital entrepreneurs and innovative software developers to compete in constructing web- and mobile-based solutions to public sector challenges facing today's cities. The winning team built a mobile application called "finger-town," which makes reporting public service problems fun and lets users know when problems are solved.[36]

To advance the public's access and use of mobile- and web-based sharing services, the city is focusing on building "smart environments," in which free Wifi, or wireless Internet, is available anywhere at any time.[37] Almost 2,000 additional wireless access points have already been established at markets, parks, and government offices.[38] Officials are targeting populations who are often underserved when it comes to smart technology by distributing second-hand smart devices to the elderly, the disabled, and low-income families, with the goal of reaching 1 million devices given out by 2015.[39] This unprecedented level of connectivity and communication uniquely positions Seoul to harness the problem-solving potential of sharing. Consistent and available communication modes also can encourage citizen participation in information sharing as well as enhance transparency in city operations.

Mayor Park Won-soon has a track record of social activism and was elected in October 2011 under the slogan "Citizens are the mayor." Creating a culture of listening has been an explicit component of the mayor's leadership platform. This pledge to listen to the people is illustrated by a large ear sculpture, placed outside of the newly constructed Seoul City Hall[40] and "installed with an interior microphone connected to the speakers placed all around the City Hall's basement."[41] The city also practices open government by providing public access to all official city documents, even those in process.[42]

Mayor Park Won-soon highlights his commitment to providing multiple venues for citizen voices to be heard. These include "the Simincheong ... in Seoul City Hall, [which] acts as a 'speaker's corner' for anyone who wants to send a video message to the city administration"; and online platforms such as Twitter, where the mayor has more than 660,000 followers who communicate suggestions and feedback to him in real time.[43] Citizen participation is also encouraged through the Seoul Innovation Planning Division, which researches how innovations from around the world might be applied in Seoul, and gathers, spreads, and systematizes creative ideas from Seoul citizens.[44] It has also been tasked to address legal obstructions to expanding the impact of sharing activities and enterprises, and to facilitate communication between startups and the Seoul Metropolitan Government.[45]

Seoul has also developed dedicated forms of co-governance to help deliver the sharing city. The city has delegated significant decision-making powers on sharing to a Sharing Promotion Committee comprising 12 members from the private sector and 3 from government, and established partnerships with both tech startups and grassroots citizen-driven organizations working to catalyze more sharing in Seoul.[46] In Dong Cho sees the city's job as providing the infrastructure for sharing:

> It is not desirable for government to directly intervene in the market to promote the sharing economy. ... The city needs to build infrastructure such as law, institution and social trust capital—the city needs to pave the way and strengthen the ecosystem for the sharing economy to thrive.[47]

The Seoul Metropolitan Government believes that public sector support and citizen participation in sharing efforts can be made stronger through appropriate collaboration with the private sector. Mayor Park Won-soon refers to this as "super-sectoral social innovation." It is evident in the way the Sharing City project fosters private sharing enterprises that use innovative approaches to tackling social problems. For instance, Dream Bank was founded in 2012 after 20 Korean banks pooled funding to create a new foundation for incubating startup companies. Dream Bank constructed D.Camp, a 1,650 square-meter (17,760 square-foot) co-working space that provides a minimum of three rent-free months of office space, networking, and educational opportunities to emerging sharing enterprises. D.Camp is also creating an online platform to connect startup entrepreneurs with investor financing.

Seoul's status as a Sharing City is attracting global attention, and the city continues to publicize this branding through hosting international conferences and events that allow other municipalities to experience a sharing city plan in action.[48] The city also plans to establish an International Sharing City Conference.[49] In November 2013, Seoul hosted the Global Social Economy Forum in its city hall. This forum was the largest event of its kind, and brought together participants from over 30 countries to share local government-led urban innovations and experience in building partnerships for social enterprises.[50]

Sharing Production: The City as Collective Commons

Cities have the capability of providing something for everybody, only because, and only when, they are created by everybody.
—Jane Jacobs

Chapter Introduction and Outline

In this chapter we pull together multiple strands of research to help us further develop the following thesis: sharing and cooperation are universal values and behaviors that are socioculturally and biologically coevolved. We then contrast this thesis with the dark side of excessive competition in our society, which can generate fraud, cheating, stress, and inequality. We examine the ways in which cities as a whole are shared domains—even their historic role as places of exchange is a product of sharing, while the essential public services, infrastructures (and the underlying resources) on which cities and their inhabitants depend are fundamentally shared. Cities like Seoul are recognizing this, and actively intervening to provide or enable provision of not only hard (physical) but also soft (social) infrastructures. We continue by outlining how such services and infrastructures can be "co-produced" utilizing mechanisms of sociocultural sharing, with particular reference to health and education.

We also explore international variations in sharing cultures through the lenses of individualism and collectivism. We look at forms of co-production emerging in commercial spheres—including peer-to-peer finance, energy, housing, and the shared production of food. We highlight the potential for cooperative forms of organization in services, production, and finance to overcome the risks of disowning responsibility and commodification that arise where sharing overlaps, respectively, public and market provision. We conclude the chapter with a discussion of how both the commercial and communal functions of cities are underpinned by a shared collective commons and on commoning (i.e., the co-production of the urban commons). By focusing on the shared collective commons, we suggest sharing can be seen as a genuine, integrative, and substantial third space, complementing both statist and market-based approaches.

The Social and Evolutionary Roots of Sharing

In chapter 1 we noted that sociocultural sharing behavior had apparently declined in modern cities with growing wealth and declining social capital. However, it is far from extinct. In households, public services, and in clubs and associations, sociocultural sharing remains pervasive in modern society in the global North as well as the global South. Often unnoticed, as Yochai Benkler says, "It sometimes substitutes for, and sometimes complements, market and state production everywhere. It is, to be fanciful, the dark matter of our economic production universe."[1]

Surveys of sharing behavior and opinion bear this out. For instance in Vancouver, Chris Diplock notes:

> For more than 60% of survey respondents, however, the term sharing was more closely associated with old models of sharing, such as communal resources and public goods. Not surprisingly, respondents reported being very active in sharing traditional physical objects and spaces, such as books and public parks.[2]

A similar picture is painted by NESTA's UK survey where 64 percent of adults surveyed reported sharing, but only 25 percent did so online.[3] Elsewhere, in Africa and indeed much of the global South, a *mediated sharing* economy is in its infancy although *sociocultural sharing* is widespread. In Latin America there is a particularly strong culture of public sharing. But also in South American cities, "co-working spaces and makerspaces are blooming, the first platforms for carpooling are being tested and free public bicycle systems are surfacing."[4]

In the Middle East, mediated sharing is also obtaining traction, especially in cities such as Cairo and Dubai, alongside traditional sociocultural sharing. According to Ouishare, "The sharing economy movement in the Arab World has seen a positive eruption in the recent few years. ... We're beginning to share more and more ... boats (*fishfishme*); skills (*Taskty*); carpooling (*Kartag*); swapping goods (*Swaphood*) or selling used goods (*krakeebegypt*)."[5]

Naturally Adapted to Share

We should not be surprised to find forms of sharing everywhere. It is, by all accounts, an evolutionary trait.[6] Shared efforts allowed our ancestors to band together to hunt, farm, and create shelter, and reciprocal forms of altruism arose naturally from repeated interactions in small groups where reputation mattered for survival.[7] Hunter-gatherer societies depended on sharing and cooperation, and this behavior instilled an instinctive preference for fair shares and broadly equal outcomes.[8]

In the modern world, as with any instinctive behavior, sharing practices are strongly influenced by cultural learning, with distinct emanations in different countries and cultures (as we explore later in this chapter). Culture, language, and cooperation likely evolved together.[9] In evolutionary terms culture can be understood as a powerful survival strategy, according to Mark Pagel, head of the Evolution Laboratory at the UK's University of Reading.[10] In symbolic and shared form, culture shaped group identity and individual behavior, affecting evolutionary selection. Culture stimulates devotion to groups, within which we reward cooperation and punish those who deviate from group norms. We are imprinted with the group culture of our birthplace by learning from and imitating parents and other teachers.

Such a view of culture fits Darwinist theories of individual selection.[11] Our genetic tendency to "help relatives" has been hijacked by culture to drive cooperation within broader yet distinct groups. Factors of group identity—such as language—effectively become markers of "reputation." These allow humans to identify those with whom to cooperate even in the absence of an immediate reciprocal benefit, and its absence indicates those from whom we should withhold cooperation.

As Pagel says, "Altruism can thrive if altruists can surround themselves with other altruists,"[12] establishing a "mutual aid" society. Within our ancestors' groups, evolutionary pressures established an "altruism arms race" as they competed to better signal status in a cooperative society. These pressures underlie the human tendency to ostentatious collective displays such as building cathedrals, or, in the modern age, hosting the world cup or Olympic Games: events that highlight a national culture's fitness in a collective gift to the world. But evolutionary pressures also underlie the daily random and ritualistic acts of kindness and civility that bind society (holding doors, helping mothers lift baby buggies onto buses), as well as all the collaborative tendencies displayed in sharing.

Over history, culture has permitted us to build ever-wider group identities, and there is no reason for that process to halt, despite its origin in *genetic* relatedness. (Indeed, Pagel notes that humans often already extend our natural altruistic tendencies not only to other humans but also to other species.) If *cultural* relatedness is now the signal for collaboration, as cultural and language barriers are reduced by globalization and cosmopolitan cities, we can rather expect the broadening of our group identities to continue.

Moreover, with the establishment of strong markers of reputation on web-based sharing platforms (through user reviews and ratings, discussed at length in chapter 5), we see the emergence of an artificial mechanism capable of rapidly extending our "mutual aid" societies across existing tribal and national boundaries. As Pagel says:

> History has demonstrated ... that we humans will get along with anyone who wishes to play the cooperative game with us. The returns of cooperation, trade and exchange that derive from that part of our nature have historically trumped our guesswork based on markers of ethnicity or other features. And they always will. ... The key is to provide or somehow create among people stronger clues of trust and common values ... and then to encourage the conditions that give people a sense of shared purpose and shared outcomes. ... Looking around the great cosmopolitan cities of our world, it is hard to avoid the conclusion that this is already happening.[13]

Cognitive and psychological science supports this. The vast majority of us are naturally reciprocators—we contribute to shared and social projects in the expectation that others will contribute too, and normally they do.[14] We are also inclined to punish non-cooperators, even at considerable personal cost, in effect policing the system.[15]

Other contemporary evolutionary theorists such as Robert Boyd and Peter Richerson reach very similar conclusions—that humans are instinctively inclined to cooperate, at least within recognized groups, and to punish non-cooperation—through the rehabilitation of ideas of "group selection."[16] As Samuel Bowles and Herbert Gintis similarly argue, groups that developed norms, mechanisms, and institutions to promote cooperation and protect the majority from exploitation by the selfish, flourished and out-competed less-cooperative groups, who could only keep up by adopting the same cultural cooperation.[17] This process of norm building is genetically predisposed, driven by the evolution of social emotions such as shame and guilt, and cemented by humans' capacity to internalize social norms in a process of "gene-culture coevolution."[18] None of these evolutionary scientists denies that humans also have evolved competitive instincts, but the consensus is that our collaborative and sharing instincts are deep-rooted, and powerful.

Exploring the same subject from a sociological perspective, Richard Sennett, Professor of Sociology at the London School of Economics, notes that "cooperation becomes a conscious activity in the fourth and fifth months of life, as babies begin to work with their mothers in breastfeeding" and that such "cued behavior ... helps the brain activate previously dormant neural pathways, so that collaboration enables the human infant's mental development."[19] Collaborative capabilities develop, even in the absence of parental collaboration, to the extent that by the third year of life the "capacity to cooperate together on a common project, like building a snowman becomes well established."[20] In these respects it appears that cooperation accompanies, or even precedes, individuation. As Sennett puts it, "We could not develop as individuals in isolation."[21] But sadly that basic truism is increasingly ignored and challenged in modern political ideology that celebrates individualism and competition without understanding their symbiotic relationship to society and cooperation.

Cooperation and sharing are closely related. Sharing a task or a resource involves cooperation, and cooperation tends to depend on a sense of fairness within the cooperative relationship. The ethics of sharing seem to emerge in childhood in parallel with the skills of cooperation, but may not be actively practiced until later. The psychologists and learning specialists Marco Schmidt and Jessica Sommerville note that "the roots of a basic sense of fairness and altruism can be found in infancy, and ... these other-regarding preferences develop in a parallel and interwoven fashion."[22]

Variable expressions of sharing in different cultures—as we saw in Seoul as opposed to San Francisco—do not invalidate the idea of genetic tendencies to share and cooperate. Modern genetic science suggests that many characteristics are expressed in response to triggers in the environment and nurture. For example, we are genetically predisposed to language: that some of us learn English and others Mandarin does not contradict this.

All the foregoing tallies well with research suggesting that human moral codes are strongly rooted in gut instincts.[23] But they are also connected with our capacity, culturally, to define and spread ethical positions and behaviors—such as the rejection of slavery, the acceptance of equal rights for women, and in recent years the rapid spread of moral inclusion of different forms of sexuality. Such shifts rely heavily on the human capacity to feel empathy for others.[24] Jeremy Rifkin argues that our developing culture and changing technology have both enabled the development of ever stronger and broader forms of empathic consciousness, with the printing press, radio, television, and most recently the Internet each accompanying new degrees of empathy.[25] This extension of empathy is not just because the groups we identify with are growing in size: empathy is also extending across group boundaries. Its growth parallels the extension of the "moral community": the group to whom we naturally assume we owe duties to treat fairly and justly.

The urban setting takes a central role in this process. Rifkin argues that empathy is rooted in urban exposure to the "alien" and different. This is supported by Sennett's view that the "arrival of a stranger can make others think about values they take for granted"—such "stranger-shock" being a feature of urban life.[26] These theoretical arguments are borne out by empirical evidence that exposure to difference in urban spaces reduces racism and discrimination.[27] Arguably, in modern cities, exposure to difference can, however, be so great as to be overwhelming, stimulating withdrawal rather than engagement and cooperation. We return to these challenges in chapter 3.

The Dark Side of Excess Competition

None of this is to argue that cooperation and sharing is inevitable because it is genetically and culturally coevolved. That would be to succumb to a naturalistic fallacy. Yet we believe it is important to emphasize the evolutionary roots of sharing, if only to challenge the more pervasive myth that such behavior runs counter to our "competitive" genetic nature, and is therefore somehow impractical or even undesirable. Nor are we arguing that competition is therefore undesirable. It, too, is part of our nature. But where it is idolized, it creates both individual and social risks.

The biologist Franjo Weissing from the University of Groningen, The Netherlands—whose research explains how competition and collaboration coexist in human and animal societies—highlights how the most competitive individuals are most wasteful. It is, he says "tempting to speculate that the external stimulation of competitiveness by societal pressure ... [could] lead to such a wastage of resources that our future survival is threatened."[28] The University of Wisconsin sociologist Erik Olin Wright argues that "competition is a powerful force for rewarding people for successfully developing their talents, and thus a certain degree of competition undoubtedly stimulates human flourishing."[29] But, he says, it also "underwrites a culture of accomplishment which evaluates people only in terms of their relative standing compared with others. ... Broadly, within systems of intense competition, most people will be relative 'failures.'"[30]

The psychological effects of such competition and inequality can be severe. The writer and entrepreneur Margaret Heffernan argues that—rather than stimulating innovation and inspiring us to do better—competition can generate fraud, cheating, stress, and inequality, and it can even suffocate creative instincts.[31] Burn outs and ethical lapses characterize the race to the top, she suggests. The problems of competition are exacerbated in today's globalized markets, cultural industries, and sporting contests, where a "winner takes all" structure awards the lion's share of the rewards to a very few elite participants who gain celebrity status and advertising endorsements that multiply their direct earnings many times over.[32] Such markets have their stars, but for every Michael Jordan or Tom Cruise there are thousands of contenders whose dreams have been crushed, despite years of economic and social self-abasement in search of success.

Cultural Variation in Sharing

Internationally, different cultures seem to place different weight on competition and cooperation, and indeed more broadly on the concepts of individualism and collectivity. This distinguishes cultures according to their sources of status and identity. Put crudely, individualistic cultures seem to value *things* and collective ones *relationships*.[33] More generally, individualistic cultures value novelty, freedom, and independence from social groupings. Collectivist cultures emphasize relationship, relatedness, and interdependence.[34] Unsurprisingly, those collective cultures also appear to exhibit higher levels of sociocultural sharing behaviors (as we saw in the Seoul case study preceding this chapter). Recent research suggests an intriguing ecological factor, suggesting that farming rice relies on interdependency, and over the generations has made rice cultures collective, while farming wheat makes for independence and individualism.[35]

Collective values systems such as Korea's *jeong* are not exclusively Asian, but do appear to be more pervasive and strongly rooted in East Asia than in the West. The psychologists Christopher Chung and Samson Cho stress that that *jeong* is effectively a "collective emotion," which appears as loyalty and commitment to the community or group concerned without obvious validation or reason. *Jeong* also underlies the prevalence of informal, assumed commitments in Korean society in contrast to the defined contractual commitments of Western culture. So Koreans "easily become members of a cohesive group at home or work, bonded by *jeong*. ... Interdependency and collectivism are highly valued, rather than autonomy, independency, privacy, and individualism."[36]

Like other communitarian forms of "bonding" social capital[37] and associated practices—such as the Chinese *guanxi* tradition of indirect reciprocal obligations[38]—*jeong* has both positive and negative implications. As Chung and Cho explain, "Warm, rich interpersonal relationships, nurturing, and caring seem common. ... Bonded by *jeong*, collective efforts toward a common goal ... are relatively frequent."[39] But *jeong* may also stimulate group discrimination. Group leaders "protect those within the circle of the in-group and discriminate against outsiders."[40] Corrupt behaviors may arise, for example in "promotions, entrance to colleges, and business contracts."[41]

Psychological research, however, highlights that individualist-collective contrasts and their implications are not set in stone. Unsurprisingly, context matters. Encouraging participants to think about themselves as part of a group, rather than as an individual, triggered more collectivist social values and judgments among Americans and Europeans. For students from Hong Kong, individualist priming triggered the opposite.[42] Swapping the lonely combine harvesters of prairie agriculture for the sociality of rice paddies, or the isolating cultures of commercial markets for collective sharing structures, could therefore be expected to generate more collective values.

Individualist—collective contrasts describe differences in individual behavior. Societies also differ in how they function at a social level. Cultures such as Japanese, Singaporean, Korean, or Pakistani can be described as "tight," with many strong norms and low tolerance of deviant behavior, whereas Ukrainian, Israeli, Brazilian, or American cultures, for example, are "loose," with weak social norms and higher tolerance of deviant behavior.[43] In figure 2.1 we loosely map the two frameworks on top of the values circumplex, developed by the social psychologist Shalom Schwartz,[44] to indicate how they might interact in a highly simplified way. Schwartz emphasizes that different cultures all display the common values he identifies, but to different degrees.

Such simple models might help us understand the emerging variability in sharing approaches internationally. Although hegemonic global culture is seeing the spread of a particular model of sharing (commercial mediated), it will have different expressions and implications depending on which local cultures it then displaces (we return to this issue in chapter 5).

Tight, collective societies can be described as traditionally "communitarian" (using the sociologist Amitai Etzioni's concept).[45] Communitarian scholars highlight the contrasts between such tight traditional, Eastern communitarianism—and responsive communitarianism, in which multiple group identities are recognized, and individuals can actively choose which "in-groups" they will join.[46] While communitarian philosophers such as Etzioni and Sandel[47] do not engage with sharing as we understand it, their work suggests that in tight societies we might expect norms to change slowly, but be powerful in establishing compliance with social goals. In looser responsive communitarianism, norms are weaker in effect, but can change more rapidly. The contrasts between San Francisco where, as our

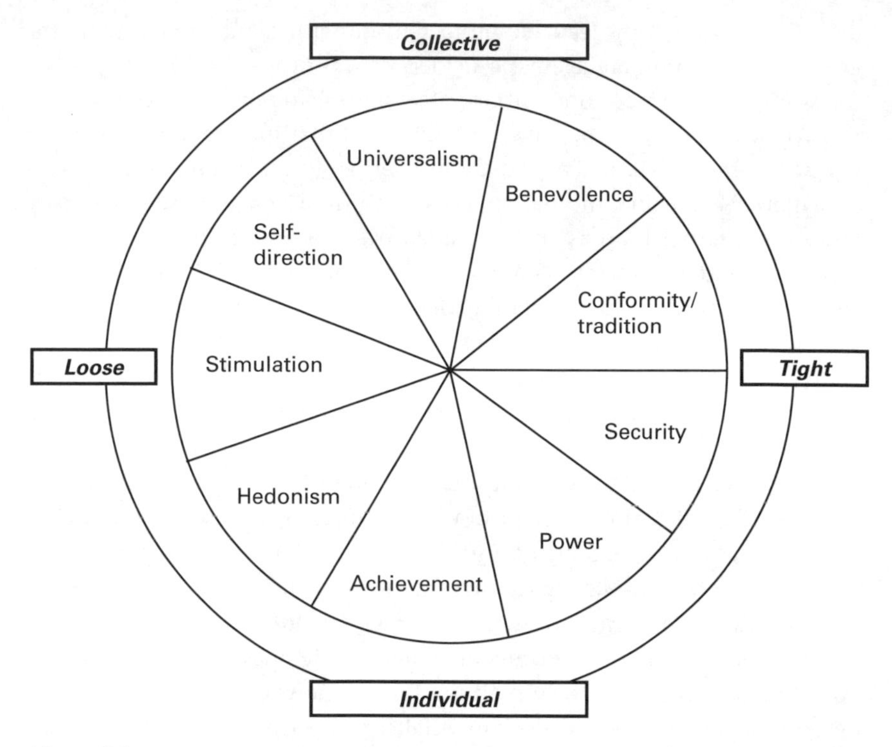

Figure 2.1
The multiple dimensions of cultural comparison. *Sources:* Circumplex after Schwartz (2006; see note 44, this chapter); "tight-loose" dimensional concept from Gelfand et al. (2011; see note 43, this chapter); and "individual-collective" from Hazel R. Markus and Shinobu Kitayama, "Culture and the Self: Implications for Cognition, Emotion, and Motivation." *Psychological Review*, 98(2) (1991): 224–253.

case study shows, mediated sharing has emerged entrepreneurially (despite some resistance from authorities), and Seoul, where the city authorities are leading the charge, may be instructive in this respect.

The apparent historical decline in sociocultural sharing (in Western societies, and in partial contrast to the examples seen in Seoul) can be associated with the development of consumer capitalism and the growth and promotion of a culture of individualism.[48] The development of advertising targeting—and recreating—an individualistic model of the self, has transformed consumption patterns.[49] Many possessions have been so privatized and individualized that Americans no longer need to share much even within their own families, still less with other members of their communities.[50] This might be, in part, a consequence of wealth. But there is a big downside to the loss of this type of sharing: loss of social capital. Those who have fewer resources—and share more—have more social contacts with their neighbors and within their community.[51]

Rami Gabriel, an associate professor of psychology at Columbia College Chicago, suggests that individualist societies like the US, where social capital is weak and there is a "thriving private sector and a starving public sector," should try to "maintain and nurture a diverse range of non-commercial relations between people and spaces for these relations to take place."[52] We would argue that such spaces are ones that can and do nurture sharing.

Reciprocity and Karmic Altruism

Overall it is indisputable that sharing and cooperation are an important part of a culturally and biologically evolved repertoire of human behaviors. Partly they arise from *direct* reciprocity between individuals who interact repeatedly[53] and thus learn that "one good turn deserves another." But in societies with identifiable groups and clear means to establish and remember reputations, humans also began to practice *indirect* reciprocity, recognizing that "what goes around comes around." We call this practice "karmic altruism." Today, humans exhibit altruism and kindness at every level, from the daily civilities that make living together possible to extreme self-sacrifice for our ideals. These form the glue for society, which in turn provides the social structures and support that are obvious to anyone who has tried to raise a child or build a business—that such tasks are ultimately collective and collaborative.

The Wharton professor Adam Grant has studied the significance of altruistic behavior in business success. He argues that the most successful executives are not those who are most competitive, and take whenever they can, nor those who are consistently altruistic regardless of personal cost, but those who collaborate or share—those who both give and receive (and not necessarily from the same people).[54] In other words, sharing is not philanthropy but rather mutual support, and it works in business settings as well as in our social lives.

There are key implications for sharing in this understanding of our evolutionary inheritance. First, our instincts for reciprocity do not stop with our family, kin, or even social grouping. Thus as the technology has enabled it, what the sociologist Juliet Schor calls "stranger sharing"—which exists "among people who do not know each other and who do not have friends or connections in common"[55]—has spread rapidly.

Second, sharing does not rely on strong and deep social relationships (nor on the total anonymity of market exchanges), but on the "strength of weak ties."[56] Nor should it be expected to create deep and enduring friendships but merely to extend and multiply our "weak ties." Like Sennett's "stranger shock," exposure to "stranger sharing" is a powerful mechanism. Conventional sociocultural sharing strengthens bonds within groups—a form of "bonding" social capital.[57] But mediated stranger sharing also promises to broaden our empathy for others, making our weak ties more diverse and robust. It increases links between social groups, and even between societies, extending the power of karmic altruism beyond the local community as a form of bridging social capital.[58]

In this section we saw how sharing is culturally and biologically evolved. We find it in all cultures and all religions. In the modern world it is extending beyond close familiars and even beyond culturally defined groups. We also saw how encouraging excessive competition harms us socially and psychologically. Enabling sharing, on the other hand, has potential to rebuild social capital. By investing in the infrastructures and practices that support sharing we argue that humanity can both strengthen the bonds *within* existing societies, and make those societies more inclusive to those who are excluded or alienated by the modern political emphasis on individualism and competition, thus strengthening the bonds *between* groups and societies. We argue in this book that cities are the best venue for such processes. Below we begin to explain why, by exploring the historic relationship of cities and sharing.

Sharing Cities in History

Nowhere in human culture is the centrality of collaboration and sharing more obvious than in the city. The city is not just a venue for the sharing of spaces, "things," and services, but is historically a shared entity in itself: a shared creation. Provision of public spaces and public goods such as sanitation, public health, and education has always been key to urban development.[59] These shared spaces, services, and goods contribute greatly to the urban commons—co-produced through collective and political action by citizens and decision makers—which in turn underpins conventional economic production in the city.

Shared urban public spaces have long served as focal places of exchange, encounter and entertainment. Cities developed in a number of places, from Mesopotamia to Asia to the Americas. Mohenjo-daro, in present day Sindh province, Pakistan, founded circa 2600 BCE, was one of the largest early cities with approximately 50,000 inhabitants at its peak. It had a marketplace, a shared well, a communal housing structure, and public baths. Egyptian cities such as Thebes, Cretan Knossos, and ancient cities in the Andes and Mesoamerica all provided shared facilities, infrastructures, or spaces. The ancient agora of Athens—literally a "gathering place" or "assembly" point developed in the sixth and seventh centuries BCE—was the locus of life in the polis: a space for feats of athleticism, artistic displays, spiritual happenings, markets, debate, argument, and political events.

In the following sections of this chapter, we highlight three ways in which cities remain fundamentally shared spaces: as places of exchange; as systems reliant on common resources, supply chains, and infrastructure; and as communities served by shared services. Of course, recognizing this does not mean that city dwellers will automatically adopt the norms and values of sharing. Nor does it deny that in practice there is conflict between shared and private uses of cities; but it does provide strong foundations to help city authorities understand the scope of sharing.

The City as a Place of Exchange

By the end of the Middle Ages, urban areas were already centers for exchange, including local, regional, and long-distance trade[60] as well as manufacturing.[61] One factor that increasingly set the urban economy apart from the rural was the transfer of goods, services, or labor in exchange for money, rather than transfer by feudal appropriation, or through informal gift exchange and barter. We discussed how such a process of commodification is affecting sharing today in chapter 1. Here we note that in some (limited) respects markets are just a more formalized way of sharing skills and resources while (more importantly) the market itself, and the rules of exchange, are collaborative, cooperative projects. As Don Tapscott and Anthony Williams emphasize, "Vibrant markets rest on robust common foundations: a shared infrastructure of rules, institutions, knowledge, standards and technologies."[62] And those shared infrastructures in turn rest on social capital produced by a "social commons … where we generate the good will that allows a society to cohere as a cultural entity."[63]

Despite the neoliberal mantra that competition is the core of economic success, there is actually a very broad consensus recognizing the entire edifice of trade, commerce, and the economy as a product of cooperation. Historians and sociologists see the roots of commerce in specialization and cooperation in prehistoric hunting tribes, modern business scholars talk of "co-opetition,"[64] politicians and regulators make and shape markets with decisions on laws, taxes, and much more. All of them understand that markets are actually a form of cooperation, as much, if not more so than a venue for competition. Yet markets, especially contemporary neoliberal ones, involve very different and limited forms of sharing, despite their common foundations.

The shared foundations of modern economies largely emerged in the new cities of the Industrial Revolution, building upon traditions of barter, commerce, and trade. At that time cities were dominated by the guild economy, but frustrated entrepreneurs developed new models of production based in the "putting out" of tasks to workshops in the countryside and smaller towns. These new methods, associated also with new technologies of water and coal power, enabled the growth of new industrial settlements, which adopted innovative forms of collaboration in factories, and rapidly grew into cities.

But new models of deliberately collaborative consumption and co-production that are often more communal than commercial, and often more sociocultural than mediated, are emerging strongly in the twenty-first-century city alongside the commercial models we highlighted in chapter 1. Initiatives to reclaim urban space for food production, especially those aimed at increasing food security among low income and minority neighborhoods; low-cost transportation options such as carsharing and bikesharing programs, as well as toolsharing, worksharing spaces and other community-use centers, are all contributing toward a redefinition of the function of cities as places for sharing and exchange. All these rely on common resources and infrastructures.

Common Resources and Infrastructure

The idea that sharing was foundational to the urban economy is one way of seeing cities as inherently shared spaces. A second, and the focus here, is that cities and their inhabitants (and their economies) rely on shared physical infrastructures, resources, and supply chains. That infrastructure—both hard and soft—represents a critical component of shared urban systems is indisputable. The physical, or "hard" infrastructures constitute roads, highways and streets; mass transit systems; electricity, gas, and district heating systems; telecommunications systems; drinking water, drainage, irrigation, and sewerage systems; public open and green spaces; and material flows, solid waste management, and materials recovery systems. Social or "soft" infrastructures (to which we return in the next section) include the institutions that provide and maintain cultural, health, and social aspects of life such as health services, schools and universities, sports and recreation facilities, and libraries.

Typically but not exclusively, even in the global North, physical infrastructures are owned by governments or some form of public utility, and are paid for by taxes, tolls, or metering. The main exceptions are energy and telecommunications networks that are usually privately owned and charge a fee for usage. Given the scale of projects, physical infrastructure maintenance and investment is often seen as a political hot potato, something to be avoided at all costs. This is economically as well as socially shortsighted: a well-functioning infrastructure is the basis of a functioning economy. As the *Economist* magazine reported: "Global spending on basic infrastructure—transport, power, water and communications—currently amounts to $2.7 trillion a year when it ought to be $3.7 trillion. ... And [the gap] is likely to grow fast."[65]

Judith Rodin, the president of the Rockefeller Foundation, makes a strong case for cities to invest in public transportation, especially bus rapid transit (BRT). She notes: "Transportation is, at its heart, about equity,"[66] and that while the NYC subway system is good in the central, most affluent area, it doesn't serve the disadvantaged residents in the outer boroughs equally well. But BRT—as proven in many Latin American cities (see our Medellín case study preceding chapter 4)—is more equitable, and in NYC it "was a successful ... answer in the days after Superstorm Sandy shut down mass transit. ... Dedicated buses ran on express lanes and transported commuters across some of the city's bridges."[67] With equity in mind too, the Barr Foundation in Boston has funded a BRT viability study by the Institute for Transportation and Development Policy (ITDP).[68]

The sharing paradigm helps reframe the debate around infrastructure in cities so that it is recognized as an investment in sustainable urban futures, and a shared resource, rather than a current financial millstone. In Seoul, pressures on hard infrastructures have driven exploration and adoption of sharing approaches—for example in transport and workplaces to alleviate congestion.

Seoul's investments in the Digital Media City and web access across the Seoul metropolitan region recognize the Internet as the modern day equivalent of roads—a non-territorial infrastructure that merits collective governance with shared norms.[69] Linus Torvalds, the founder of Linux, also sees open source software and technology as a public utility, like transport infrastructure.[70]

More recently, the concept of "green infrastructure" has entered the lexicon, focusing on interconnectivity, sustainability, and the "natural" environment in cities and the shared use of environmental and "life support" resources.

Shared Use of Environmental Resources

Natural common resources such as clean air and water are necessarily shared everywhere, but in cities the sharing process is more intensive than elsewhere. It requires close planning, intervention, and management, for which, fortunately, we can learn from the management of natural commons elsewhere.

Land is also intensively shared in the form of dense and high-rise development, in multi-purpose streets (see "Streetlife: Sharing Streets" in chapter 3) and public spaces—especially parks and other green spaces (see "Intercultural Public Spaces," also in chapter 3), where, of course, clean (or not so clean) air is equally a shared resource. The less visible sharing of water is epitomized by the (perhaps urban mythical) claim that tap water in London has on average already passed through seven people before being drunk. Such reprocessing and reuse of water will become ever more critical in the face of climate change.

Cities also share land and natural resources located outside their boundaries. The ecological footprints of cities are huge and their supply chains for food, water and materials can span continents.[71] As we will see later in the chapter, forms of urban agriculture are increasing, but most food eaten in cities is grown or raised elsewhere, often on land dedicated to intensive agriculture. Learning from the principles of the sharing paradigm, that land could be better managed as shared "multifunctional mosaics" enhancing food production, water management, soil protection, and biodiversity simultaneously.[72]

Materials used in all sorts of products from buildings to newspapers are also shared. Advocates of the sharing economy rarely talk about recycling and recovery of materials from waste. But this is sharing in turns just as much as car sharing or bikesharing. Some materials (e.g., surplus building materials) are recycled in a peer-to-peer fashion through Freecycle, but most recycling in the global North is conducted by state intermediaries, or their contractors, as part of the services of waste management. Even here, as we saw in Seoul, waste management can be undertaken as co-production between the city and its citizens (we examine co-production of services in more detail later in this chapter). In the global South recycling is dominated by informal "waste pickers." In cities such as Buenos Aires waste pickers are coming together in cooperatives, renaming themselves as *cartoneros profesionales* (professional recyclers) or *ingenieros callejeros* (street engineers) and collaborating to simultaneously obtain recognition, improve their conditions, and boost recycling rates.[73]

"Redistribution markets" such as Freecycle and Craigslist focus on the efficient reuse, in the sharing economy, of functional products that can still meet consumer needs. Recycling and recovery focus in parallel on the efficient use of the useful materials or components in no longer functional products. In this respect, sharing through redistribution markets is environmentally preferable to recycling and recovery, but the latter are forms of sharing nonetheless.

Recycling makes more efficient use of the resources such as paper, plastics, and metals that are extracted from our environment—and typically transported to cities through globally extensive supply chains—by sharing their functionality "in turns" between different products. Cities have become key players in recycling and recovery markets, as they are places where collection of recyclable material is very practical, *and* where the scale of supply and demand can sustain efficient reprocessing facilities. Following a series of campaigns and demonstration projects in Recycling Cities, such as Sheffield circa 1990, recycling rates have grown rapidly in the UK, and cities now face legal duties to provide curbside collection (as we saw in Seoul). Curbside collection improves participation rates and ensures higher quality recyclate. It also raises awareness of recycling, helping to increase purchase of recycled products and thus closing the participation loop of sharing by recycling. Local reprocessing is becoming common in cities: for example in New York City recycled paper is processed at the Visy Paper Mill on Staten Island.[74] Closing the loop locally not only offers environmental benefits, it can also prevent the exploitation of lower environmental standards elsewhere in "dumping" of hazardous or toxic wastes on disadvantaged communities elsewhere in the country or overseas.[75]

By raising recycling rates, cities cut their environmental footprints and reduce the impacts of the urban metabolism on the natural environment.[76] The higher the proportion of material recycled, and the greater the number of cycles, the more overall impact is reduced. The ideal is summed up in the visions of industrial ecology and the circular economy.[77] Here all wastes are seen as potential inputs; all end-of-life products are recovered for components and materials; all products are designed for subsequent disassembly and recovery; and innovation focuses on ways to minimize virgin material input. Consumers play a role not just in the co-production of new inputs by recycling but also in the co-production of new lifestyles that preferentially utilize the products of the circular economy. The Danish environment minister Ida Auken suggests: "Everything needs to be redesigned when we move from a linear to a circular economy. ... This will include new leasing and sharing models and creative ways for businesses to engage with their customers"[78]—and, we might add, new forms of sharing cities.

Since their earliest times, cities have been places of sharing. But they are not only places where sharing happens, cities also rely critically on shared resources, supply chains and infrastructures to function. Even the commercial functions of cities rely on the collaborative rules and institutions, and social trust that underpin markets. And shared infrastructures are typically provided by state and city authorities as public goods, funded by taxation. We saw above how hard infrastructures like transit are critical to modern cities—but soft infrastructures are no less important.

Public Services as Sharing

The third way in which cities are inherently shared spaces is in the social or "soft" infrastructure of shared services and facilities that make them so attractive as places to live. In this section we look both at the history of shared services (and "soft" infrastructures) and at ways in which they are currently shared.

Public services were integral to early urban centers. Even in the time of Aristotle, the city authorities of Athens intervened in food production to ensure that all citizens were provided with fair rations. In the Middle Ages, city governments ensured food supply for their residents through city-owned mills and granaries. Today, ask any incomer what are the "pull factors" which draw them to a particular city and—besides job opportunities—they will likely mention things like healthcare, childcare, education and libraries, and public transport; all services which are typically shared, and all services which are harder to provide at scale in rural areas, especially in less-developed countries.

In many countries such services are provided by national or municipal public authorities, although the specific remit of city authorities varies between countries and services. As we saw in the case study, Seoul's vision of sharing embraces such services. Government has even been described as the "ultimate level of sharing" in which we practice "collaborative consumption through societal organization of public services."[79] Rajesh Makwana, the director of the UK NGO Share the World's Resources agrees:

> Building on ... perennial human values and principles, modern systems of welfare are arguably the most advanced sharing economies ever established, and represent an important extension of the human capacity to share what we have in order to protect the least well off in our societies. Through the process of progressive taxation

and redistribution, we share a portion of the nation's financial resources … so that it can benefit society as a whole. In this way, sharing underpins the functioning of entire economies by ensuring that members of society take collective responsibility for securing basic human needs and rights for all citizens.[80]

The history of childcare, healthcare, and education as elements of the welfare state highlights not only the efficiencies of sharing at both macro and micro scales, but also massive potential for enhanced equity and participation. For example, kindergartens are more efficient than everyone having their own nanny; and universal access to shared services such as education enables wider participation in society. Access to childcare is particularly important for enabling women's participation. This is an example in which displacement of sociocultural sharing can be a clear step toward freedom and genuine development: when organized within networks of family, kin, and friends, childcare duties fall almost exclusively on women. This is why, within the US, New York City mayor Bill de Blasio's initiative to provide pre-kindergarten access to every 4-year-old in the city is an important signal. It still falls well short of provision in countries like Sweden, where all children are guaranteed heavily subsidized kindergarten places from turning one until starting school, and parents are funded to share a full year's parental leave at around 90 percent of full salary to cover that first year of a child's life. It also implies reliance on market models of childcare, where profits can take priority over children's needs, but de Blasio's move is still a step in the right direction.

Yet it is also controversial. In the US there is much less of a public consensus about the desirability of publically funded services in education, healthcare, and childcare than in most European nations. In the US, shared services are widely perceived as less desirable than individual private alternatives, and as an inappropriate reason for taxation, regardless of whether shared provision is more efficient and therefore cheaper than private. And with a much greater share of taxation locally based, shared provision in poorer districts is likely to appear much worse than private provision in rich districts. As we will see in chapter 4, greater tax sharing might also improve the prospects for shared public services in the US.

Yet, for all that the continuing successes of some Scandinavian nations in delivering high-quality education, childcare, and healthcare suggest—that such public services can be both effective and efficient—such sharing doesn't necessarily require direct "state provision." Recognizing the potential of public delivery to offer both efficiency and inclusion is important, but we can still acknowledge the potential downsides of bureaucratic approaches and look for ways in which service users (and taxpayers) can be involved in the governance and delivery of public services as we saw in the Seoul case study. In Quebec, for example, the majority of childcare is provided by nonprofit cooperative daycare centers run jointly by daycare workers and parent volunteers and financed by a combination of parent fees and state subsidies.[81]

In some ways our arguments might imply a larger state, with higher tax rates and higher spending, but at the same time they suggest a more localized state (virtually and physically) in which autonomy is distributed and citizens closely involved.[82] Participation is a critical step on this road. Seoul is one of many cities—now including Paris, France,[83] and Boston, Massachusetts[84]—imitating Porto Alegre's rightly celebrated participatory budgeting process in Brazil (see "Collective Governance" in chapter 3).

Even deeper participation—in planning and delivery as well as governance—is embedded in the idea of service co-production.[85] In the public sector, co-production takes a somewhat different form to the more recent visions of open sourcing, prosumers, and wikinomics discussed later in this chapter. The late Nobel Prize–winning political economist Elinor Ostrom is credited with coining the term "co-production" in 1970 to explain how public safety is jointly produced by the police and the community—a process which was being undermined at that time in Chicago by the removal of police from the beat, into patrol cars.[86] Healthcare similarly is necessarily an outcome of co-production between health professionals, their patients, and the families and communities that support those patients. These applications of co-production refer to what Marxists might call the "reproductive" economy (reproducing labor), which advocates of co-production have more recently termed the "core economy."[87]

Even in countries with strong public services, over-specialization has often combined with the erosion of social capital to raise costs and trigger a crisis of confidence in service delivery. A weakening core economy of care and support in families and communities has left specialist services uncoordinated and unable to care properly for the needs of the whole person, whether in education, healthcare, or social services. All too often in contemporary healthcare, patients and families are left reliant and subservient to doctors' expertise and although—with no other option—they are involved in delivering care, they are not actively co-producing the service (in the center left and bottom left cells of figure 2.2 below).

Figure 2.2 summarizes the variety of user-professional interfaces in the design, planning, and delivery of services. Typically, while professional service provision has not involved users or community members (top left cell), the other cells show differing levels from a variety of co-working styles,

		Responsibility for design of services		
		Professionals as sole service planners	Professionals and service users/ community as co-planners	No professional input into service planning
Responsibility for delivery of services	Professionals as sole service deliverers	Traditional professional service provision	Professional service provision but users/communities involved in planning and design	Professionals as sole service deliverers
	Professionals and users/communities as co-deliverers	User co-delivery of professionally designed services	Full co-production	User/community delivery of services with little formal/ professional
	Users/communities as sole deliverers	User/community delivery of professionally planned services	User/community delivery of co-planned or co-designed services	Self-organized community provision

Figure 2.2
User and professional roles in the design and delivery of services.
Source: David Boyle and Michael Harris, *The Challenge of Coproduction*. London: NEF, The Lab and NESTA, 2009, http://www.nesta.org.uk/sites/default/files/the_challenge_of_co-production.pdf; Bovaird (2007; note 85, this chapter).

with users, other community members, or with both, to full co-production (center cell).

Some aspects of co-production have emerged where administrations have emphasized "citizen centricity," that is, "where citizens themselves play a more active and ongoing role in defining and even assembling the basket of services they need."[88] This would fall in the top center cell of figure 2.2. To this end, Canada established a "single agency focusing on development, management, and delivery of all social services for all citizens."[89]

But co-production at its best means more than a focus on the citizen as service user: it means understanding citizens, as the philosopher Immanuel Kant argued, as "ends" in themselves, not means, and building their capabilities. In other words, it means citizens and public bodies not only co-producing healthcare services, but also co-producing the outcome of good health. Full-blown service co-production initiatives (those in the center cell in figure 2.2) share seven characteristics that illustrate their commonality with the sharing paradigm as we have described it. As outlined by Lucie

Stephens, Josh Ryan-Collins, and David Boyle, writing for the New Economics Foundation, they: provide opportunities for personal growth, treating people as assets, not burdens; invest in building empathy and capacity in local communities; use peer support networks to share knowledge and capabilities; blur the distinction between producers and consumers of services; turn providers into catalysts and facilitators; devolve authority and encourage self-organization; and offer incentives which help to embed reciprocity and mutuality.[90]

Stephens and her colleagues emphasize the preventive nature of co-production: its capacity to provide mutual support systems and social networks that prevent problems arising or becoming acute, and that help people continue to thrive when they no longer qualify for all-round professional support. Whether we call this "social capital" or "community solidarity," it is the same inclusive fabric that the sharing paradigm seeks to weave and repair.

Co-producing Good Health

Lehigh Hospital near Philadelphia offers a simple, if limited, example of co-production in practice. On release from the hospital, patients are "told that someone will visit you at home, make sure you're OK, if you have heating and food in the house."[91] That visitor will be a former patient, sharing their knowledge and experience; and the new release is expected to pay the same favor forward. The program has dramatically cut re-admission rates and built a support community in which many former patients "come back time after time to help out in the neighborhood. They stay in touch, not just with the hospital, but with the patients they visited."[92]

In Stockholm co-production is being harnessed to save lives in emergency situations. The SMSLifesaver program enrolls people with training in cardiopulmonary resuscitation (CPR) as volunteers. When a call comes in to emergency services, as well as summoning an ambulance, a text message is sent to all volunteers registered within 500 meters of the person in need. SMSLifesaver volunteers reach victims before ambulances in 54 percent of cases. This has contributed to a near quadrupling in survival rates after cardiac arrest, from 3 percent to nearly 11 percent, over the last decade.[93]

In Greece, drastic austerity cuts have left many without free access to healthcare. In response, volunteer doctors have established thirty "solidarity clinics" in a communal co-production model. These

> not only provide free health care to those who have lost access, but … have a different way of thinking about health and heath care—basing it in horizontal assemblies,

breaking down hierarchies between patients and medical professionals, and seeing health as an overall question, not a sum of body parts. All decisions related to the running of the clinics are made by the assemblies, comprised of those who volunteer in the clinics and ... those receiving care from them.[94]

Online, collaborative health care is particularly useful for rare conditions. Web platforms like WeAre.Us and Patients Like Me enable sharing between sufferers and their caregivers, and can provide a much richer source of experience and information than local doctors who rarely encounter more than one patient with the condition.[95] Such online shared platforms ensure access to "a second opinion" and also permit anonymous or depersonalized access to advice, overcoming the shame or embarrassment that might be experienced in face-to-face consultation. Such groups also provide valuable support and reinforcement for preventative health activities, and even offer direct psychosocial benefits to health. By helping patients feel more in control of their lives and by overcoming isolation,[96] they turn patients into agents, both building their capabilities,[97] and directly reducing illness rates. In a definitive review for the World Health Organization, the public health epidemiologists Richard Wilkinson and Michael Marmot make clear:

As social beings ... we need friends, we need more sociable societies, we need to feel useful, and we need to exercise a significant degree of control over meaningful work. Without these we become more prone to depression, drug use, anxiety, hostility and feelings of hopelessness, which all rebound on physical health.[98]

Orsi and her colleagues also identify the preventive health potential in sharing, and suggest an overlap between the sharing economy and co-production. They propose: "Instead of relying upon emergency rooms, preventative eldercare can be delivered through a peer-to-peer marketplace or a time dollar program."[99] Time dollars are the medium of exchange of time banks, in which participants build up credits by helping others. (We return to time-banking and discuss the role of complementary currencies more generally in chapter 4.)

Learning Together: Co-producing Education

Education is another space where co-production appears to have massive potential. All around the world, contemporary education is increasingly dominated by competitive approaches with frequent testing and ranking of students. Teaching methods are primarily one-to-many, where one teacher "educates" many students. And although participation in governance is widespread, with parent-governors and parent-teacher associations common, often the only hint of co-production in delivery is the unfunded delegation of anything outside the core curriculum—such as school sports clubs—to parents or the wider community.

The residual core educational process is proving singularly ineffective at producing flourishing individuals. Such systems stifle creativity and innovation, says the international education advisor Ken Robinson.[100] The geographer Danny Dorling is even more scathing. Testing, ranking, and the tendency to exclude disruptive students from school conspire to reproduce inequality, he suggests.[101] These actions establish elitism and ideas of inherent inferiority as norms, and they damage not only the development of those labeled as inferior but that of society as a whole, discouraging cooperation at all scales. Collaborative learning is less uncommon in higher education, not only in conventional educational institutions, but also in the emerging ecosystem of massive open online courses (MOOCs), and P2P skillsharing enterprises on shared web platforms. For instance, Skilio is a mediated commercial peer-to-peer skillsharing marketplace (currently in beta), where people can share their experience on any subject—from brewing to programming—via a web conferencing platform.[102] Members create biographical profiles with details of the skills they offer and what they would like to learn. "Educators" and "learners" jointly agree the fees and schedule for their online skill sharing. Skilio provides the platform and handles the payment, for a 15 percent commission. FutureLearn, the UK's first MOOC platform, seeks to trigger more student collaboration, rather than one-to-one interactions. It "aims to build a community around its courses, and group discussion is used as a tool to support participatory, collaborative learning."[103]

Genuinely student-led collaborative peer-to-peer learning, however, is rare at any level. In such approaches the students ask questions and explain things to each other and lead group discussions while the teacher facilitates and observes. This helps students see their peers both as equals and as resources, reveals their capabilities for collective problem solving, and promotes a sense of shared responsibility.[104] Access to information and learning resources is made easier by Internet access, and even participation can be enhanced in online learning environments, where real-life inhibitions can be alleviated.

Whether face-to-face or online the methods are powerful: students learn to be proactive yet also to depend on their peers. The education blogger Matt Davis notes how "students are encouraged to share and to listen to each other's individual interpretations ... underscoring the notion that there can be multiple right answers."[105] This approach teaches "dialogics"

–an expanded understanding achievable even without finding "common ground"—rather than a dialectic pursuit of consensus or one right answer.[106] In this way it better prepares students for life outside the academy, while also enhancing their academic performance.

Self-organized education takes such ideas even further, building on sociocultural sharing traditions. Sugata Mitra, a professor of educational technology at Newcastle University in the UK, has conducted a series of experiments providing access to computers for children in India (the so-called hole in the wall experiments). The results were stunning: without any instruction or supervision, they were able to teach themselves a surprising variety of things, from DNA replication to English. The methods have proved transferable into classrooms in the form of self-organized learning environments (SOLEs). Here the educator poses a question, relevant to the course, and the students engage with it in self-organized groups, with access to all the resources of the web. Both teachers and students report learning more, greater retention of knowledge, and more enjoyable learning experiences.

In SOLEs, learning is

> distributed and democratized rather than individualized. This frustrates the evaluation of children in terms of more or less "ability" ... since the children's movement between groups, taking knowledge with them, "stealing" knowledge through sharing or building from one group to another, leads to a more uniform learning across the class.[107]

In this form SOLEs can be supported within "both a human capital model of education within the neoliberal paradigm and a progressive child-centered model focusing on the importance of creativity and transformation."[108] They also have the potential—with greater student autonomy in the setting of questions—to become a "radical transformative pedagogy, working in antagonism to the dominant framework of largely individualized learning ... without a need for teachers to position themselves as contesting the curriculum."[109]

Such approaches are used in the shack dwellers University of Abahlali baseMjondolo, in Durban, South Africa, established as "a space for the creation of knowledge about survival, hope and transformation, where the shack dwellers themselves are the scholars, the professors and the teachers."[110] Here the knowledge of "marginalized and excluded people" is co-created, valued, and shared. The philosophy stands in direct opposition to "knowledge enclosure" whereby elites "acquire knowledge for the purposes of gaining political, spiritual, and cultural leadership and power."[111]

The principles of self-organized learning apply outside the classroom, too. Children who engage in more free play have more highly developed cognitive skills such as organization, planning, self-regulation, initiative, and ability to switch between activities.[112]

Self-organized education reflects a proud tradition of cultural and social theory. The philosopher and social critic Ivan Illich advocated the "deschooling" of society in 1971,[113] emphasizing his belief that conventional education was geared to creating "good" consumers, whereas people *learn* better by themselves, at their own pace, and outside the constraints of an institutional environment. More recently the author and activist Charles Eisenstein has taken up the baton, arguing a need to undo classroom "habits like looking to authority for answers and instructions, following a program determined by someone else, needing to be right, addiction to meaningless praise [and] conditioning to dull, trivial work to obtain external rewards."[114] Such habits powerfully influence the development of children's identities and sense of self, and reproduce the cultural meanings of dominant elites as the basis for knowledge in the education system and behavior in society, what the late sociologist and philosopher Pierre Bourdieu called "habitus."[115]

It is therefore in the rejection of both the habits and habitus of the classroom in techniques of popular education,[116] life-long education,[117] and self-organized education that we might anticipate the development of more independent selves (or identities), exhibiting the evolved intrinsic values and empathy that particularly enable effective collaboration and sharing.

Co-production: Disowning Responsibility?

Just as we are alert to the downsides of the sharing economy, however, we must acknowledge the potential downsides of co-production of public services like education and healthcare. Too often it is seen—under the pressures of neoliberalism (and the "post-political trap" mentioned in the introduction)—not as a complement to professional services, but simply as an opportunity to cut costs and enable the closure of collective facilities. In particular, the individualization of budgets can allow privatization of care services by stealth and, at worst, risks abuse of vulnerable individuals who need a long-term responsible caring relationship, not a service bought on the market on a day-to-day basis.[118]

This is not just a hypothetical concern. Under Margaret Thatcher in the UK and Ronald Reagan in the US, mentally ill people were released from institutions to so-called Care in the Community, but often enough the alleged beneficiaries of this policy became its victims, and ended up living on the streets. Ironically the healthcare and policing costs of homelessness have become so great that some progressive city authorities are acting on the realization that it is cheaper to provide free housing. In Utah, for example, city resources are used to provide houses and social-worker support for the homeless, resulting in a net saving to the public purse.[119] Yet such approaches persist. Prime Minister David Cameron in the UK has attempted to encourage small groups, charities, and business to play a role in a range of welfare services under the rubric of the "Big Society." The policy was rolled out in a period of fiscal austerity and significant spending cuts, largely demolishing any credibility it might have had. Sennett was scathing: "It's fair to liken the 'Big Society' idea of David Cameron to economic colonialism ... the local community, like the colony, is stripped of wealth, then told to make up for that lack by its own efforts."[120]

While public spending cuts are a poor way to encourage co-production, individualized budgets may have a place. Stephens and her colleagues argue: "Clients often know best what priorities they have and how the money allocated to them should be spent. ... [individual budgets are] also a way that service users can play a role in their own development."[121] But, they stress that "'self-directed support,' in which money is only one of many assets on which people can draw,"[122] would be preferable, or even "mutual support networks, backed by 'community budgets' ... [or] mechanisms that allowed people to pool their budgets collectively when they chose."[123] These arguments over individualized budgets echo those over a citizen's income (discussed later in this chapter), which political liberals like for the freedoms it implies, and which libertarians may support as a route to rolling back the state.

Tapscott and Williams highlight further practical issues that governments encouraging co-production will need to address:

> Among other things governments will need to guard against threats to consumer privacy, data security and the potential misrepresentation of government content and services. ... If companies and nonprofits build new interfaces to popular services, who's accountable if something goes wrong? If a nonprofit or business folds, who will ensure service continuity or take custody of personal data? Will governments guarantee that ... services remain accessible to citizens irrespective of their income bracket?[124]

As in the more commercial end of the sharing economy, the specific measures are perhaps less important than the structures that determine the ethos of the sector. There are real limitations of collaborative healthcare—especially in market systems. Patients, it is said, "make poor shoppers."[125]

Institutional reform must enable the same culture of collaboration among the professionals tasked with delivering the support—whether they are in public employ or in private or cooperative enterprises—as it does among patients. If the driving ideology is one of saving taxpayers' money, or rolling back the state, co-production is unlikely to flourish. Yet with commitment to co-production as a preventative, supportive, and participative whole, it is likely that the costs of public services will fall—or at least not continue to escalate.

In Quebec, for example, social economy services that "provide the kind of ongoing practical support that makes it possible for the elderly to stay in their own homes ... like house-cleaning, meal preparation, shopping assistance, and odd jobs"[126] offer a practical alternative to residential care. A network of nonprofit and cooperative homecare businesses supports over 70,000 elders, employing around 8000 caregivers. Its existence and vibrancy owes much to the Chantier de l'économie social, a "network of networks" promoting the social economy.

So, as with so many of the examples and initiatives we have reported in these pages, co-production of care services is enabled and enhanced at the intersection of urban space and cyberspace. All the elderly support services provided in Quebec are available in the sharing economy in almost any city today, but without coordination and support, it would be wrong to expect elderly people to rely on them. Yet, simply providing face-to-face contact with care staff by using services such as Skype on a robot platform can go a long way to enabling independent living for some ill, elderly, or disabled people.[127] In Barcelona, the city is establishing a collaborative care system for all elderly residents, employing digital and low-tech strategies to make sure every citizen over 65 has someone who checks in on them regularly.[128] More ambitiously, city agencies could act as online service aggregators to package together sharing economy services for the elderly, disabled, or vulnerable. We are not suggesting that all caring services will be delivered online—only that such services can be better coordinated through the enhanced connectivity and social networks of cyberspace, and more easily delivered (and seamlessly integrated with formal care such as hospital visits) in the urban arena.

In this section we have seen how shared services—or "soft" infrastructures—are a fundamental plank of urban living. The wiring up of modern cities offers new opportunities for involving citizens more fully in the design and delivery (or co-production) of services such as health and education. But, as so often we have seen, careful planning and management will be critical if the new models are to be forces for democracy and justice, rather than tokenistic excuses for public authorities to disown responsibility while cutting service funding. Next we ask: If co-production can be a force for equality and democratization in public services, can the same principles be applied in the conventional economy?

Co-production in the Conventional Economy

So far in this chapter we have discussed market rules, infrastructure, resources, and public services—which are all part of the *settings* for a productive urban economy. Here we turn to explore forms of co-production that are also emerging within the "productive" or conventional economy of consumer goods and services, including open source software, domestic microgeneration of energy, and self-build homes.

The trends involved range massively in scale and significance. Self-assembly of furniture purchased from the increasingly ubiquitous IKEA stores is clearly co-production, but in a minimalist and indeed unhealthy form: the consumer (or some hired help or maybe a TaskRabbit) transports the furniture home and physically assembles it. The model has allowed IKEA to cut costs, not just in store space and transport, but also by shifting production of these no-longer so bulky goods to low-wage, nonunionized locations. In this particular emanation of co-production the consumer has become complicit in the exploitation of their own labor. But in other spheres, co-production involves people reclaiming and reinventing work. Even when commercial in motivation, in many cases it is conducted without formal intermediaries in contemporary expressions of sociocultural sharing behavior. Open source software is perhaps the most widely cited example. According to Tapscott and Williams, Linux is the "quintessential example of how self-organizing, egalitarian communities of individuals and organizations come together—sometimes for fun and sometimes for profit—to produce a shared outcome."[129]

Many contributors to Linux code are paid by existing employers who use Linux as a whole in their business, and benefit from having in-house experts in the software. Such approaches are communal and self-stimulating, for example as new programmers, documenters, and debuggers are recruited from among the user community. Participants may benefit financially through skill development and enhanced reputation. Open source approaches are increasingly seen as commercially competitive. The former *Wired* editor in chief Chris Andersen believes that open source "community" business models will ultimately win in most markets.[130] In South Korea the open source software sector is seen as so important that public administrations, including the Seoul Metropolitan Government, are officially expected to use open source to boost the country's skill base in the sector.[131]

But the exercise of "consumer labor" in the open source movement can also represent active rejection of the logic of capital. Even in commercial settings, co-production often engages people's individual and collective capacities to invent, create, shape, and cooperate, without direct monetary incentive. Intrinsic motivations for contributing to open source software can include intellectual stimulation, fairness, a sense of an obligation to "give something back" (indirect reciprocity), and—particularly—group identity.[132] In one survey of contributors to open source software, 80 percent of respondents said the hacker community was their primary source of identity.[133] The outcomes are also potentially socially useful, for example, helping reduce barriers to "digital inclusion" and disrupting monopolistic corporate power. Clive Thompson suggests that it is actually easier for community purposes to benefit from the "collective intelligence" such cooperation generates. "Motivation is a problem," he says. "Few people think profitable companies deserve their free work,"[134] and commercial incentives therefore crowd out public spirit.

The international development policy advisor James Quilligan notes:

> When [commons] users are directly involved in the production and delivery of goods and services, they develop cooperative skills, knowledge and wealth beyond the constraints of extractive profits, patents, trademarks, copyrights, traditional ownership, paid work, commodity values and other value-added measures. Social [co-]production thus entails not only new forms of property management, but also a different measure of value.[135]

Some spaces of commercial co-production also offer major potential social and environmental benefits. Domestic or community scale energy generation—for example—can be both liberating and transformative. However, the capital costs are currently such that its spread is only genuinely socially inclusive in countries with high subsidy or support regimes, such as the German or (until recently) Spanish feed-in-tariffs. Here millions of residents have benefited, not only relatively wealthy homeowners. Elsewhere incumbent energy utilities have actively made the transition harder, despite the potential for city-scale aggregation. They have resisted measures designed to ease connection, and offered tariffs well below market rates for the electricity they buy from microgenerators. In Arizona, for example, the state utility has imposed a charge on household solar owners for supplying energy back to the grid, and promoted an initiative to increase property taxes to include the value of leased solar systems.[136]

Sharing models—both cooperative and rental—have helped accelerate and widen uptake of solar PV (photovoltaic) technology in US cities. Consumer cooperative models are spreading, with some—like DC Solar United Neighborhoods—aggregating householders who want solar on their own roofs, and others—such as the Clean Energy Collective, based in Carbondale, Colorado—offering the opportunity to collectively invest in solar gardens, which are good for those who rent their home, live in an apartment, or have too much shade on their roof.[137] Solar rental companies like Elon Musk's Solar City, or Sunrun, treat solar panels as a product service system. Such models appeal to those who cannot afford to buy (or cannot obtain a loan to buy) their own panels or to those who cannot face the administrative hassle involved. Less ethical companies will "rent" roof space to install solar panels, leaving the householder with a second-class service. In the UK some installers using this model have promised householders free electricity—but only when production is high and grid demand is low—meaning at inconvenient times.

Nonetheless, rental models allow many more households to join the market—including those without access to capital to invest upfront. In the US this group is disproportionately comprised of ethnic minorities—something the incumbent companies have been quick to highlight in their campaigns to hinder the spread of rooftop solar.[138] From our perspective, recognizing that ethnic minorities and indigenous groups also suffer disproportionately from the pollution and risks caused by conventional fossil and nuclear power,[139] the solution is not to resist domestic energy co-production and storage, but to subsidize and support it in ways that mean no one is left unable to participate, nor with a second-class service.

To move microgeneration and community generation of energy center stage everywhere will require a smarter, more open grid system. The conventional energy grid is a centralized broadcast model—one-way, one-to-many.[140] The future shared grid will need to be like the Internet, and indeed will rely on the technologies of the web, providing a platform for the delivery of energy services—including convenient demand management and distributed energy storage facilities—as well as being an aggregator and balancer for the produce of many small-scale generators. Working with smart appliances and switching systems, a decentralized smart grid can better share peak generating capacity, marginally reducing unnecessary consumption in millions of fridges and freezers for example, or releasing previously stored energy—from hot-water tanks, or electric vehicles—at peak times.

The realization of a smart, shared grid is beginning at the city scale. Some countries with city-level energy utilities—like Germany—have already shown that microgeneration can be a major part of the energy system. The German model has also delivered innovation and economies of scale in local area-based programs for energy-demand reduction. The proKlima cooperative in Hannover is a leading example, promoting a city-based approach, having pioneered the achievement of the Passivhaus Institut standard (which requires virtually no energy inputs for heating or cooling) in retrofitting existing homes to cut energy use.[141] Funded jointly by the local authorities and the local energy company, proKlima has an advisory board including workers, suppliers, customers, and environmental and consumer protection organizations. It also organizes courses for unemployed people with technical backgrounds to become energy efficiency advisers.[142]

In Germany as a whole, "citizens, cooperatives, and communities own more than half of German renewable capacity, vs. two percent in the US."[143] This level of co-production has helped push remarkable levels of renewable generation. Renewables accounted for 27 percent of German electricity generation over the first quarter of 2014, for example. Moreover, according to the UK think tank Respublica, approximately "half of the capital borrowed by co-operative energy groups comes from co-operative banks, [which] … have a much greater understanding of the models and risks involved, and a greater level of trust in the businesses that they lend to."[144]

As smart grid technologies and smart meters are rolled out in electricity networks across the world, the potential for householder participation in the energy system will grow exponentially. However, smart meters—like much of the Internet of things—also raise concerns over data privacy. Consumer resistance has emphasized not only fears that they might be used inappropriately to control household appliances, but also that they could reveal too much about personal habits. In a surveillance society such concerns may seem unavoidable, but they should be treated as another serious reason to resist the spread of unnecessary surveillance—and to reverse the slippery slope where more surveillance leads to less public trust and declining social capital, which in turn appears to demand more surveillance.

Co-production extends from generating energy to building houses. Self-building is a growing niche in the rich world, and an essential necessity for many in the global South. In the global North, self-build often delivers higher energy efficiency and lower environmental footprint than the commercial construction industry, and high efficiency is a key selling point for many of the self-build kits on the market. In the UK, self-build has grown to around 20,000 homes per year (about 8 percent of new construction) and is reported to cut the costs of housing by up to 50 percent.[145] However, statistics published in 2012 indicate that only around 1 percent of those who would like to self-build actually get the opportunity.[146]

In Northern countries like the UK, despite historical traditions of hutting[147] and plot-lands,[148] self-build has long been something of an elite privilege. The majority of new self-build projects, in the UK at least, are detached homes in rural or suburban settings. However there is a new wave of cooperative cohousing self-build projects such as LILAC in Leeds, which has imported a cooperative mutual ownership model from Scandinavia. This makes the development affordable even to those on low incomes.[149] The emergence of a trend to microdwellings—more in tune with the minimalist aesthetic seen among Millennials—is also extending the affordability of self-build.

In the global South informal self-build is critical, but there is wide variation in the support provided. Some cities support the formalization of informal settlements. Elsewhere such developments remain precarious, more vulnerable to natural disasters and at constant risk of clearance on political whim or in the interests of development. Some self-build is directly supported by NGOs. Local Habitat for Humanity (HfH) affiliates for example, build and renovate houses in partnership with people in need, who then can purchase the home by repaying a "no-profit" mortgage. Mortgage payments contribute to a revolving fund that provides capital to build more houses.[150] Shack/Slum Dwellers International (SDI) is a bottom-up movement, which has grown from modest beginnings among "pavement dwellers" in central Mumbai in the 1980s to operate in more than 20 countries today.[151] SDI uses residents' knowledge attained through building houses and infrastructures to press for access to land and resources for poor people to build their own developments; it employs strategies—such as full-scale self-built house "modeling" as a means to negotiate with local authorities—replicated through an informal peer-learning network.[152] Today SDI affiliates in hundreds of informal settlements are involved in profiling facilities, collaborating with academics to co-produce influential information, establishing city-wide community managed revolving funds accessible to the poor, improving public spaces in collaboration with street vendors and waste pickers, and in-situ upgrading of settlements and building new homes.[153]

Models like these actively involve the community in both implementation and decision making, and help build a wide range of capabilities, say housing researchers Ivette Arroyo and Johnny Åstrand from Sweden's Lund University. On the other hand, typical top-down models of "aided self-help housing," such as the "sites and services" approach supported by the World

Bank, don't fully involve users in design as well as delivery, generate urban sprawl, and fail to provide the finance and capacities for subsequent incremental improvement.[154]

In energy and housing, co-production is clearly significant in both commercial and communal forms, and there is potential for it to spread significantly.

Peer-to-Peer Finance and Crowdfunding

Co-production models in finance, on the other hand, are already booming, stimulated by plummeting levels of trust in the conventional finance industry following the global financial crisis and bank bailouts.[155] P2P lending is predominantly mediated rather than sociocultural in form, but once again we find a tension between commercial and communal models. In P2P models web intermediaries bring together borrowers and lenders, who agree on rates, often via a "bid and ask" process analogous to eBay. Some have mechanisms to spread risks across multiple investors, but most loans are unsecured and the model generally relies on building a sense of community to achieve low default rates. In theory, by cutting out the conventional intermediaries and complexities of bankers, credit card operators, and predatory pay-day loan companies, P2P lending can deliver both higher rates to lenders and lower rates to borrowers.

Peer-to-peer lending in the West is forecast to grow rapidly from 500 platforms sharing $3 billion in 2013 to over 1,000 platforms lending $17 billion in 2015[156] benefitting from tighter reserve requirements on conventional lenders and new rules in the US Jumpstart Our Business Startups (JOBS) Act of 2012, which substantially loosened crowdfunding rules for businesses.[157] The new rules have shifted the center of gravity of the sector away from personal projects to business purposes, and leading companies in this sector appear to be swiftly commercializing. San Francisco–based Lending Club, the largest P2P lending company, is valued at $3.76 billion and appears set for a stock market initial public offering. It recently acquired the more conventional finance company Springstone, which specializes in financing elective medical procedures and private school education.[158] Web-based P2P lending arrived in Europe before it did in the US with Zopa founded in 2005.

P2P finance is also now widespread in East Asian countries like Korea and China. China already has the largest P2P finance sector in the world. Demand is growing fast, partly because banks have tightened credit restrictions. And conventional finance companies are also competing to offer

P2P services. By June 2014, there were 944 predominantly commercial P2P lending platforms, lending $3 billion a month on average. By the end of 2014, monthly lending was expected to reach $5 billion on 1,300 platforms.[159] The strength of China's P2P sector might seem to reflect the benefits of the *guanxi* tradition, which sustains a net of mutual obligations throughout Chinese communities. *Guanxi* would suggest that borrowers naturally recognize their obligation to the lenders, while lenders accept the possibility that such an obligation might be repaid in other ways if necessary, or by other members of the extended *guanxi* community. But this does not appear to be the case in this commercially dominated sector, with 74 platforms declaring bankruptcy in the last quarter of 2013 alone.[160]

Some platforms elsewhere, however, seem to be making more effort to help lenders align their lending with their values and maintain the long-term sustainability that comes from a community base. Although VenCorps sees itself as an "American Idol" for entrepreneurs, it is seeking to extend P2P values to venture capital—crowdsourcing its evaluations as well as investments, using "reward points" (a form of alternative currency) to pay its crowd community, and targeting some of its funding at solving specific social problems. For example, one VenCorps competition targeting traffic congestion led to it funding a ride-sharing web platform—iCarPool. Similar values are emerging in the idea of "hybrid finance" for social entrepreneurs. The marketing entrepreneur Ellinor Dienst and the former McKinsey consultant Markus Freiburg started the Financing Agency for Social Entrepreneurship in 2013 to aggregate funds from private investors prepared to accept lower rates of return than venture capitalists, and deliver them to social enterprises in need of patient (i.e., long-term) capital.[161] Many sharing enterprises could clearly benefit from such finance, rather than having to dump their social values to secure venture funding.

Among the most inspiring current models is Kiva City—a P2P lending platform launched in the US in 2013. It makes interest-free loans to small businesses in disadvantaged areas of cities like Pittsburgh and Oakland. In participating cities, the city authority, community organizations, and local financial institutions such as credit unions work together to support the scheme. It helps entrepreneurs who might struggle to obtain conventional credit, but who can win endorsement from local community organizations. In Oakland the program has supported a local store that provides supplies for seed-to-table farming and gardening, and in Pittsburgh an indoor mountain biking and BMX park. Community interest and the web platform intersect to create shared obligations that have kept repayment failure rates to just 12 percent. Another Pittsburgh beneficiary describes the moral imperative to repay the loan:

> I feel obligated to make sure that payment is made when it's supposed to be ... Ninety-five total strangers lent me the money to get started; to me that speaks volumes. You don't want to let anybody down who believes in you enough to contribute to your funding.[162]

Kiva City utilizes KivaZip, a direct interest-free loan mechanism using PayPal and mobile payments that is also used by Kiva internationally. Kiva began as a crowdfunded way to support international development, enabling microfinance online. Kiva channels most of its lending through existing microfinance field partners whose high interest rates (often over 30 percent) are presumably reduced by the availability of funding at zero percent interest, but is developing KivaZip as a lower cost way to provide finance directly. Since its founding in 2005, Kiva has recruited over one million donors and lent more than $500 million at a repayment rate better than 98 percent,[163] successfully extending sharing links across international and cultural boundaries. Eighty percent of Kiva's international loans support initiatives by women, who are more often excluded from conventional finance. Kiva believes its investors are more engaged with the borrowers than typical charity donors; the loans pay no interest and the majority of investors recycle repaid loans into new lending, which suggests that they see their involvement as a form of charity, rather than as investment. Indeed Kiva itself is charitably funded, rather than taking a cut from the loans or repayments.[164]

Kiva was inspired by the Grameen Bank and other offline microfinanciers who have practiced peer-to-peer lending in developing (and some developed) countries for many years. Tapscott and Williams bullishly claim microfinance as a success: "The aggregate results, notably 100 million customers with a repayment rate in the high ninetieth percentile, have proven that a networked and largely self-organized, system of peer-to-peer lending not only can work, it provides a sustainable way to lift millions of people out of poverty."[165]

But closer examination finds no robust evidence for either poverty alleviation or women's empowerment,[166] and suggests that high repayment rates may come at harsh cost to both borrowers and their wider communities, while many loans are used, not productively, but to pay for healthcare or immediate consumption needs.[167]

The Cambridge economist Ha-Joon Chang argues that the weaknesses of microfinance are partly because its promoters have insisted on an individual, rather than collective, understanding of entrepreneurship.[168] Despite high levels of individual entrepreneurship in the global South (75 percent of all Bangladeshis, for example), the lack of institutions that could allow these self-employed individuals to build large enterprises is a much bigger barrier. Rather than helping, microfinance exacerbates these problems because of its focus on individuals, rather than on collectives, such as cooperatives.

Yet microfinance has not been an unmitigated failure. Despite his concerns, policy consultant David Roodman argues: "Sustainably extending the financial system to poor people *is* development. … Poor people deserve access to financial infrastructure just as they deserve access to clean water, sanitation, and electricity."[169] But the critiques highlight the underlying problems of the now-dominant model of microfinance, in which commercial financiers lend small sums at (extremely high) market interest rates to individual borrowers who would otherwise struggle to access them. Kamal Munir of the Judge Business School at Cambridge University describes this model pointedly as "minimalist," and highlights how it came to dominate, as mainstream international financial institutions got more involved in microfinance

> attracted by … tales of helping the poor and making a buck at the same time. With international capital, however, came unprecedented pressure for growth and quarterly profits. Those providers who tapped into the equity markets responded by seeking out more borrowers, and then when defaults loomed they tightened the screws to keep things on track. They devised elaborate public shaming rituals and used these ruthlessly to destroy borrowers' social capital.[170]

There are clear parallels here with the domination of commercial models of sharing by venture capital and their co-option by neoliberalism (described in chapter 1). The problem is not the mediated model, or the sharing principles, but the way it is commercialized in attempts to scale up swiftly.

Like Roodman, Munir supports positive alternative models of microfinance, such as "socially embedded microfinance institutions that organize entrepreneurs, provide them with training and then deploy them in larger ventures."[171] These, he says, "are much more effective though high-cost propositions. Interest-free microfinance, based on charity, similarly offers much greater relief." Roodman also sees opportunities in phone banking to extend financial inclusion, taking advantage of the same technologies Kiva has used to develop KivaZip.

Alongside microfinance, the other progenitor of peer-to-peer lending is crowdfunding. Famous for helping fund Obama's presidential campaigns, web-based crowdfunding has become increasingly common in media (as well as in politics). In music and film, crowdfunding—using sites like Kickstarter—is a reaction to the same web-enabled trends that have undermined conventional models of production and marketing, but it is largely communal, rather than commercial, in motivation and structure. High-profile examples, such as *The Age of Stupid* film that pioneered crowdfunding in 2004 and the music of Amanda Palmer (supported by 25,000 Kickstarter investors), show how crowdfunding shares much more than money. Supporters typically exhibit communal values and play a part in co-marketing the end product. As Palmer says, "There's just something magical about Kickstarter. ... You immediately feel like you're part of a larger club of art-supporting fanatics."[172]

In the five years from 2009 to 2014, Kickstarter accumulated over $1 billion in pledges by more than 6 million people, funding more than 60,000 creative projects—a success rate of around 44 percent of all projects posted on the site.[173] Kickstarter uses Amazon's payment platform and takes a 5 percent fee from successfully funded projects, but backers don't get a financial share in the project. "Instead, project creators offer rewards to thank backers for their support. Backers of an effort to make a book or film, for example, often get a copy of the finished work. A bigger pledge to a film project might get you into the premiere—or a private screening for you and your friends."[174] In these ways crowdfunding best reflects the claim that peer-to-peer lending platforms enable lenders to reflect their values in their investments. Yet even Kickstarter, unable to maintain its original ban on business startup funding, has come under fire for losing its soul to commerce.[175] Yet many of the commercial products promoted through Kickstarter—such as the Kano computer kit—themselves embody ideas of co-production. Kickstarter has responded by clarifying its guiding philosophy: now it simply demands that "projects must create something to share with others."[176]

As in other sectors, online mediated collaborative finance is experiencing tensions between social and commercial purposes, in which questions of ownership, participation, and accountability will loom large. We return to the prospects for cooperative financial organizations later in this chapter, but first we consider another booming—yet contentious—segment of co-production: growing food.

Growing Together ... or Growing Apart?

In chapter 1 we saw how food consumption is a boom area of mediated sharing. But co-production in the food system is perhaps even more common.

Many cities provide allotments, or shared community gardens, where residents can grow their own food. These are typically sites of sociocultural, informal sharing, and in many cities represent venues for deliberate efforts at intercultural mixing. Community gardens deliver benefits for both the individual and the community, including an increased sense of attachment to the neighborhood, and the added benefits of increased physical activity and better nutrition.[177] But they are also contested spaces, and sadly often temporary as developers dominate the competition for urban land.

Mediated sharing has a foothold in this space too, enhancing the efficiency with which remaining garden land is used. For example, in the UK, Landshare was founded in 2009 by the celebrity chef Hugh Fearnley-Whittingstall to bring together "people who have a passion for home-grown food, connecting those who have land to share with those who need land for cultivating food." It now links over 70,000 growers, sharers, and helpers.[178] Tuintjedelen (Sharegarden) is a similar not-for-profit Dutch initiative, funded by charitable foundations.[179] Such approaches are clearly valuable for increasing local food production and building social interaction in cities without a good allocation of land for allotments or other forms of communal gardens (i.e., most cities).

Making good use of productive city land is also the aim of urban orchards. These sites take many forms, ranging from Seattle's community-led, city-funded, seven-acre edible forest on public land just 2.5 miles from downtown[180]—echoing the common rights to harvest berries and mushrooms on forested land that countries like Sweden still enjoy—to the UK's growing number of cider cooperatives, in which garden owners contribute their apples and get back a share of the resulting cider.[181] Such initiatives also help preserve varietal diversity, a valuable resource in the face of climate change. Not Far From the Tree, founded in Toronto in 2008, is representative of a range of urban fruit tree projects run by charities in cities in Canada and elsewhere. It puts food-bearing trees to good use, splitting the bounty three ways: tree owners keep one-third, volunteer pickers share another, and the final third is delivered to food banks, shelters, and community kitchens by bicycle.[182]

Effective sharing of gardens and orchard produce can be encouraged where city authorities permit use of home kitchens for food production—such as baking or jam making—under "cottage food laws" and support provision of shared commercial kitchens.[183] Also referred to as "kitchen incubators" or "community kitchens," shared kitchens provide a licensed, equipped commercial kitchen available for rent on an as-needed basis, typically with access to onsite cold and dry storage, and often with accounts with wholesale distributors for food and supply needs. Like farmers' markets, cottage food and community kitchens form part of a movement that seeks to develop closer links between food producers and consumers, hoping thereby to enhance sustainability and justice.

In the same spirit of establishing transparent and local connections along the supply chain, makers on Etsy—an online marketplace designed for P2P selling of handmade goods—seek to establish a sense of connection and community with their buyers. Both are fulfilling people's desires for distinctive products and for individual connection in normally anonymous marketplaces. Buyers like to know the history and story of the goods they use—as Juliet Schor puts it—in a more considered and careful form of materialism.[184] Shared goods can offer the same benefits of transparency, traceability, and connection.

But in both P2P markets and farmers' markets there can be a downside in which minorities are misrecognized and framed out, and in which transparency can mean little if immoral labor practices continue. Due to the dominant framing of the food discourse from a white, middle-class perspective,[185] much community food security work also reflects white cultures of food and white histories that may be culturally insensitive to those being "served," as the geographer Julie Guthman notes.[186] Observing a food swap event, for instance, Schor noted how cultural capital or class privilege shaped and directed trades: "Only participants with the "right" offerings, packaging, appearance, or "taste" received offers or, in some cases, even felt comfortable returning to the event."[187]

Guthman has noted that many urban farms or community gardens targeted at underserved minority communities are met with less enthusiasm and participation than the organizers had anticipated. Such initiatives often encourage community residents to grow their own produce. Yet sometimes these residents share specific cultural histories that associate farming and growing food as instances of past oppression. This can be the case especially with African American communities whose ancestors have had a tormented past of slavery and sharecropping in the US. An act seen as positive and empowering from a local food perspective can be perceived as an unwanted reminder of past injustices from a black cultural perspective.

Similarly, "buy local" has become a mantra for the Alternative Food Movement (AFM) but what happens if that construction of "the local" by the AFM is seen by some members of the local community as exclusive rather than inclusive? In increasingly diverse societies, there are many ideas on what "the local" means. For example, the Filipino immigrants in San Diego, California, interviewed by the researcher Jiminiz Valiente-Neighbors demonstrate “translocalism”: “Filipino immigrants carry with them the idea that Filipino food is local food, which they cook at home or eat in restaurants. They also exercise this translocalism when they tend their fruit and vegetable gardens.”[188] The same could be argued of African immigrants in Washington, DC, who travel out to farms in Maryland to buy fresh African “garden eggs”—tiny green African eggplants—and chocolate habanero peppers. Is local to be defined geographically, as some in the AFM would argue, or in cultural terms, as many immigrants define it, *trans*locally?

“The local” has been imbued with multiple connotations, many of which are not necessarily deserved. As the University of Washington researchers Branden Born and Mark Purcell have pointed out, scale is socially constructed and there is nothing inherent about any scale, “the local” included.[189] Injustices and inequality can be perpetrated at any scale, and acting on the local level does not alone guarantee a more sustainable or just result.

The framing of the local food movement has confused the *ends* with the *means*. In other words, the goal has become the creation of a local food system, rather than the creation of a more sustainable and just food system using localization as the means. Born and Purcell call this “the local trap,” and explain: “No matter what its scale, the outcomes of a food system are contextual: they depend on the actors and agendas that are empowered by the social relations in a given food system.”[190] As one reviewer of Margaret Gray’s study of *Labor and the Locavore* suggests, the local trap is also a labor trap where “*local* and *moral* are not synonyms. ... Our society’s tendency to idealize local food allows small farmers to pay workers substandard wages, house them in shoddy labor camps, and quash their ability to unionize to demand better working conditions.”[191]

An altogether more optimistic vision of an integrated local food economy is nevertheless emerging in some of Boston’s poorest minority neighborhoods. Since 1994, City Fresh Foods, a catering company with 100 employees based in Boston’s Roxbury district, has served fresh, locally sourced, nutritious, and culturally appropriate food to a range of community institutions. In 2009, Glynn Lloyd, the founder of City Fresh Foods, developed City Growers, which helped stimulate

> an emerging network of urban food enterprises in Roxbury and neighboring Dorchester. From a community land trust that preserves land for growing, to kitchens and retailers who buy and sell locally grown food, to a new waste management co-op that will return compost to the land, a crop of new businesses and nonprofits are building an integrated food economy. It’s about local people keeping the wealth of their land and labor in the community.[192]

Co-production, Power, and Ownership

Our discussions of finance, energy, and food have highlighted risks of injustice and exclusion in many co-production and P2P models, which arise where power remains in the hands of a commercial intermediary, or is retained within a culturally hegemonic group.

On the other hand, co-production has great potential to reduce some of the inequalities endemic to the conventional capitalist economy. For example, it reduces labor specialization, which otherwise tends to result in excess leisure for some—namely the unemployed, with all the lack of purpose and identity that label brings—and overwork for the rest.

To many sharing economy and co-production boosters, new technology is central to its social potential. Tapscott and Williams highlighted the role of participatory web-based networks—the eponymous "wiki," emerging first in fields such as software and cultural products, and extending with the development of modular design and decentralized fabrication technologies to many other sectors, including industrial products.[193] Rifkin eulogizes the role of 3-D printers, describing the Maker Movement as a transformative hybrid of the "appropriate technology" and "free software" movements.[194] As the technologies have developed, "prosumers" have become ever more closely involved in the design and production of the goods they then consume. For example, the customers of Local Motors in the US not only participate in the design of the company's products—made from off-the-shelf motor industry components—but can even help build their own car at one of the company's "microfactories."[195]

Tapscott and Williams also highlight how "hacking" norms have combined with demands for customization of products to stimulate suppliers like Apple and Amazon to see their products as a platform for crowdsourcing customization and innovation, rather than as an end-product used by a passive consumer.[196] The key issue here—as with concerns over the exploitation of drivers by Uber, or householders by unscrupulous rooftop solar installers—is not about the technology per se, but about the distribution of power, as Gorenflo highlights in his advocacy of transformational modes of sharing.[197]

As we saw with the co-production of public services, a lack of power can mean that participants are simply treated as cheap labor when public authorities cut budgets and disown responsibility. There we argued for genuine co-production in which participants are involved at all levels. In the conventional economy, Tapscott and Williams argue in similar terms: "Exploiting crowdsourcing to get services on the cheap is not sustainable" and success involves "members sharing in ownership and the fruits of their creation."[198] The cooperative movement (discussed at greater length later in this chapter) is a long-standing model of co-production that involves shared ownership. The critical definition of a cooperative is in the collective *ownership* of the enterprise, by either producers or consumers. However, shared ownership can also be facilitated by technology. "Fab-labs" or "maker workshops," with shared facilities including 3-D printers can offer greater independence from corporate networks. Some—like the Fab Lab Seoul set up by the TIDE foundation[199]—are nonprofit foundations or established through universities or other public research facilities, providing opportunities for localizing production and repair.

But others are commercial, contributing to cultural consequences we discuss later. (See "The Contested Power of Sharing" in chapter 5.) For example, TechShop has spread from its first facility in Menlo Park, California, to provide shared workshops in 10 US locations where members can access the latest 3-D printers, laser cutters, engraving systems, and tools for electronics, woodworking, metalworking, and almost all imaginable craft. The centers also offer classes to help members use the equipment, and the chain plans to expand to the UK in 2015. At present, fab-labs are more typically oriented to the production of prototypes or one-off products, but as the costs of 3-D printing fall and its scope widens, the prospects for local production of a wide range of products and parts will grow. Rifkin highlights the potential for self-replicating 3-D printers (powered by microgeneration of electricity and using waste-derived feedstocks), as well as open-sourced designs shared over the Internet, to liberate prosumers from many conventional markets. Unfortunately, so far even the best 3-D printers can currently replicate only about half of their parts[200]—presumably the simpler, lower value half. And barriers of cost and skill may remain to broad use even with higher levels of self-replication.

Still, fab-labs can be seen as an extension and transformation of the local tool-bank, building local capacities and independence from centralized markets. Like domestic and community renewable energy plans, and also local food production and distribution programs, they suggest ways co-production could build local capacities and also increase freedoms by enhancing security from unstable and insecure global markets.

Many incumbent businesses might understandably resist the spread of co-production in the conventional economy. Yet it is controversial for other groups, too. In the market sector, for instance, co-production might imply not just the exploitation of participants as cheap labor, but also the commodification of leisure. Whether in the commercialization of home cooking, "making" as an expensive hobby, or the involvement of prosumers in designing new Lego toys,[201] co-production threatens to bring even more of life into market spheres. In our view, the mechanisms and institutions designed to stimulate and support co-production in these areas will be critical if it is to support the growth of community and sustainability, rather than being co-opted into a new cycle of conventional economic development. Yet again, culture is the critical issue.

Sennett emphasizes the potential role of craft skills in the building of cooperation and community culture.[202] Clearly this effect could be diluted if co-production is commodified. The musician and cultural commentator Pat Kane also highlights "the importance of craft—the personal construction of objects and services, as a route to meaning, mastery and autonomy," but he goes much further. Play, he argues, "can help redirect our passions from consumption to craft, from lifestyle narcissism to joyful participation, and thus live lighter (though just as richly) on the planet." He highlights especially "the power of festivity and carnival—forms of collective, organized behavior whose end is experiential pleasure, and whose means is participatory involvement."[203]

Co-production is a central part of the sharing paradigm, enabling cooperative activity and sharing experiences and capabilities. It is the potential of co-production to meet needs—not only the desire for novelty and entertainment and freedom, but also the needs for security, community and solidarity and identity—while stepping off the treadmill of growth and consumption, that makes its potential so exciting. With widespread, contagious uptake no longer could waged jobs be assumed to define people, and no longer could they be the key basis for politics. Nor could consumerism hold such powerful sway over politics if greater levels of well-being were generated by such participatory activity, rather than by consumption of the end products. But all this relies on recognizing and avoiding the threat of commodification.

Avoiding Commodification through Communal Sharing

The sharing paradigm clearly encompasses more than economic cooperation. It recognizes the centrality of fairness to individual motivation and to social solidarity. So while we recognize, and in many respects, welcome the fact that co-production and sharing can generate innovations and create business opportunities in the conventional economy, we also question whether focusing on such opportunities is using the wrong lens, or even the wrong frame altogether.

Cities can indeed, as Orsi and colleagues argue, "lower the cost of starting businesses by supporting innovations like shared workspaces, shared commercial kitchens, community-financed startups, community-owned commercial centers, and spaces for 'pop-up' businesses."[204] But the value in this is surely because it enables useful and valuable work to be done with underused resources, including under-used labor and skills. Yet again, we are forced to ask the questions, "What is the nature of these sharing businesses?" and, in particular, "Who owns them?" Orsi offers cooperatives as one "right answer"[205]—businesses owned by their members and in which profits are shared on basis of patronage, not capital share. We pick up this idea in the next section.

We saw also in chapter 1 that there are many ways in which cities might help citizens supplement their income by participating in the sharing economy. But we saw that commodifying sharing in this way can crowd out other motivations for realizing the environmental and social benefits of sharing. Would it therefore be better for cities to facilitate cash-less exchanges—in the form of gifts and barter? Would more social capital be built by providing the facilities, infrastructure, and, if necessary, insurance and guarantees to enable sharing, while actively seeking to prevent it being commodified? Using the abstraction of money as a medium of exchange has several problems. It makes it easier to ignore the *people* behind the goods and services we use and weakens empathic bonds.[206] It reduces the need for trust in reciprocal transactions, which might oil the wheels of the economy, but at the cost of social capital. It sidelines the rich diversity of cultural forms of trust in gift and non-money exchange.[207]

But it's not as simple as "gifting good, money bad." Not even Christianity claims that money itself is the root of all evil, but rather that love of money is to blame. Medieval Christianity, like modern day Islam, largely prohibited the practice of usury—lending money for interest—while the iniquities of financial debt have been widely condemned.[208] As the anthropologist and sociologist Marcel Mauss long ago highlighted, gift economies impose cultural obligations, too.[209] Money-based exchanges and markets can free people—especially women—from unjust cultural obligations, as for example with unpaid domestic labor. And the use of money as a medium of exchange and a store of value can be useful even within a gift economy.

In healthy forms of gift economies, people give away goods and services "without any explicit agreement for immediate or future reward."[210] Instead the model is fuelled by indirect reciprocity, what we earlier called "karmic altruism," sometimes called "paying forward." Paying forward is perhaps doubly valuable when the gifts we offer are the product of self-provisioning, rather than purchased on the open market, but its essence rests not in whether the assistance or kindness rendered requires payment in money. It lies instead in the request that the beneficiary, rather than repaying the donor in some way, pays forward by helping another member of the wider community.

The idea is not new. Benkler cites a letter written in 1784 from Ben Franklin, providing financial aid to a young man in need, which neatly overcame any stigma the recipient might have felt in accepting a gift. Franklin wrote: "I do not pretend to give such a sum, I only lend it to you." But the loan "was of an unusual kind." Franklin continued, "When you meet with another honest man in similar distress, you must pay me by lending this sum to him."[211]

Paying forward and other indirect forms of reciprocity raise a key question: How broad is the community concerned? By promoting such communal forms of sharing, do we risk reinforcing tight and potentially socially exclusive communities where members are obliged to follow narrow cultural values and social norms? We are alert to this concern, but do not see it as a major issue. We do not only advocate a single-minded pursuit of communal models. And even within communal models of sharing there is broad scope to build bridging social capital, too. For example, paying forward seeks to build bonds of community in ways that resonate with the complexity and anonymity of modern urban life. It builds on the simple courtesies that lubricate "living together,"[212] and is intrinsically a communal form of sharing what we have with the diverse set of citizens around us.

Gift models of communal sharing are also compatible with mediated web platforms that extend to "stranger sharing," although funding platforms in such models may be challenging. Freecycle, which started in Arizona in 2003, now claims over 5,000 groups worldwide, and operates on a charitable nonprofit basis using simple email groups.[213] Freecycle links those wishing to gift still functional goods—rather than dumping or trying to sell them—with potential users. Peerby, which started in Amsterdam in 2012, links potential lenders and borrowers of virtually any item or service. Rather than seeking to fund the platform through a share of rental fees, exchanges are totally free, although Peerby plans to offer a paid insurance option, in which lenders can cover what they lend against damage or loss. Peerby's startup phase was funded by charity and social enterprise finance, although it has also now obtained venture capital backing.[214] These services work not because individual gifts or loans are marketized, or even directly reciprocated, but because the system as a whole is one of mutual support.

Streetclub in the UK, which connects neighbors with a private online community noticeboard for posting both offers and needs for shared tools and equipment, and community events such as bring and buy sales, also depends on such mutual support. Streetclub was started in the UK by home improvement retailer B&Q for corporate social responsibility reasons: "We believe that when neighbours [start] talking, they also help each other with DIY (Do It Yourself) projects, and in this way, by sharing help, a ladder or offering local advice, they will help improve the nation's houses and thus support our company's ambition of 'Better Homes, Better Lives.'"[215]

Of course B&Q might also anticipate that enabling more home improvement by sharing tools will generate more sales of consumables such as paint and wallpaper. They have nevertheless committed to avoid any marketing to Streetclub members (either from B&Q or third parties), and take great care to ensure effective data privacy within the clubs.

These communal sharing platforms promise to help reverse the effects of neoliberalism on "the scope of community ... [which] tends to be narrowed to the level of personal relations and local settings rather than extended to broader circles of social interaction."[216] Or as Tonkinwise notes, attached in divisive ways to nation or religion—stimulating conflict and division between different communities of interest and identity groups. He castigates the "neoliberal ideology [that] insists 'the most important thing is love, which money can't buy, so just tolerate your precarious lot because all that really matters is family or nation or religion'"[217]

So our argument here is not against money per se—nor against sharing businesses, but simply that it may be better not to monetize everything, and to seek different ways to incentivize sharing.

In contemporary Western societies there is, however, a tendency toward monetization: public services are privatized, commons are enclosed, even carbon emissions are priced. Neoliberals see markets as enhancing freedom, while those of a more left-leaning bent grudgingly accept that without a monetary value, resources such as clean air and wild nature will be treated as if they are worthless. The tendency to commercialize sharing, as with so many other things, is perhaps part of our general contemporary failure to see alternatives to capitalism, rather than alternatives within it. As Žižek argues, it has become "easier to imagine the end of civilization than the end of capitalism."[218] But all our imaginations suffer as a result.

So in sharing cities, the authorities should work to enable sharing within commercial markets, but also sharing outside of markets, both in the form of communal, peer-to-peer barter and gifting, and in the form of public services and public infrastructures paid for through taxation or insurance.

A pragmatic approach would choose the mix of means best suited to local circumstances—which will vary from city to city and from resource to resource—rather than take an ideological bias toward market provision or state provision, but recognize that sharing offers a real "third way" where direct peer-to-peer exchanges can predominate. These sentiments can be translated into the economic realm through adaptation of cooperative and commons governance models. We consider cooperatives in the next section, and commons governance at greater length in the next chapter.

Cooperatives as a Catalyst for Co-production

We focus here on cooperatives, not only as sharing organizations themselves, but also because they appear to offer a sound organization model for sharing economy enterprises with a social purpose.

Cooperative models of farming, finance, and craft production date back at least to ancient Babylonia and early China. But it wasn't until the mid-19th century in Britain that the cooperative as a structure for modern business originated. A combination of industrialization, poverty, poor working conditions, and the general hardships faced by many pressed the case for cooperation. In 1844, a group of 28 workers in Lancashire formed the Rochdale Society of Equitable Pioneers, a consumer cooperative selling basic items that the members could not afford individually. As their group grew, they drafted rules for the operation of a cooperative, in which members would have democratic control and need to pay only limited interest on capital. The policies developed in Rochdale quickly became the standard for other cooperative ventures and gave birth to the modern cooperative movement. An international association was formed in 1895 and cooperatives are now widespread globally.

Cooperative forms vary depending on what is being shared. They include retailer cooperatives (sharing manufacturer discounts), worker cooperatives (shares in the enterprise); consumer cooperatives (sharing retail and financial services) and housing cooperatives (various forms of housing shares, membership, or occupancy rights). The delivery of shared services through cooperatives is also growing for example, through community cooperative ownership of small-scale renewable generation capacity.

Cooperatives need not always be small. The world's biggest workers cooperative—the Mondragon Cooperative in Spain—acts as the parent company to 111 small, medium-sized, and larger cooperatives and has global sales of €15 billion ($17.1 billion).[219] Spain's struggle through the double-dip recession, with its desperate austerity measures and 26 percent unemployment, has hit its people hard. However, Mondragon has built-in solidarity structures—the cooperative members contribute a portion of profits to the functions of the collective.[220] These have enabled it to share the pain between owners and workers—in contrast to the typical approach of private corporations—and its workforce of around 84,000 people worldwide (around half of them full co-op members, and the remainder employees) has remained broadly constant.

The African American historian Jessica Gordon Nembhard is a great fan of cooperatives. She says, "Cooperatives solve economic problems in different ways than conventional for-profit businesses. They operate on the values and principles of democratic participation, inclusion, solidarity, sharing, and 'for need' rather than 'for profit.'"[221]

Nembhard places cooperatives firmly in the "solidarity economy," arguing that they "develop—and survive—as a response to market failure and economic marginalization. … [They] fill gaps that other private businesses and the public sector ignore." Moreover they address critical issues such as "the pooling of resources and profit sharing in communities where capital is scarce and incomes low, … [and] they, like other elements in the solidarity economy, start where people are and build from the ground up."[222]

Some cities have already begun to heed the call to support cooperative growth. New York City has established incubators to help cooperatives with startup challenges, and Cleveland has catalyzed funding for the Evergreen Cooperatives in Cleveland's low-income neighborhoods. According to Orsi and her colleagues, the Evergreen cooperatives

> are models in urban wealth-building. They provide services to anchor institutions, like local hospitals and universities, and include a green industrial laundry, a solar installation firm, and the largest urban greenhouse in the US. The Mayor's Office connected the Cleveland Foundation and other Evergreen partners to Cleveland's Department of Economic Development for help finding innovative sources of funding. The city's Sustainability Office helped identify energy incentives like Solar Tax Credits.[223]

Both Nembhard and Orsi's group also suggest that cities might use their procurement spend and contracts preferentially to support cooperatives. Smart use of procurement spend is an important tool for city authorities, although it might prove problematic, particularly in the EU, where it might be seen as a restriction on free trade. The benefits would be manifold, however, so cities could clearly gain from pushing the boundary and making the case for reform of such trade rules.

The most ambitious modern cooperatives challenge not only economic, but also governance models. In Quebec cooperative development is promoted by le Chantier de l'economie social. Its membership includes social economy enterprises, regional associations, community development centers, and social movement organizations such as labor unions and environmental NGOs.[224] It also includes a network of First Nations representing Canada's indigenous peoples. Each category elects representatives to the Chantier's board. The associational democracy of the Chantier is one of four functional foundations of the Quebec social economy, according to Wright, alongside social economy investment funds, targeted state subsidies, and participatory organizational forms.[225]

In Catalunya, Spain, even more ambitious models are emerging. The Cooperativa Integral Catalana (CIC) functions as a political project linking consumer and labor cooperative initiatives with alternative currencies and efforts to establish a basic income, among other goals.[226] The CIC has established an alternative currency called the eco, which helps pay a basic income to some members, buying centers (where bulk purchases can be stored, cutting out retail intermediaries) and a collective bus, and is working on plans for cohousing in a "post-industrial, post-capitalist eco-colony."[227]

Like the best examples of the sharing paradigm, the CIC explicitly seeks to build social relationships through regular fairs and markets, bringing together members of the different *ecoxarxas* (eco-networks) to exchange products, skills, and entertainment, and primarily using the eco as currency. The model is being explored and replicated in the Basque country, Madrid, and Valencia.

The CIC works in "three concentric economic spaces."[228] At the center lies "a gift economy ... based on mutual aid between individuals." Surrounding and protecting this from the conventional economy—in which even CIC members continue to use the euro—is an intermediate space of "direct and indirect exchanges" based on reciprocity using a social currency. In this respect the CIC bridges and integrates all the various domains of the sharing economy described earlier, and suggests a model within which both communal and commercial sharing can be promoted.

The cooperative principle is also thriving in the world of finance and banking. According to the World Council of Credit Unions:

> Credit unions ... are member-owned, not-for-profit financial cooperatives that provide savings, credit and other financial services to their members. Credit union membership is based on a common bond, a linkage shared by savers and borrowers who belong to a specific community, organization, religion or place of employment. Credit unions pool their members' savings deposits and shares to finance their own loan portfolios rather than rely on outside capital. Members benefit from higher returns on savings, lower rates on loans and fewer fees on average.[229]

Credit unions date back to 1852 in the Kingdom of Saxony in present-day Germany, where they are now widespread and vary from multibillion dollar enterprises, such as the US Navy Federal Credit Union with assets of around $50 billion and over 3 million members, to small groups of volunteers. The number of credit union accounts in the US increased by as much as 650,000 in the fall of 2011, in the wake of the global financial crisis and a call from the Occupy movement for people to move from Wall Street banks to local financial institutions, especially credit unions.[230] Run on the principles outlined above, credit unions consistently report greater customer satisfaction than banks. In the UK and a number of former colonial nations, such as Jamaica, Australia, and New Zealand, there is another variant on the cooperative financial institution, the Building Society. These are mutual societies, set up especially for mortgage lending. Community Banks are locally owned and managed, in some countries run by local or regional authorities. Jim Blasingame, president of the media company Small Business Network, notes how US community banks[231] make almost 60 percent of small business loans—despite holding only 20 percent of all bank assets:

> Small business owners don't care much about a bank's asset size. But they care very much about ... relationship banking. To a small business owner a community bank ... is locally owned and managed. ... [It] takes into account a business owner's character when making loan decisions ... [and] decides small business loans by a local committee, not credit scoring by a computer.[232]

Clearly, between them, mutual financial institutions—credit unions, building societies, and community banks—are major players in the financial lives of many people who want a different way of conducting their financial and banking affairs. But if they are to underpin a sharing economy that does not fall victim to the drivers of misplaced commercial motivations, they may need reinvention in the model of the contemporary sharing paradigm, adopting the best features of P2P finance and crowdfunding discussed earlier in this chapter. City authorities should look for opportunities to support the further development of mutual financial institutions in their localities as part of the urban commons.

Sharing Work, Sharing Income

It would be absurd to discuss the sharing of productive equipment like 3-D printers, capital funds, or facilities such as gardens or kitchens, and to ignore the sharing of jobs. In times of economic hardship, sharing jobs can be a key way of sharing scarce resources. Those involved in job sharing typically report better work-life balance. Formal job sharing (two people jointly holding a single full-time position) is particularly valued by parents, and also offers benefits in terms of creativity and collaborative working. Countries that practice work sharing (cutting hours rather than jobs) in response to economic recession not only minimize the hardship triggered by unemployment, but also better maintain workforce skills.[233]

Anna Coote of the New Economics Foundation proposes an alternative to existing workplace values:

> A slow but steady move toward a 30-hour week for all workers. This will help solve a lot of connected problems: overwork, unemployment, overconsumption, high carbon emissions, low well-being, entrenched inequalities and the lack of time to live sustainably, to care for each other and simply to enjoy life.[234]

Long hours may be one product of the vicious cycle of consumerism driving the economy.[235] To Joseph Stiglitz of Columbia University, this implies that "individuals can learn—they learn how to consume by consuming, they learn how to enjoy leisure by enjoying leisure."[236] But, he says, education and advertising provide strong incentives to shift preferences toward consumption. So shifting preferences instead toward leisure and toward sharing should be possible, especially if collaboration is encouraged and enabled in education and in the broader public realm as well.

The relationship between shorter hours and sharing should be virtuous, with shorter hours enabling people to engage more in collaborative consumption, production, leisure, and politics. More sharing would potentially reduce the high costs of living (especially for housing), enable people to choose to work less and also to resist the commodification of leisure discussed above. For so many people the greatest regret is the lack of time to spend with family, friends, and community, but as leisure is increasingly commodified, not only do we need to work more to fund our leisure activities, but also their nature is transformed in ways that undermine their psychological and cultural benefits. Time rescued from the dictates of markets can instead be spent simply being, or connecting with others.

We noted above how job sharing can promote equality in a stable or shrinking economy. A system of redistributive taxation can do the same, as long as people are allowed or enabled to do informal work, in the charitable, volunteer, shared, or gift economies, rather than deterred from doing so in order to maintain eligibility for benefits by remaining "available for work." The idea of a basic income paid to all citizens regardless of whether they are in work could have the same effect. Switzerland will hold a referendum on a citizen's income in 2016, while the Dutch city of Utrecht plans to trial replacing benefits with a citizen's income starting in 2015. In theory a citizen's income offers an opportunity to reduce bureaucratic welfare systems and enhance equality, without the stigma of means-testing, particularly by improving the negotiating position of those on the lowest incomes.[237] It also properly values informal labor in households and communities (which is mainly women's work), and frees people to undertake work and collective projects that they themselves have reason to value. And it clearly has scope to change norms and values: if the state treats people as citizens, rather than wage slaves, then they are more likely to behave as citizens.

Yet there is a Catch 22: all efforts to persuade people to trade work for leisure—even if conceived as regulatory—imply winning a public political debate. In some countries that debate is live—in others it is unlikely without a revival of the political public square. Moreover, winning it depends critically on reducing the power that consumerism holds over our identity in the modern world, through the role consumer goods play in defining our extended selves.[238] While the nexus of consumerism and growth-oriented politics persists unquestioned, trading income for leisure will likely continue to appear undesirable and even impossible.

In this chapter so far, we have seen how infrastructures, public services, and their role as marketplaces make cities inherently shared places. We have also shown how co-production is transforming economic practices in cities in sectors as diverse as energy, housing, food, and finance, as the opportunities enabled by new Internet technologies intersect with the dynamics of urban space. We have explored the tensions emerging as commercial sharing models threaten to transform gifted social goods into commodities and further concentrate economic power; and some of the models of communal and cooperative sharing that could mitigate these tensions. Such models include approaches that would share work or income, with profound implications for consumerism, identity, and the public square. We will explore those implications in some detail in chapter 3, but first we discuss one final way in which cities are inherently shared spaces: as co-created urban commons.

Commoning: Co-producing the Urban Commons

Clearly the economies, services, and infrastructures of cities are spaces of sharing and co-production that can be designed to promote social justice and sustainability. But cities themselves are also co-created. As does David Harvey, we see urban areas, and their physical and social infrastructures, resources, and institutions as "commons" co-produced by their citizens,[239] not just a collection of buildings and facilities produced by the construction and development industry in line with (or more often despite) city government planning rules or ordinances. This understanding has two important consequences for city governance and management.

First, at least at relatively small scales, with communication and the construction of cultural rules for sharing, effective self-governance of commons is entirely plausible.[240] We will see some experiments with commons governance in chapter 3, when we look at social movements in the city, and explore its potential for cities further in chapter 5. Harvey, however, questions the capacity of self-governance to scale up. He argues—somewhat counterintuitively—for the "enclosure of non-commodified spaces in a ruthlessly commodifying world"[241]—for instance to protect rainforests from further logging and exploitation; or urban districts from gentrification. But it seems Harvey has chosen the word "enclosure" simply for shock value. From our perspective the concept he describes is active "exclusion of neoliberalism," which might be pursued by deliberative polycentric governance focused on cities as commons. Harvey is right, however, to warn us that exclusionary enclosure can also run counter to social justice, citing the habit of the rich of "sealing themselves off in gated communities within which an exclusionary commons becomes defined."[242] In this respect, cities and citizens would do well to remember the origin of the term "beating the bounds," in which commoners collectively removed illegal fences to maintain the commons against enclosure.[243]

Second, like natural resources, we conventionally think of common land as a limited shared resource that is depleted or degraded by overuse. But most urban commons are in fact social or cultural, and like knowledge and some technological commons, they are not depleted in use. Such commons are extended or enhanced by investment, for example, in infrastructure or in research. Providing adequate incentives or rewards for such contributions, while maintaining access and benefit for all is the governance challenge here.

Simply supporting cooperation, as Richard Sennett argues,[244] is a critical and often overlooked step. Take, for example, urban public space, which we consider at greater length in chapter 3. Its value as a commons in part depends on those sharing it using it properly—not littering, or more extremely, not mugging passers-by. In part it also depends on the private yet collective investment of surrounding property owners and occupiers. In practicing and developing the skills of 'living together' we create an urban commons—sociocultural as well as physical—that enhances all our lives in myriad ways.

The public goods and services—like education—provided by city authorities also contribute to the quality of the urban commons, but only—as Harvey points out—through political actions which maintain their funding and direct them to a common purpose. In this way the production of the urban commons is a result of a "social practice" of "commoning," that is, creating collective and non-commodified cultural, social, and physical spaces in the city.[245] Such spaces might, incidentally, contribute to market exchange—for example, a community garden from which food is sold. But as Harvey points out, such a garden is a good thing in itself, no matter what food may be produced there. We might further emphasize the activity of gardening, and the practice of cooperative skills.[246] In other words, as we will see again in "The Crucible of "Democracy" in chapter 3, with the experience of protest movements in urban space, process and place are as important as outputs.

Gardens are a particularly important venue for the practice of living together in diverse neighborhoods and cities. Placemaking through sharing of produce, seeds, knowledge and recipes—in gardens and across garden walls and kitchen tables—is one of the most effective bridges between cultures, rooted in shared practice and experience. For example, predominantly Latino/a urban community garden projects in Los Angeles and Seattle connect growers to local and extra-local landscapes, creating an "autotopography" that links their life experiences to a deep sense of place.[247] In effect, users are writing their cultural stories on the land- or cityscape. This is a type of immigrant "cultural commoning" or placemaking through the growth and celebration of culturally appropriate foods. As Teresa Mares and Devon Peña explain, "One gardener, ... a thirty year old Zapotec woman, described her involvement at the farm in the following way":

> I planted this garden because it is a little space like home. I grow the same plants that I had back in my garden in Oaxaca. We can eat like we ate at home and this makes us feel like ourselves. It allows us to keep a part of who we are after coming to the United States.[248]

Above we noted the importance of incentivizing investment in cultural or knowledge commons. Public investment is one way, particularly relevant to urban commons. But all too often in cities the problem is not how to incentivize private companies to invest in commons, but governing how they exploit commons produced by the community, so as to maintain the community benefit. Harvey highlights, for example, how the "ambience and attractiveness of a city ... is a collective product of its citizens, but it is the tourist trade that commercially capitalizes upon that common,"[249] while simultaneously displacing poorer residents from enjoying the city by forcing up housing and living costs. Gentrification may begin with attractive, lively, and diverse streets, but the very process debases the cultural commons. Air-conditioned and burglar-alarmed houses with private rooftop decks may overtake the streets, while the original residents are forced out by rising rents. Harvey argues: "The better the common qualities a social group creates, the more likely it is to be raided and appropriated by private profit maximizing interests."[250]

Cultural activity is one of key co-producers of the urban commons—and one of the first victims of this process of appropriation. Artists and musicians come together in creative clusters, in cultural and countercultural cycles typical of the sector. A new artistic or musical movement is as likely to begin in poverty and oppression, in shared accommodation and shared facilities, and even in squats and informal communities, as it is in wealth.

For example, the former curator and novelist Nicola White highlights "the collective, egalitarian feel of Glasgow, the multitude of practices and groupings, the respect for hard work, the "now.'"[251] In genuine co-creation, taking advantage of relatively low rents, "the do-it-yourself culture of the city's artists, who built their own institutions rather than rely on established ones, has been crucial." The authorities have played a role too: "the city council, which once appeared wrongfooted by the riches on its own doorstep, has now invested hugely in studio complexes."[252] More generally:

> Artists are not simply attracted by cheap rents alone, but by places that appeal to the "artistic habitus" or a lifestyle rooted in the aesthetic of older often industrial neighborhoods that contain buildings with historic architecture and adaptable, open floor plans and which are typically found in walkable, mixed use central city locations.[253]

These are also typically areas of mixed ethnicity and higher than average crime rates vulnerable to gentrification. And indeed these artistic districts of studios, galleries, cafés, and venues can rapidly attract redevelopment capital, in extreme cases expropriating not only the physical spaces but the very identities of the previous occupants—in marketing "artists lofts" and "artists quarters" to those eager to adopt the cachet of hip or cool, but with the wealth and income to afford to live in these newly gentrified districts.

In contemporary cities this has become a deliberate strategy—alongside the creation of cultural clusters of new artistic venues—largely oriented at attracting wealth-creating "creative classes" as consumers of culture, rather than supporting clusters of artists as co-creators of culture. Examples such as the "Westergasfabriek in Amsterdam (a former gas factory, and now a conglomeration of cultural activities such as concerts and exhibitions) and the well-established film and video cluster Film in Soho in London,"[254] suggest economic benefits. But such clusters might be less successful artistically, and socially.

Miguel Martinez, an assistant professor at the City University of Hong Kong, notes how in cities like Paris, "Authorities praise the artistic squats over others, and they are more prone to tolerating or subsidizing their continuity because they are conceived as city landmarks for the so-called creative class. They also appeal to tourists."[255] But, says Martinez, the authorities "forget that low-paid and precarious artists need an accessible place to live, too. As a consequence, the housing question is often ignored."[256]

As one anonymous blogger living informally in a disused warehouse noted in February 2014: "The landlord of the so-called 'warehouse community' where I live in Tottenham, London, recently adorned our homes with signs that read 'Artists Village—for creative people.'"[257] This was not a symbol of recognition and formalization for the existing occupiers, but "the start of a drive to shrink-wrap the diverse lives of his tenants into a neat marketing package, a 'creative hub' commodity, to be made available to Wharf-strutting City types at sharply increased rents."[258]

This is co-option and commodification of the very idea of co-creation, not simply its products.

Similar processes are co-opting parts of the sustainable economy movement such as local food and artisanal products, exploiting them in "cultural capital myth creation" for wealthy consumers who ignore the other precepts of sustainable economies (such as downshifting to trade work and income for leisure).[259] In both cases, the investments of countercultures are used to market a distinct and valuable "local culture," among other consequences, raising rents in property markets.

The artist and musician David Byrne describes the potential end-point of such a process:

> Most of Manhattan and many parts of Brooklyn are virtual walled communities, pleasure domes for the rich. ... There is no room for fresh creative types. Middle-class people can barely afford to live here anymore, so forget about emerging artists, musicians, actors, dancers, writers, journalists and small business people. Bit by bit, the resources that keep the city vibrant are being eliminated.[260]

"Cities may have mercantile exchange as one of their reasons for being" Byrne accepts:

> But once people are lured to a place for work, they need more than offices, gyms and strip clubs to really live. ... The city ... generates its energy from the human interactions that take place in it. Unfortunately, we're getting to a point where many

of New York's citizens have been excluded from this equation for too long. The physical part of our city—the body—has been improved immeasurably ... [with] bike lanes and the bikeshare program, the new public plazas, the waterfront parks and the functional public transportation system. But the cultural part of the city—the mind—has been usurped by the top 1%.[261]

He acknowledges that the wealthy elite—who, he says, do not pay their taxes—habitually contribute to support parks, museums, and symphony halls. "But it's like funding your own clubhouse," he complains. "It doesn't exactly do much for the rest of us or for the general health of the city."[262]

Harvey argues, however, that some hope remains.[263] To capture the monopoly rent on such places, cities must maintain difference and uniqueness, rather than simply joining the indistinguishable globalized city whose emergence Saskia Sassen of Colombia University documents.[264] Harvey argues that cities must therefore allow, and even support "divergent and to some degree uncontrollable local cultural developments" and "spaces for transformational politics."[265] Spaces for co-creation are thus reproduced even in the face of capitalist redevelopment. The challenge is to expand and replicate those spaces and communities as shared commons going forward. The revival of sharing is one potential tool to help in that task.

Cultural producers are likely to play a powerful role. As Harvey argues, artists are typically "transgressive about sexuality, religion, social mores, and artistic and architectural conventions."[266] It is has never been unthinkable for them to be transgressive toward commodification and market domination either. And today the relationship between cultural producers and their "consumers," especially in music, has been shaken by the online sharing models that now dominate the sector. Moreover, the new countercultural norms toward sharing adopted here are now disrupting the wider relationship between consumers and conventional markets. (We return to artistic countercultures in chapter 3, and to the scope for countercultures to transform the mainstream in chapter 5.)

We have outlined here a process in which city dwellers co-create a sociocultural, and—in some respects—physical urban commons. In this shared environment of the city, security, stimulation, and cultural experience are co-produced. Yet the urban commons is not governed as such, rather it is typically enclosed and commodified for profit by the property development industry—abetted by local governments desperate for higher tax revenues—in ways that can exclude those original producers. At heart this is the same process as happens in the global South when the social commons of communication and exchange in poor communities is marketized for corporate profit.[267](We return to this at length in "Commodification: Displacing Gift Economies and Social Sharing" in chapter 4.)

The commodification and enclosure of the public realm—the urban commons—is as much an injustice as the annexation of native lands by colonists in the US, Australia, Canada, and elsewhere, with no account taken of the use rights of the former occupiers, nor of the collective investments in land management that made those territories productive and sustainable. These are failures of recognition—communal and sociocultural sharing are simply not perceived as of value—as well as expressions of inequalities in wealth and power. (The term "recognition" is used here in the sense of the justice concept explained in more detail in chapter 4.) The Right to the City needs to be underpinned by formal recognition of the urban commons and the role of citizens in its creation. We need to recapture the commons as the "theatre within which the life of the community was enacted."[268]

Summary

Like San Francisco, Seoul is proactively realizing its role in nurturing the sharing paradigm with Mayor Park Won-soon driving "Sharing City, Seoul" throughout the Seoul Metropolitan Government. The city-funded project has led development of an impressive array of mediated sharing services, and seed-funded startups and partnerships, such as Creative Commons Korea, whose role is to increase accessibility to and awareness of Sharing City Seoul. In doing this, cities like Seoul are rediscovering their roots as spaces of sharing, which can be traced all the way back to the most ancient cities.

We saw how shared infrastructures and services have underpinned urban development and quality of life in cities throughout history. In various guises (hard, soft, social, green, etc.), infrastructures as shared assets or commons underpin markets as well as urban society. We saw how co-production can help deliver services like healthcare and education, as well as products like housing, energy, food, and manufactured goods. We saw that urban markets and urban density make the processes of co-production more efficient, as well as help the norms that drive them spread through real-world, face-to-face collaboration. We also saw how norms of sharing and cooperation are a natural product of evolution, although they are expressed differently in different cultures.

In this chapter we have built the case for understanding the productive realm of the city as a commons—shared physical and social resources which support our ability to flourish all the more, the more we invest collectively in them, as Jane Jacobs so presciently suggested. From health and education, to open source software, cooperatives, and fab-labs, these emanations of the sharing paradigm build an arena that is productive yet potentially frugal in its use of scarce resources. We have also noted some of the injustices arising from abuse of the urban commons, such as co-production being used as a shortcut for service cuts, or problems around the fetishization of "the local" in urban agriculture. We will delve more deeply into equity and justice issues surrounding the sharing city in chapter 4. But the city as commons is not just an economic space—indeed it is not even primarily an economic space, but a sociocultural and political one, which is the topic of the next chapter.

3 Case Study: Copenhagen

Copenhagen—lying on the shores of the Øresund strait—is the capital and most populous city in Denmark. Denmark is home to the happiest people in the world, if one is to believe the United Nation's 2013 World Happiness Report.[1] In Copenhagen, people are at the heart of the city's renowned urban space design and planning efforts. According to Jan Gehl, the famed Danish architect and international urban design consultant, "Cultures and climates differ all over the world, but people are the same. They will gather in public if you give them a good place to do it."[2] As professor emeritus of urban design at the Danish Royal Academy of Fine Arts, Gehl has been studying and documenting the evolution of Copenhagen's public spaces since the Central Pedestrian District was first conceived in 1962 as a strategy to breathe life back into the city center. Over the years, many of the city's former parking spaces have been reclaimed for shared public space.

But despite Copenhagen's current star status, Danes have not always been keen sharers. Apparently, when the Danes colonized Greenland after 1721, they taught the Inuit "that communal living—shared food, shared hunting trips, shared wives—was sinful."[3] Yet now Denmark has embraced sharing in its many forms despite having a very private national culture. Clearly, sharing norms can change over time.

Tina Såby, Copenhagen's city architect, highlights how planning in Copenhagen pays particular attention to the areas where public and private spaces meet.[4] In a dramatic contrast with the controversial use of "pavement spikes" and other "defensive urban architecture" in many cities to prevent homeless people sitting or sleeping there,[5] Såby explains that the city council encourages building owners to put tables, chairs, and planters on the sidewalks near the building as a way to make these edge zones more inviting. In addition to creating more opportunities for public interaction, the city uses planning to actively promote eye contact as a way to build community. For example, 75 percent of a building's ground floor walls must be glass so that people can easily see in and out.

The city of Copenhagen lays out its vision for becoming the worlds' most livable city in an action plan, titled *A Metropolis for People.*[6] To encourage support for the principles set out here and elsewhere, Sustainia has created a highly visual *Guide to Copenhagen 2025.*[7] The city's vision frames the idea of livability around sustainability and how urban space can create opportunities for people to partake in unique and varied urban activities. The report describes three concrete goals the city will pursue for year 2015: 80 percent of residents will be satisfied with the opportunities they have for taking part in urban life; the amount of pedestrian traffic will increase by 20 percent; and Copenhageners will spend 20 percent more time in urban space. The city urges a shared responsibility for these goals, stating that the municipality alone cannot *create* urban life, but "together with citizens, site-owners, business, ... and experts we can create a city which invites people to an urban life."[8]

Copenhagen is encouraging people to spend longer in its public spaces by encouraging playful use of the urban landscape and its features. Architects like Gehl, designers, artists, and city residents are challenging the traditional view of urban space as something to be passively observed from a distance.[9] From public trampolines along the city waterfront to sculptures that double as informal playgrounds, the people of Copenhagen are invited to share the cityscape as playscape. The recently installed "off-ground" fixtures on the city's boardwalk offer another example. The design and purpose of the fixtures challenges the common assumptions that seating facilities in public space should be rigid benches and all playing facilities in public space should be in kids' sizes. Brightly colored adult-sized hammock slings, made from recycled fire hoses, aim to transport people back to their childhood, and beckon visitors and locals alike to lounge, swing, and linger.[10]

In addition to recreation and social interaction, the design of shared public space in Copenhagen is intended to facilitate democracy—and not just within Copenhagen. In 2013 the Danish Agency for Culture, the Danish Centre for Culture and Development, and the Danish Egyptian Dialogue Institute invited Danish and Middle Eastern urban planners, architects, artists, and designers to Copenhagen to visit innovative public spaces and explore the cultural significance of reclaiming public space for bottom-up democracy.[11] The program included site visits and dialogue on topics like cultural expression and history in urban structures, city space for the socially marginalized, and street art in public spaces.

One of the sites explored by Copenhagen's Middle Eastern visitors was Superkilen, a park in Nørrebro, which is the most ethnically diverse district of the city, with more than 50 nationalities represented. The park celebrates this intercultural diversity by incorporating objects from around the world in a sort of global exposition for the local residents, who have been invited to contribute their own ideas and personal artifacts to the park. In one public space, this diverse community can gather on Iraqi swings, Turkish and Brazilian benches, around a Moroccan fountain, and under Japanese cherry trees, all while sharing space for meals, games, and conversation.[12]

Another destination was *The Wall*, curated by the Museum of Copenhagen, "a twelve meter long mobile structure with a built-in, interactive multi-touch screen that provides access to 20,000 pictures depicting Copenhagen's history."[13] Photographs have been solicited from citizens as well as provided by the museum. This project has been selected to serve as a model for a similar wall in Cairo to help celebrate its cultural and natural heritage and history.

Superkilen and *The Wall* might now be well known in Cairo, but the Christiania district of Copenhagen is famous, or perhaps notorious, the world over for its illegal but open cannabis trade. For many years practically autonomous, the "Freetown" of Christiania is situated in a disused military district in central Copenhagen squatted since 1971. Initial motivations for the squat included the lack of affordable housing elsewhere in the city. Tom Freston describes his excitement on visiting Christiania in 1972:

> Artists, feminists, hippies, anarchists … had actually conquered a part of town, were holding it, and were living there for free. … Christiania even had a mission statement: "to be a self-governing society … self-sustaining … and aspiring to avert psychological and physical destitution." The possession of private property was thought to be immoral.[14]

Since 1994, Christiania's 900 or so residents have paid taxes and fees for utility services and in 2012 struck a deal—as a resident's collective—with the Danish government to purchase the site for substantially below market value. In the intervening years residents established an alternative local currency, restored the buildings, built new homes and regulated the district according to collective anarchist governance models.[15] Today, concludes Freston:

> Christiania's survival is a good bet. The Danes are proud of it now. After all, these are people who built their own homes, who stood up to the government and criminal elements for decades, who took in the poor and disadvantaged, who were

eco-friendly and racially diverse before anyone else, and who sent the world a strong image about the creativity and tolerance of Denmark.[16]

While Copenhagen has served as a laboratory for different forms of shared politics, the city has probably received more recognition for two further sharing innovations: Cohousing, and its cycling culture and infrastructure.

The modern cohousing movement originated in the greater Copenhagen area in the 1970s and has now spread internationally. Cohousing communities consist of several families living separately with extensive communal space, in a neighborhood designed specifically for social interaction. The model emerged from Nordic feminist community project models with a goal of creating a "just society in which children's and women's needs and the social reproduction of all peoples and natures are valued as central motives for action."[17] In practice, social interaction and shared experience are the major catalyst in the founding of cohousing communities: "When you do things together," says Anna Jorgensen, cofounder of Jernstoberiet, a cohousing community in Roskilde, Denmark, "you create history together."[18] Soren Fredericksen, an active cohousing community member, noted the impact on his own personal interactions with people outside the community, as well as the impact on his teenagers. "Living so closely with people, I am much more open minded and able to put myself in someone's shoes in a situation. My children have learned teamwork, responsibility, and I see the effect it has had on their upbringing."[19]

Copenhagen has been dubbed "City of Cyclists" because more than one-third of its population uses a bicycle each day to get to and from work or education destinations around the city. The city places great value on citizen satisfaction and feedback in improving cycling conditions, as is demonstrated through its Bicycle Account, a biannual assessment and public survey of cycling development in the city. In its Bicycle Account 2012, Copenhagen sets ambitious and concrete goals to elevate its status as the City of Cyclists.[20] One such goal is to have four out of five cycling Copenhageners feel safe while cycling. A second objective is that 50 percent of the city's modal share will be made up of bicycle trips by 2015, up from 36 percent in 2011. The city is allocating significant public funds to improve cycling conditions in order to reach these goals.

To encourage suburban commuters to opt for cycling over driving, the city is partnering with surrounding communities to construct a network of 26 new bike routes, dubbed "cycling superhighways," to better connect surrounding suburbs with the city center.[21] The superhighway project aims to ensure that there are standardized, continuous bike routes into the city for distances reaching out up to 14 miles.[22] The Capital Region of Denmark, a political body responsible for public hospitals as well as regional development, has provided $1.6 million for the superhighway project. One regional councilor said that "anything we can do to get less pollution and less traffic is going to mean healthier, maybe happier, people."[23]

The first bicycle superhighway is an 11-mile link connecting Albertslund with Copenhagen. It features several innovations to move more people through the city quicker and optimize cycling conditions. The "green wave" technology times traffic lights to suit bikers in a way that allows cyclists maintaining a certain pace to ride straight through the city without having to stop. Tilted footrest bars greet cyclists at traffic intersections to allow for resting without dismounting, and bike pumps can be found at every mile mark in the event of a flat tire.[24]

More generally, Copenhagen's cycle lanes are slightly raised above the street and wide enough to allow bikers to feel buffered from vehicle traffic.[25] They have been expanded from 2.5 meters to 3 meters (roughly 8 to 10 feet) in width, which has optimized capacity, allowed for varied speeds and enhanced the sense of safety among cyclists.[26] Many cycle tracks run between on-street parking places and the sidewalk to shield cyclists from street traffic.[27]

Rather than a simple bicycle rack installed at popular destinations like supermarkets, there are entire bicycle parking sections that rival the size of space available for cars. Also when it snows, bike lanes get plowed as soon, if not sooner, than streets.[28] The city is even constructing three new bicycle bridges to connect Nyhavn with Christianshavn and Holmen so cyclists and pedestrians have more options when traveling around Copenhagen's harbor.[29]

To complement its impressive cycling infrastructure, a new, high-tech bikeshare system, Cykel DK, has been introduced, with 1,850 bikes installed during summer 2014.[30] The share bikes feature an electrical assistance system, a built-in Android tablet mounted on the handlebars (offering a GPS guidance system and real-time information on station location and availability), as well as public transit schedules and ticketing information.[31] To facilitate the convenience and connectivity of this system even further, riders will be able to reserve their bikes ahead of time. All of these initiatives are working toward enticing city residents out of their cars and into active and public modes of transportation.

Yet the #1 rule for biking in Copenhagen, as displayed on the city's bicycle map, is not environmental or economic—it is cultural: "Spread positive karma. It doesn't take much to spread good karma in traffic so everyone can get out and about."[32]

The City of Cyclists aspires to be the world's first carbon-neutral capital by 2025. This is truly a shared ambition among city government, businesses, grassroots groups, and citizens alike, involving ambitious action on both transport and energy. Copenhagen lord mayor Frank Jensen explains that city residents are even investing their own money to help reach the city's sustainability goals, with half of the turbines in the Middelgrunden harbor wind farm being funded by individual Copenhagen shareholders.[33]

In addition to the impressive cycling infrastructure, the city is increasing the reach and capacity of its subway, with the new City Circle Line, due to be completed by 2018, which will bring 85 percent of city residents within 650 yards of a metro station.[34] Since energy consumption of buildings makes up 75 percent of Copenhagen's total carbon emissions, the city is also upgrading its shared district heating and cooling infrastructure through projects like the Adelgade cooling plant, which draws in chilled seawater from the harbor and distributes it to buildings via insulated underground pipes.[35]

Adding to its rising fame, Copenhagen has been awarded the prestigious European Green Capital 2014 Award. The city's efforts to encourage more citizens to choose cycling as a primary mode of transport, as well as its goal to be carbon neutral by 2025 underpinned its nomination. Copenhagen's proactive efforts to share its sustainable solutions with other municipalities from around the globe were also a factor. For example, early in 2014, Copenhagenize.com publicly shared a translated summary of the city's urban landscape Design Manual, complete with mandated design principles for its cycle tracks and bike parking infrastructure, as well as urban space planning and design. In fact, "sharing," in this context of Copenhagen sharing its experiences with the rest of the world, has been the focal point of the city for celebrating this award in 2014. The program is called Sharing Copenhagen, and transformed the city into a giant showcase inviting the world to explore solutions together.

Program Manager for Sharing Copenhagen, Casper Harboe, says the city's invitation was "come and see, and come and share with us."[36] Shining a spotlight on some of the initiatives that helped Copenhagen earn the award provided visitors and locals alike with the chance to learn more and be inspired about what is being done, ranging from cycling infrastructure to resource efficiency, to climate adaptation. Harboe also explains that the city is just as excited by the opportunity to learn from those who visit and participate in Sharing Copenhagen.

Over the course of 2014, together with 75 partner organizations, the city hosted numerous events, conferences, workshops, and guided tours.[37] The city's shared challenges and recognition for its progress make for a collective culture that holds both great pride, and great potential to reach its ambitious goals.

Sharing Politics: The City as Public Realm

We are caught in an inescapable network of mutuality, tied in a single garment of destiny.
—Martin Luther King Jr.

Chapter Introduction and Outline

In this chapter we explore the city as public realm through the seemingly eclectic political and cultural dimensions of the sharing paradigm, under three broad and interrelated themes. First we consider questions of space and place: in particular how urban spaces and places have been central to political movements, and especially insurgent, participatory, countercultural movements—such as Las Indignadas and Occupy—which have the capacity to transform societies; and how such movements now equally depend on public cyberspace. Second we turn to the ways in which sharing can underpin democracy in practice, building social capital, supporting a healthy public realm, and challenging the hold of consumerism on our individual identities. In doing so we explore the role of shared or "collaborative" leisure, and take a close look at streetlife, as well as at the notion of Complete and Incomplete Streets, the former being a largely design-led, prescriptive approach critiqued by the latter as excluding the social narrative of the street. We also explore the notion of interculturalism: the acknowledgment that increasing difference and diversity in our cities needs to be met by a pluralist transformation of public space and place, institutions and civic culture, and a proactive engagement between cultures. Third, we examine ways in which sharing is emerging in the practices of urban governance, from cryptostates to co-production, also giving attention to the key challenges to urban democracy in land ownership and taxation.

The Crucible of Democracy

We introduced the economic and exchange aspects of cities in chapter 2. But the urban geographer Paul Wheatley argues convincingly that cities arose more as centers of political power, combining temple and city hall.[1] Subsequently, cities have been a cradle of democracy—not only in ancient Greece, but more generally: as the Germans say, "Stadluft macht frei" (city air makes [one] free). During the Renaissance cities became sites for gatherings, discovery, expression, and political participation,[2] activities that continue to define the role of cities to this day. Cities enable also a sense of citizenship and belonging. Sandel points out that many

> of our public institutions—public libraries, public transportation, public parks and recreation centers—are only partly for the sake of looking after those who couldn't afford those services left on their own. They are also traditionally sites for the cultivation of a common citizenship, so that people from different walks of life encounter one another and so acquire enough of a shared ... sense of a shared life that we can meaningfully think of one another as citizens in a common venture.[3]

In this chapter we focus on public space as the crucible of democracy. The historic and continuing role of public space in cultural and political progress is critical to the potential of sharing cities all around the world. Yet the specific cultures and histories of different places also matter. Latin Americans adopted colonial Spanish plazas,[4] but in West African culture, less formal spaces such as streets form the "living tissue" of the public sphere, offering gathering spaces, shade, and forums for communication through "oral history, verbal navigation" and now mobile technologies.[5] In the nineteenth century, open spaces in England and the US were controlled by the bourgeois, a dynamic that once again threatens to dominate.[6] In the early twentieth century, however, many parks and open spaces were created for the poor in response to congested living conditions and resulting health concerns.

Today public urban spaces can be used for *security*, for *resistance*, or for *possibility*.[7] Security is promoted in contrasting ways by "inclusive" or "exclusive" urban spaces.[8] Inclusive spaces are the aim of "the New Urbanists, Urban Villagers and 24 Hour City people who want to 'crowd out crime' through mixed use and maximizing activity in public areas."[9] Exclusive spaces, in contrast, are the domain of "the 'designing out crime' proselytizers who seek closure and limitation of use of spaces." Beijing's Tiananmen Square and Cairo's Tahrir Square were spaces of resistance as were the streets, plazas, and squares in Libya, Bahrain, Syria, and Yemen, among other Middle Eastern countries, where citizens gathered to protest against long-standing repressive regimes. In 2011, a gendered focus took hold in Toronto, as the Slut Walk phenomenon grew out of a careless and crass comment by a Toronto police officer to students at York University. Possibility is where the fullest expression of the human spirit lies. Through various methods of urbanism—guerilla, DIY, tactical, pop-up, and open source to name a few—city dwellers are reimagining and redefining their environments. In Copenhagen, a good example is *Happy Wall*, a 30-meter-long sculpture of around 2,000 colored wooden boards that anyone can flip to effect a color change.[10] Patterns, creatures, and words or larger statements are morphed at different times of the day on the whim of the curious passersby. By their nature, such movements interpret the collective urban commons as a shared space, but rather than accepting places as they *are*, they are redefining what they can *become*.[11] Indeed, "space as possibility" is another way of framing a "right to the city," in recognition of Harvey's conclusion that such a right should be seen not as "an exclusive individual right, but a focused collective right."[12] These uses are not exclusive: as we will see later in this chapter, the strength of the Occupy movement arose in part because it successfully used urban space for both resistance and possibility.

Some of these "insurgent tactics" are centuries old, and some brand new. Physical activities that create public space include digging, dancing, selling, building, and sitting, among many others. These movements are embodied in urban visionary Jane Jacobs's description of the "sidewalk ballet," wherein people shape the street through a choreographed chaos.[13] Tactical urbanism is working to revitalize the art of the public sidewalk ballet, which has been lost in many places throughout the world to private automobiles, suburbs, indoor malls, and restrictive laws. The sidewalk ballet has echoes in Doreen Massey's concept of "throwntogetherness,"[14] of which the University of Cambridge geographer Ash Amin notes:

> The ethics of the situation ... are neither uniform nor positive in every setting. ... Social pathologies of avoidance, self-preservation, intolerance and harm [can arise], especially when the space is under-girded by uneven power dynamics and exclusionary practices. ... The compulsion of civic virtue in urban public space stems from a particular kind of spatial arrangement, when streets, markets, parks, buses, town halls are marked by non-hierarchical relations, openness to new influence and change, and a surfeit of diversity.[15]

In 1958, Jacobs wrote: "Designing a dream city is easy, rebuilding a living one takes imagination."[16] And it is with imagination, this sense of possibility, that self-proclaimed urbanists and neighborhood elders are erecting parks from parking spaces, benches from shipping pallets, and gardens from rubble. Acts of "insurgent public space" can be fleeting everyday occurrences that loosen the constraints of public space. These actions include skateboarders who take advantage of the curvatures of a freeway underpass, or who have (re)claimed parts of London's South Bank, and Yangee dancers who use the streets of Beijing on a daily basis to practice their dance ritual.[17] Although not their intention, these everyday performances and actions can shift the meaning of public spaces.

Other interventions are intended to challenge people's conceptions of their built environments. In 2005, San Francisco–based design collective Rebar transformed a downtown parking space into a park, an act of experimental public space creation that laid the foundation for many of the insurgent projects that have followed. The collective reprogrammed the space by adding street furniture, trees, and grass—fixtures denoting a traditional Western park. A small park island in a sea of concrete, the installation attracted curious passersby and promoted social interactions that a parking space fulfilling its intended purpose would not have done. The concept tapped into a far-reaching discontent, or curiosity, or excitement among fellow urbanites seeking to challenge their auto-dominated concrete cities. Rebar launched an annual event, called PARK(ing) Day, which is now celebrated internationally.[18] Such new forms of public space intervention have been enabled, in part, by communication technology.

While technology has threatened the importance of place-based public space, it has also allowed mass mobilization and the open-source dispersion of new urbanist tactics. The Street Plans Collaborative's *Tactical Urbanism* guide includes suggestions for urban gardens, informal street furniture, and food vendors in co-production between citizens and local government. For city governments such experimentation can be cost effective. More generally in a faltering economy, the responsibility of managing public spaces may fall increasingly on co-production. In the UK there is already a crowdfunding platform dedicated to projects that improve local public spaces and facilities that are freely accessible to local communities. Spacehive allows organizations to propose projects, which are independently assessed for viability before being opened to funding from online supporters. Kathleen Stokes and colleagues report that: "To date, 51 projects have been successfully funded, including … a project to transform a flyover into an urban park in Liverpool."[19]

Well-designed and managed public spaces are particularly important for gender as well as ethnic inclusion: mixed-use spaces and porous design—in which home and work are less rigidly separated—enable women to use city spaces safely. Yet historically urban designs have favored gender-segregated suburbs and garden towns. Part of the problem is of course, that disciplines like architecture, urban design and planning remain heavily dominated by white males.[20]

Since 2011, multiple uprisings around the world have seen citizens reclaim public space as a symbolic means to revolt against unjust power dynamics.[21] These uprisings included those in Cairo's Tahrir Square in 2011 and more recently in 2013, and that in Gezi Park in Istanbul in 2013. The sociologist and urbanist Manuel Castells highlights common patterns in transformational social movements that swept the globe—from Iceland to the US, via southern Europe and much of the Arab world between 2009 and 2011.[22] These patterns emphasize the importance of shared communication and activism in both cyberspace and urban space.

In the Arab Spring revolutions, and in the activities of Las Indignadas in Spain and Occupy in many countries, the movements typically converged on the occupation or "liberation" of a symbolic urban territory—such as Tahrir Square, or more metaphorically, Wall Street. Here activists "escaped the authority of the state and experimented with forms of self-management and solidarity."[23] The existence of "an occupied territory that anchored the new public space in the dynamic interaction between cyberspace and urban space"[24] made possible powerful connections between digital social media, real-world social networks, and mainstream media (notably Al Jazeera). Legal and police antagonism to Occupy Democracy's attempt to occupy London's Parliament Square in 2014 suggests that the powers-that-be recognize the importance of such symbolic spaces, too.[25]

Originating in a "structural economic crisis and ... a deepening crisis of legitimacy,"[26] the movements studied by Castells were all multiply networked (on- and offline), spontaneous, decentralized, viral, and leaderless. They created "spaces of conviviality and autonomy" at the intersection of urban- and cyberspace. Occupy groups constructed community kitchens and assembled libraries. The libraries were especially symbolic for the movement as shared resources, freely available without fees or interest, contributing to a knowledge commons which could grow and expand with use.[27] In the occupied spaces activists deliberated and reflected, achieving a transition from "outrage to hope," shifting public values, and stimulating "new forms of political deliberation, representation and decision-making."[28] Researchers at the New England Complex Systems Institute (NECSI) in Cambridge, Massachusetts, have confirmed in statistical analysis and systems modeling studies that such movements emerge from stress and inequality, and spread in contagious, self-organized fashion.[29]

The movements were—and remain—remarkably creative and politically countercultural, shifting wider social norms by winning broad sympathy from the public. They represent a powerful emanation and example of the possibilities of a healthy shared public realm. And they have impacted on conventional politics. Iceland has a new constitution. Tunisia has its first directly elected president. And in Spain, new political parties have carried forward the spirit of Las Indignadas, Podemos won five seats in the European Parliament just months after its founding; and activist Ada Colau of Barcelona en Comu (Barcelona in Common) won in the city's 2015 mayoral elections—one of several such upsets across Spain.

These movements and insurgencies overlap real and online spaces. The high density of wired and wireless connections to virtual space in cities reinforces the more conventional opportunities for political assembly in the physical public spaces of the city, effectively multiplying the challenges to power and injustice that such social uprisings create. The state is confronted by virtual assembly and physical assembly simultaneously, and despite growing powers of cyber-surveillance and collaboration by corporate providers, the very multidimensionality of these new spaces makes them more durable venues for protest.

The Arts and Counterculture

The density and diversity of cities also creates the spaces in which artistic or musical—as well as political—countercultures can flourish. Artists "locate near concentrations of artistic venues and specialized institutions (e.g., nightclubs, art spaces, design schools) to gain access to their consumer base, industry gatekeepers, and potential employment and contract opportunities."[30] The resulting physical clustering and interaction, and sharing of ideas and agents for example, also helps stimulate new artistic and cultural movements.[31]

In common language, "culture" is taken to imply the arts, music, theater, and other forms of entertainment. But as noted in the introduction we understand culture as a social and political concept. To avoid confusion, here we use the term "the arts" as a collective noun for those various cultural activities. In chapter 2 we introduced the arts as a key contributor to the co-created urban milieu or shared urban commons. Here we explore their role as a political agent of change in cities, through the role of cultural cycles and their potential to spread wider changes in social norms.

Rami Gabriel sees the arts "as a means for articulating emotional—and rational visions of meaning as they pertain to universal human experiences."[32] Artistic participation therefore offers a key outlet for the expressivism that is a core element of contemporary individualism—an alternative to consumerism and reliance on brands to express our identity. Artistic communication offers a means to strengthen empathy and the complex emotional bonds of society, helping us identify with our fellow humans. Moreover, the arts can articulate a public voice, while artistic movements, and even individual works of art or performances, can help mobilize dissent in the public realm.

The arts offer a venue for the establishment of countercultures, which permit the imagination (and expression) of alternate futures, and contain the seeds of potential transformations to realize those futures. As Wilson Sherwin and colleagues note, "Historically, creative 'areas' such as New York's SoHo in the 1980s, have evolved organically and provided important space for counter culture and innovation to develop."[33] At the same time, they are, as we noted in chapter 2, highly vulnerable to co-option and commodification at multiple levels. So, as Sherwin comments, "Today's purposely built creative clusters, often supported by local government and/or private businesses, to varying degrees, are accused of reducing the arts to the state of a commodity, potentially undermining opportunities for true innovation."[34]

Artistic countercultures have formed a vibrant part of public political activity in many countries and periods including the Situationist International's art of spectacle, the counterculture of New York's SoHo, Berlin's squatted artistic center the Kunsthaus Tacheles[35], Christiania in Copenhagen, and even the Occupy movement.[36] As Harvey puts it, effectively bringing together identity politics, anticapitalism and "the Right to the City" movement, "striving for a certain kind of cultural autonomy and support for cultural creativity and differentiation is a powerful constitutive element in these political movements."[37] Of course, as Harvey recognizes, and Gabriel too,[38] these countercultures are typically assimilated into or co-opted by the mainstream of consumer capitalism—which has also mobilized more formal public art and museums as a reformist institution[39] and as an economic development tool to attract the so-called creative classes; but without the sort of cultural spaces the city offers, the hope of truly transformative political movements is lost.

The debates and contests over the commodification of artistic movements echo those we see today in the sharing movement. Musicians have long been accused of "selling out" and "going commercial" when they follow the dictates of the record industry rather than their artistic consciences in order to "sell product," especially if the products they sell are not even their own, as when they allow their music to be used for advertising. On the other hand, commercial channels broadcast the emotional and rational content of the artistic creation far beyond the reach of most individual artists. Sadly, too often the content of the musical product promoted by the industry is unremittingly superficial and reinforcing of taken-for-granted elements of mainstream culture: consumerism and brand identities.

Some argue that in the last 25 years the commodification of music has become irresistible as the urban spaces in which musical countercultures form have been gentrified. The music journalist Taylor Parkes for example, suggests that the commercialization of Britpop

> was about the end of … whatever was left of a counterculture. … About … the dumbing-down and depoliticisation of what used to be known as "alternative" culture … About the beginning of self-righteous privilege, the demonization of the working class, the full assimilation of mass art into neoliberalism … while … Camden Town—Britpop's spiritual home— abandoned its radical bohemian past to become a millionaire-owned, tourist-oriented pop-cultural charnel house.[40]

Going Underground, Online

In this context—where the urban spaces for counterculture have been eroded, the battle over the equivalent cyberspaces is of great interest. Online, musical creation and dissemination have been transformed by online music sharing. Back in 2004, Yochai Benkler noted that "in displacing industrial distribution, peer-to-peer distribution is thought both by its critics and by some of its adherents to be likely to undermine the very possibility of industrial production of music."[41] Artists concerned that artistic expression was being constrained by the industrial mode of production might welcome that prospect, while those within "the machine," as the rock band Pink Floyd called it, may well resist. As Benkler pointed out, "Certainly there are recording industry executives whose roles have no existence outside of the industrial organization of music production and distribution."[42]

At the same time, arguing strongly against the criminalization of file-sharing, Benkler dismissed the view that the new modes would somehow halt artistic creativity:

> As for creation, it would be silly to think that music, a cultural form without which no human society has existed, will cease to be in our world because we … abandon the industrial form it took for the blink of a historical eye that was the twentieth century. Music was not born with the phonograph, nor will it die with the peer-to-peer network.[43]

In fact it appears that online music sharing has shifted the economic balance of the industry back toward artists, with higher demand for live performances and higher ticket prices. These are of course, genuine shared experiences, as well as being increasingly important to musicians' livelihoods.[44] Nor indeed will the cycle of culture and countercultural movements within the arts die also as long as spaces remain for new movements to be born in our cities and online. For instance, one of today's hottest musical movements is pop-up gigs. For example, Sofar Sounds "curates secret, intimate gigs in living rooms around the world,"[45] spotlighting emerging artists in 85 cities across around 40 countries. Sofar describes itself as is an international collective of fans, artists, and music professionals. But in a sad echo of the key challenge of commodification and consumerization of urban space, its business model appears to rely on leveraging its network of fans to enable it—through its "talent and licensing division" Sofar Creative—to discover new talent and license their music for advertising.[46]

The examples of collective protests, insurgent urbanism and countercultural arts so far in this chapter all foreground the city as a venue, as a shared creative space—as opposed to a place where gentrification and an unequal economy exclude both countercultural and political art and the artists themselves. The public and common spaces of sharing are thus as contested as the commercial ones we focused on in preceding chapters. Artists and activists alike must therefore also share a right to the city: not just as an entitlement to share in the life, facilities and resources of the city; but also as a right to collectively change and reinvent the city, its citizens' identities, and their politics.

The Power of the Web: Politics in an Internet Age

Art and protest are but two emanations of the increasingly blurred nexus between urban space(s) and cyberspace, which hold promise for the rediscovery of a collective politics now and in the future. These spaces are fundamentally important for forms of participation invented and controlled by the people. And the Internet is far more valuable than it might appear from the prevalence of celebrity gossip and lolcats.

Clive Thompson highlights how fears of "dumbing down," trivialization, and social atomization are nothing new to the Internet age: similar concerns have been raised, over the ages by writing (as opposed to remembering knowledge), the telegraph, the telephone, coffee houses, and even mass-market novels; against which one 1835 essay fulminated: "Perpetual reading ... inevitably operates to exclude thought. ... It is apt either to exclude social enjoyment, or render the conversation frivolous."[47]

Thompson suggests rather that the Internet offers cognitive benefits in the form of collective intelligence and the broadening of our "ambient awareness" of what is going on in—and around—our social networks. Ambient awareness allows us to exploit our many "weak ties" in intelligence gathering, and potential collaboration—ties which we would often not maintain in the days before Facebook and Twitter.

This effect is most powerful if we manage to build diverse networks, rather than linking only to people like us, and allowing network algorithms to filter out anything unusual. In a diverse network, tools like Twitter allow us to "think out loud" and benefit from the collective conversation. Like "positive encounter spaces" in the urban setting (see "Intercultural Public Space" later in this chapter), online discussion boards for hobbies and cultural pursuits expose young people to a diversity of views, and the process gets them comfortable dealing with strangers.[48] Thompson notes further: "This wasn't true for kids on Facebook, because Facebook doesn't encourage you to interact with strangers."[49]

Castells suggests also that social networking sites on the web form "living spaces connecting all dimensions of people's lives," and their expansion "transforms culture by inducing the culture of sharing."[50] He cites World Values Survey data which suggests that "Internet use empowers people by increasing their feelings of security, personal freedom and influence" especially strongly for "people with lower income and less qualifications, for people in the developing world and for women."[51] A Pew Foundation survey confirms that "the more socially active people are online, the more civically active they are offline too."[52]

Online connections are proving particularly valuable for members of culturally oppressed groups, such as lesbian, gay, bisexual, and transgender (LGBT) youth or women facing oppression in fundamentalist religious groups,[53] because of the scope for anonymous participation in collective political action, and the possibility of making effective long-distance bonds linked by a common cause. Social media were also critical throughout the transformative movements studied by Castells. Not only did they facilitate advance discussion of grievances and demands online, but also, he says, "In every single case, the inciting incidents of the Arab Spring were digitally mediated in some way."[54] The Internet creates "the conditions for a form of shared practice that allows a leaderless movement to survive, deliberate, coordinate and expand," as well as being a decisive tool for mobilizing and organizing. It also provided the communication needed to protect "the movement against the repression of their liberated physical spaces."[55] Moreover, social media also facilitate artistic political creativity, changing culture (in its narrow sense) "as a tool of changing politics."[56]

However, at the same time that neoliberalism is forcing a reduction in public space, it is also threatening the same in cyberspace. They might seem far from the struggles of the Arab Spring, but ongoing battles over net neutrality are an expression of the same confrontation. Genuine net neutrality is based on the principle of equality: that Internet service providers (ISPs) and governments should not discriminate on the basis of (legal) content or introduce different charges, whether by type of user, particular content, particular site, platform, application, type of equipment, or mode of communication. Some prefer the term "open internet" where there are no restrictions and no extra charges. Genuine net neutrality is just one of a suite of measures needed to sustain the sharing paradigm, alongside rights to a secure online identity and privacy regarding personal data (see "Cybertrust and Identity" in chapter 5.)

Tapscott and Williams see open access to cyberspace as a critical contributor to the public square, fearing continued decline in civil liberties and press freedoms especially in "authoritarian regimes."[57] Of course efforts to limit web freedoms in countries such as Turkey, Azerbaijan, and China are of grave concern, but so should be the idea that the "free for business" cyberspace pursued by the US administration, dominated by market forces and advertising, is genuinely a space of democratic freedom. The Harvard law professor Jonathan Zittrain highlights how easily a digital social network platform could influence electoral results simply by reminding people in particular groups or places to vote on the day—in what he describes as "digital gerrymandering".[58]

For Thompson the "the biggest conundrum for politics in the digital age"[59] is that the largest online spaces, such as Facebook, are technically private, and can act to prevent political speech or endanger those involved in activism or protest, for example, through rules against the use of pseudonyms. He suggests we might address this through a charter of Internet rights—as proposed by Tim Berners Lee, the "inventor" of the "World Wide Web," and actively campaigned for by the international Internet Rights and Principles Coalition[60]—but also by recognizing that these corporate spaces are functioning as utilities and regulating them accordingly.

Net neutrality is clearly desirable for the effective functioning of the multiplicity of web-based sharing platforms mentioned elsewhere in our book, but it is of more than instrumental interest. It is easy to imagine incumbents seeking to resist sharing approaches by limiting web access to competitors, but it can become a matter of life and death where governments seek to close down sharing of political activism in cyberspace.

In doing so, governments have in part relied on creating a climate of fear of cyberterrorism, part of wider narratives designed to bolster political control by elites through the creation or exaggeration of fear. This approach characterized the post 9/11 decade and continues today. The conservative political agenda pursued by George W. Bush and Tony Blair (and others) elevated fear of terrorism, crime, and racial and cultural difference above any objective assessment of the real risks involved. In doing so they created political space for oppressive measures, and an economic "shock therapy" that predated the financial collapse of 2008 but was further boosted by the narratives of "austerity," which have dominated ever since.

In the use of narratives of fear to justify greater control of the Internet, governments have embarked on something of a race to the bottom.[61] Castells is perhaps over optimistic in his view that the freedoms of cyberspace are "beyond reach of usual methods of corporate and political control."[62] The great firewall of China, and the reported 300,000 Chinese "Internet police" monitoring the web for material against the Communist Party's definition of national interests,[63] are matched by the NSA's blanket monitoring of social media, for example. Turkish president Erdogan's efforts to strangle tweeting might be motivated by his fear of it being used for political organizing by his opponents (such as that around the Gezi Park protests mentioned above), but such measures threaten not only political sharing online, but also social and economic sharing. By casting a dark shadow of surveillance and control across even social networking tools like Facebook and Twitter, governments can only exacerbate the breakdown of social capital, and increase the withdrawal of citizens from society. However, such efforts are rarely effective in bolstering control in the short term, as they drive innovation in evasion and avoidance of Internet controls. For example, in January 2011, the Egyptian government blocked almost all Internet access, as well as text and Blackberry messaging services. In response activists and hackers around the world helped create work-arounds mainly based on landline telephone connections. The blockade was so effectively circumvented by these means—and by mobilization in physical spaces—that within four days the government gave up trying.[64] Such efforts are also utterly shortsighted—both for the economic and social future of societies so emasculated, and for the long-term political survival of the governments wielding the knives.

In contrast to narratives of fear, sharing could underpin a more collective political response in terms of the narratives it inspires. The sharing paradigm lends itself to political narratives of personal benefit, of human flourishing, of security and independence (from remote institutions and corporations). Insofar as such narratives shape politics they confront and challenge those based on fear, insecurity, and conflict, and can be reinforced by sharing politics online.

Tapscott and Williams argue that, "Unlike the older hub and spoke architecture of the mass media, the peer-to-peer architecture of the blogosphere is more resistant to capture or control by the state,"[65] with broad implications. Writing more than a year before the Arab Spring they noted that "against all odds, the transnational Arabic blogosphere is exploding and liberating new channels for uncensored dialogue," citing 35,000 Arabic language blogs in 18 countries where "arguably the blogger has become the new freedom fighter." They concluded at that time, however, that it was not yet clear that the Arab blogosphere had "the critical mass needed to change the course of events in Arab countries."[66] With hindsight—looking at the various conditions of Syria (civil war), Egypt (military in power) and Tunisia (free elections) at this writing in 2015—we can see that activism in the virtual realm can break through into the material world, but alone is not sufficient to deliver sustained transformative change. That only becomes possible with the effective interlinkage of virtual and physical activism.[67]

A Danish member of parliament (MP) in Copenhagen, Uffe Elbæk, and Neal Lawson, chair of the UK's Compass think tank, see the Internet, the smart phone, and social media as flattening the world, empowering collaborative rather than hierarchical politics, delivering "informed, enabled and empowered citizens."[68] They argue:

> The new rules of this epochal shift go with the grain of a good society precisely because ... we talk and participate as equals. ... The post–1945 social settlement could never hold, because it was built on well-meaning but hierarchical institutions. And so the counter-revolution of neoliberalism took hold in the late 1970s. ... This doesn't deny the need for struggle. The big corporations will try to commercialize these new flat planes, and the threat of authoritarianism is real. But here at last is a terrain that can be genuinely contested by radicals because democracy and equality are now what we struggle with, and what we struggle for.[69]

An important theme in both online and urban radical activism is anonymity. This is not just a question of protection from arrest and retributions by oppressive states but a much more general condition of spaces for political activism that do not privilege existing elites. Anonymity allows participants to shed cultural preconceptions and to engage freely, on equal terms. Eurig Scandrett, a sociologist at Queen Margaret University in Edinburgh, in a paper for the UK NGO Friends of the Earth, notes how, online not only is it possible to hide the age, gender, occupation, accent, geographical location etc. and all the other visible signs or symbols of power in stratified societies, but it is also possible to obfuscate the influence of that power by deliberately adopting false or fictitious persona.[70]

This helps empower what Scandrett calls "invented participation," "where affected people take collective action to find new ways to intervene in decision making processes on their own terms,"[71] in both urban and cyberspace.

Scandrett also reminds us that to deliver justice, political engagement must extend into economic activity—for example in the realms of cooperatives and co-production: "So long as participation is focused on the allocation and distribution of resources alone, whilst the production of those resources is left to the market, then inequalities in power and access to resources will persist."[72]

We see the power of the sharing paradigm, in its reach across economic, sociocultural and political spaces—both on- and offline—as offering opportunities to move beyond such circumstances, not only by redefining our roles in production and consumption, but also, as we discuss in the following sections, by redefining our identities in more collective and democratic societies.

Sharing and Democracy

It should already be clear that we see civic engagement and activist protest as equally valid and equally collective forms of democratic politics. This section argues that both forms depend on a healthy public realm both of physical and virtual spaces. Yet consumerism—driving commodification and individualism—not only directly undermines collective political action, it also degrades the public realm. Moreover it promotes an ideology in which markets replace politics, political interests are fragmented and personalized, and individual behavior change replaces community organizing. We argue that thoughtful investment in sharing can reverse these trends, dampening consumerism, invigorating the public realm, rebuilding both bonding and bridging forms of social capital, stimulating solidarity among diverse cultural and identity groups, and encouraging collective action.

We begin with a brief outline of some of the ways sharing builds social capital. Then we look at the various ways that, in parallel, the public realm is being commercialized, and our identities commodified. Finally we explore how lively and inclusive public spaces can be restored as venues for public debate and democracy, with new forms of political engagement.

Building Social Capital by Sharing

Sharing is a communal act that links us to others in networks and organizations, creating feelings of solidarity and trust.[73] These "bonding" and "bridging" interconnections and sentiments within and between groups are the building blocks and infrastructures of social capital. Sharing businesses, public services, nonprofits, and activist movements all promise to strengthen social capital, with multiple benefits. As Robert Putnam, a professor of public policy at Harvard, notes:

> Life is easier in a community blessed with a substantial stock of social capital. … Networks of civic engagement foster sturdy norms of generalized reciprocity and encourage the emergence of social trust, … facilitate coordination and communication, amplify reputations, and thus allow … collective action.[74]

Michael Woolcock and Deepa Narayan of the World Bank highlight important distinctions between bonding and bridging capital: "The poor, for example, may have a close-knit and intensive stock of 'bonding' social capital that they can leverage to 'get by' … but they lack the more diffuse and extensive 'bridging' social capital deployed by the non-poor to get ahead.'"[75] Both forms need to be valued. As Woolcock and Narayan say, "Without weak intercommunity ties, such as those that cross various social divides based on religion, class, ethnicity, gender, and socioeconomic status, strong horizontal ties can become a basis for the pursuit of narrow sectarian interests."[76] However, "the vitality of community networks and civil society" is arguably, say Woolcock and Narayan "largely the product of the political, legal, and institutional environment."[77] So, an integrated or "synergy" conception of social capital is needed in which co-production of social capital between community networks and state institutions comes to the fore, while policymakers still recognize social capital of the poor, build social bridges, and enforce civil liberties as well as institutional transparency and accountability.

Yet social capital can be threatened by excess pubic intervention. Because bridging social capital relies on interpersonal relations between strangers or weakly related people, strong state provision can damage it as much as powerful markets do. Both replace trust in other people with trust in institutions rather than complementing it. This is another reason we are concerned about over-commercialization of sharing, as it relies similarly on alienating institutions of reputation management and insurance.

Cities and districts with strong social capital typically have strong collaborative institutions, public services, cultural facilities, and voluntary organizations that serve to strengthen both bonding and bridging capital.[78]

Shared spaces and facilities, such as community gardens, increase relationship building and social interaction between neighbors and strangers, including different ethnic groups, and decrease the isolation of vulnerable populations.[79] Shared public spaces not only increase social interaction and social capital but also offer places for physical activity and personal reflection, leading to improvements in physical and mental health. For example, in City Repair's "intersection repair" projects in Portland, Oregon, streets are painted and sharing nodes—including mini-libraries, public seating, or self-serve cafés—set up for the public to enjoy. Residents within a two-block radius report increased social interaction, improvements in mental health, a stronger sense of community, and increased social capital.[80]

Research into the social fabric of urban neighborhoods has uncovered how livability and social capital can be delivered by taking streets back from cars, for people. Jane Jacobs's seminal work *The Death and Life of American Cities*, published in 1961, challenged the modernist, Le Corbusier–inspired, form-over-function understanding of cities. Donald Appleyard's *Livable Streets*, based on research conducted in the early 1970s in San Francisco, showed that traffic volume on streets correlated with livability and social inclusion. Residents on a street with 2,000 vehicles per day had three times the number of friends and twice the number of acquaintances than those living on a street with 16,000. Subsequent research has largely confirmed similar findings in Basel, Switzerland,[81] and Bristol, UK.[82] More walkable neighborhoods and streets correlate with happier people who are more socially involved and connected.[83] The existence, significance, and enhancement of the social fabric of streets and neighborhoods are now seen as "placemaking"—a dominant goal in urban planning (as we saw in the Copenhagen case study preceding this chapter).

"Isolating" technologies and habits, such as person-to-person communication and social interaction via the Internet, may also be reducing local social interaction and diminishing community "bonding" social capital,[84] even as they enable the building of "bridging" social capital with remote others. Most likely the physical and virtual factors both combine with changing lifestyles, with worrying results. According to recent surveys, 75 percent of people in the US and 60 percent in the UK admit to not knowing their neighbors.[85] But better designed virtual spaces might help increase local connections, in the same way as street design can support social interaction.

For example new forms of web-based communities are emerging—focused virtually at the street level. Sites like Streetlife, RightMovePlaces, Your Square Mile, and WeCommune aim to enable local community and face-to-face interaction by providing simple tools to help with coordination of community activities and to share local information.[86] Low-tech Pumpipumpe provides simple stickers so residents can advertise things they have to share with neighbors on their mailboxes or front doors.[87] On the web, Streetlife uses zip/postal code identifiers to help users find people with common interests nearby, promote local events such as craft fairs and community clean-ups, lend skills and belongings, and find out about local clubs, services, and facilities.[88] Following a successful pilot in Wandsworth, London, Streetlife is spreading throughout the UK—at the time of writing, claiming close to 100,000 users in more than 500 communities—backed by Archant Digital Ventures, the incubator/investment arm of a large regional media company. The investor sees possibilities in the sale of targeted advertising, and for synergies with its publishing activities, as a result of access to local stories on the site.[89] Even acknowledging the risks of a funding model like this, such sites sit right at the "intersection" of urban- and cyberspace and can succeed only by rebuilding social capital through communal mediated sharing.

The ongoing generational shift may also help rebuild local social capital. Millennials experience the public realm differently through personal and digital technologies, increasingly preferring public transit as a means through which to remain plugged-in both technologically and socially.[90] Fewer own cars: more car share, take public transit, walk, or cycle. More perceive the experience of the street as one in which they have rights as individual users to interact in myriad ways. This generational shift should help a transition from car domination to shared streetscapes that are more democratized and just.

The nature and quality of online spaces matters more widely, too. We saw the value of online communities and discussion spaces in the last section. But certain types of sharing platforms are proving better for social capital. For example, online exchange communities such as Freecycle foster greater community spirit among their members than do equivalent cash-for-goods websites such as Craigslist.[91]

Deron Beal started Freecycle in 2003 because in his day job, he found it hard to "rehome" excess office supplies and equipment, simply because of the practical difficulties in coordinating donors and potential charitable recipients, and even though plenty of each existed. In just seven years, Freecycle grew to over 7 million members in more than 95 countries,[92] successfully rehousing an almost infinite range of products including many things that even charity shops will not take (such as broken appliances or left-over building material). Freecycle also triggers personal relationships and helps build community feeling. As Botsman and Rogers say, "The social capital generated by Freecycle grows every time something is passed on."[93]

Research confirms that different types of groups and associations have different implications for well-being and, we would suggest, for politics, too. Like community groups, church organizations, sports clubs, art and literature clubs, fraternal groups, and youth associations, Freecycle is the sort of group—emphasized by Putnam—in which people participate primarily for intrinsic motivations. Intrinsically motivated groups build community engagement and underpin collective political participation. They also raise reported happiness. For example, being a member of three or more such groups is about half as influential on well-being as being employed rather than unemployed.

Such groups contrast with those that offer instrumental incentives for membership—for example, fraternity associations, unions, and professional organizations.[94] Membership of the "instrumental" groups "goes, if anything, with lower reported happiness."[95] Politically, such groups are more likely to be involved through lobbying and advocacy for particular interests.

In other words engagement in collaborative group activities can stimulate involvement in both the practice and content of collective politics. But if sharing platforms recruit members through instrumental extrinsic financial motivations they may do little for happiness. On the other hand, where they stimulate intrinsic motivations—such as meeting new people or gaining a sense of community—they can be expected to increase happiness as well as material welfare. Cities keen to maximize the benefits of sharing should look to prioritize their support for the latter sort of initiatives.

Yet, designed and managed well, sharing of all types can contribute to mutual respect and solidarity. A shared public realm can contribute to social capital further by reducing crime.[96] Crime and the fear of crime are critical to the desirability and quality of urban life. It is often assumed that disintegration of trust, widespread withdrawal from social interaction, and decline in social capital contribute to growing crime rates. But in recent decades much of the problem has been exaggerated fear of crime—in most cities and countries crime has been falling.

While both structural cultural changes,[97] and environmental factors—such as reduced levels of lead pollution[98]—have probably contributed to falling crime rates, it seems unlikely that trust and social capital are irrelevant. In a vicious cycle, where people withdraw ever further from the public realm, falling trust and rising crime can go together.[99] Yet a virtuous cycle of engagement, rising trust and falling crime is equally possible in the right context, culture and environment. (We come back to vicious and virtuous cycles of cultural change in chapter 5.) Put simply, people respect the public realm, property, and the police more the more they perceive that others do so, too. Where citizens are cooperating to co-create the urban commons we can expect lower crime. Kahan highlights the apparent successes of the Chicago Alternative Policing Strategy (CAPS) that actively engages communities in co-production, increasingly using digital communication such as Twitter as well as direct community liaison. Although it may be hard to disentangle its positive effects from others driving a long-term decline in crime, the 10-year evaluation (published in 2004) suggested a correlation between awareness and involvement in CAPS and reduced crime and fear of crime, with predominantly Latino/a districts exhibiting lowest involvement and rising crime, bucking the general trend.[100]

While reviving a shared public realm and rebuilding social capital may help reduce the negative impacts of crime, an even bigger benefit might be found in the potential for sharing to contribute to proper recognition for minority groups. Even though actual crime levels have fallen, the US prison population has never been higher, especially among African Americans. Worse the prison population among African Americans of college age is larger than that in college, despite the fact that it would be cheaper to send these young offenders to an Ivy League school like Princeton rather than prison.[101] We return to measures that can promote recognition and inclusion later in this section.

All of this suggests ways in which more sharing can help to revive communities that are well placed to engage in collective politics, with strong networks, a healthy shared public realm, and social capital providing practical and normative foundations for public debate over the best ways of supporting society as a whole. But in too many cities the public realm is not healthy. For every Copenhagen with lively streets and diverse cultural spaces, there are many cities where fear of crime, and austerity budgets have decimated the public realm, leaving privatized commercial malls and entertainment complexes in its place. Citizens have been replaced by consumers.

Putting the "Citizen" Back in the City

In the last century most cities experienced increasing privatization and commercialization of public spaces. The transformation of public spaces into a commercial realm in the postwar era of consumerism and individualism[102] was accelerated by the "neoliberal onslaught" of the 1980s, which brought a "trenchant reregulation and redaction of public space."[103] Public space is no longer primarily the domain of shared citizen politics, or collective leisure activities, but of commercial marketing and consumption. This transformation has intertwined with dramatic shifts toward more individualist consumer identities in modern societies.[104]

Cities are often portrayed as melting pots in which such conventional identities are disassembled and individualism thrives. There is truth in this: migration to cities weakens conventional ties and creates opportunities for otherwise suppressed identities (such as LGBT ones) to flourish. More generally, in contemporary societies, increasing mobility, fluidity, and insecurity have eroded longstanding fundamentals of identity such as geographic community, occupation, and religion.[105] Consumerism has leapt into the void, preying on our inclination to construct our own identities as ongoing narratives. Consumer goods and brands are marketed as symbols of group identity and as ways of meeting our psychological needs for identity, stability, and belonging.[106] The dominance of consumer identities is closely linked with the processes of suburbanization and the establishment of mass markets for consumer goods—especially cars and household appliances—in consumer capitalism.

Yet cities are, perhaps paradoxically, also crucibles for empathy and cooperation, where our exposure to diverse others allows our empathic bonds to extend.[107] But the scope for cities to play this role has been stripped back by the commodification of the urban commons and the domination of public spaces and discourses by advertising and commercial or corporate interests.[108] Harvey similarly critiques the commodification of urban quality of life, in which "intense possessive individualism" is increasing "isolation, anxiety and neurosis in the midst of one of the greatest social achievements … ever constructed in human history."[109]

The leisure functions of public urban spaces are also being commercialized and commodified. Much sharing activity in leisure is still unmediated and sociocultural. From sport to food and drink, leisure activities are marked by informal sharing rituals such as buying rounds of drinks or sharing sporting equipment and transport, to and from matches. Many leisure activities are also naturally collaborative undertakings, shared with family or friends.

Yet leisure is also a battleground of commodification. Storytelling, sport, music, dance, and theater all long predate capitalism. But in the modern world, leisure time has been reduced, and leisure as an activity has been commodified: we pay for holidays and tourist activities, to attend artistic and sporting events, and even for gym memberships. And although for children most leisure and play remains informal, even here, the intrusion of market relationships seeks to carve out an ever-larger space for commercial providers. Childcare too is increasingly commercialized. All this commodification of the experiences of childhood is a particularly powerful influence challenging the natural cooperative instincts of children, instead weakening social capital and establishing consumer identities.

The example of toy sharing is illustrative of the opportunities and risks. *Swap Shop*, a 1980s UK Saturday morning kids TV program, introduced a generation to swapping toys by combining high-profile televised swaps with a roadshow of local events where children exchanged everything from soft toys to model airplanes. But it also helped open a door for commercially mediated swapping. Toy and game manufacturers have leapt in, creating and marketing a veritable horde of games that involve the swapping of cards or other pieces to build sets or enhance decks, but at the same time institutionalize repeated purchases and consumerist identities.

Toy libraries, on the other hand, provide a communal mediated version of toy sharing. They are not new, having become widely established in the 1960s, at least in the US. But modern web-based versions appear to be erasing the stigma that used to be associated with relying on borrowed and hand-me-down toys.[110] Moreover, in the UK, toy libraries have been "found to be successful in engaging isolated families in areas of social deprivation, and redressing part of the imbalance between the supply of play equipment available to children from affluent areas and those growing up in poverty."[111]

They also appeared to serve "as a hub for the community, provide opportunities for volunteering and the learning of new skills, create jobs, and build community capacity."[112] Moreover the "toy library experience 'stretched' children in ways parents had not anticipated, with boys trying toys that are typically associated with girls, and less active children becoming more physical in their play, for instance."[113] In other words sharing models of toy provision, as opposed to commercial ones, not only alleviate inequality but also help children experiment with different—nonconsumer—cultural identities and build social capital.

Research in New Zealand confirmed that toy libraries build social capital in the form of a sense of community among a large majority of users. It also found that toy library use particularly supports parents who share values of frugality, sustainability, anti-consumption and generosity.[114] The research concluded that "sharing is an alternative market structure that can be adopted by anti-consumption consumers"[115] in ways that successfully reduce consumption and its impacts. So in the single case of toys we see sharing partly co-opted by commercial culture; partly adopted into mainstream culture—shifting values toward community and also helping strengthen a counter-cultural movement among consumers.

In adult leisure, whether watching or participating, sports and performance arts remain by definition cooperative or shared activities, and depend on a well-maintained public infrastructure. Yet, as commercial spectacles backed by corporate sponsorship, and with the repeated sale of branded team uniforms reinforcing an aggressive tribalism, it can be hard to see the potential for rebuilding community identity in professional sport. And art galleries and performance groups are increasingly forced to rely on corporate sponsorship rather than public support, further disconnecting them from the urban community. Michael Sandel bemoans the effects of commercial branding of sports infrastructure: "In recent decades," he says, "the money in sports has been crowding out the community."[116] He explains:

> Sports stadiums are the cathedrals of our civil religion, public spaces that gather people from different walks of life in rituals of loss and hope, profanity and prayer. ... The public character of the setting imparts a civic teaching—that we are all in this together, that for a few hours at least, we share a sense of place and civic pride. As stadiums become less like landmarks and more like billboards, their public character fades. So, perhaps, do the social bonds and civic sentiments they inspire.[117]

Nonetheless, leisure remains an essential part of the core economy. Public authorities almost everywhere subsidize the arts and sporting facilities. Copenhagen even installs "adult play" facilities in public spaces. Charities invest in encouraging participation in sports by disadvantaged communities. Community groups and clubs coalesce around almost any conceivable leisure interest, from shooting to sailing, from stamp collecting to singing. All this helps resist further commercialization and commodification.

Such resistance is needed on multiple fronts to defend the wider public realm from the onslaught of commercial marketing. Children are routinely exposed to advertising in schools, much of it only loosely disguised as educational material. Advertising increasingly swamps our social media feeds. Marketing samples are virtually forced on women who have just given birth in UK hospitals, in exchange for personal details. Public spaces are converted into commercial malls to save on maintenance costs. Billboards and shop-front advertising dominate city centers.

Some cities are pushing back radically. For example, Bergen (Norway) and São Paulo (Brazil) have banned advertising in outdoor public spaces. In a 2011 survey conducted in São Paulo, 70 percent of residents thought the ban was beneficial.[118] The removal of billboards revealed a rich urban beauty. As local reporter Vinicius Galvao said, "You start getting new references in the city. The city's now got new language, a new identity."[119]

In this contested space the impact of advertising *within* sharing models is a serious concern. It's not only that deals like Barclays sponsorship of London's city cycle program were greenwashing. "It's obvious what was in it for the bank" says the *Guardian*'s Anne Perkins:

> For a mere £50m, which is a tiny fraction of the actual cost of running the scheme [program], its brand was to be—pun alert—pedalled around the capital for at least five years. A kind of redemption-through-association, a bank in need of a better image hitched to a clean, green transport policy.[120]

But such sponsorship and business models reliant on selling advertising space to fund online sharing platforms share a further problem: the values and norms communicated both by the presence and content of advertising risk contradicting those that sharing advocates hope to stimulate through sharing. Even though some advertisers are withdrawing from the "social web"—car giant GM, for example, recently pulled all Facebook advertising[121]—that does not signal an end to commercial marketing in online "public spaces."

We are not arguing however, that only countercultural creative activities such as subvertising and graffiti can help co-produce a citizen-oriented public realm (helpful as these can be). Even commercial advertising can be regulated to communicate more sustainable norms.[122] Like creative activities anywhere, co-production in the commercial economy, motivated by desires to share knowledge and skills, can also offer new foundations for personal identity, disentangled from consumerism. Also, online social networks make it easier for us to show "status, group affiliation and belonging" without material consumption,[123] and to begin to turn those expressions of identity into collective political activity as citizens. But as we have seen in this section, coordinated action—for instance to reduce the intrusion of consumerism and advertising into shared spaces, and particularly into children's shared environments—can help support a transformation in how people build their identities by relying less on consumer goods, ownership, and the narratives of marketing, and more on relationships, communities, and places.

Making Space for Public Deliberation

A healthy public realm is both an expression of successful participatory democracy and a venue for its practice. In a working democracy, public discourse and public reasoning influences political decisions. But, as Amartya Sen, a Harvard professor of economics and philosophy, argues,[124] such participative deliberation is in itself central to the development of a good society. It is a fundamental human capability, without which we lack a critical freedom. It is of course also an essential means of discussing what we think is good, rather than relying on consumer markets to define well-being.

Yet, as we have seen, crime and consumerism alike are putting urban public spaces' democratic functions under threat. Does this really matter? Ash Amin is critical of "urban essentialists" who see urban public spaces as politically critical. He argues that in the modern era

> of organized, representative, and increasingly centralized and also veiled politics ... sites of political formation have proliferated, to include the micro-politics of work, school, community and neighborhood, and the workings of states, constitutions, assemblies, political parties and social movements. Urban public space has become one component, arguably of secondary importance, in a variegated field of civic and political formation.[125]

Civic practices, too, argues Amin, are shaped not only in public spaces, but also "in circuits of flow and association that are not reducible to the urban (e.g., books, magazines, television, music, national curricula, transnational associations)."[126] Yet it remains possible to believe both that the importance of urban spaces has declined, and also that they could still be essential, especially as they overlap with public cyberspace. Tim Stonor of the Academy of Urbanism argues that

> urban policymakers should be careful not to abandon the historically successful form of dense, compact and continuously connected cities. The two processes—online and on-land—should be made to work together, to create an effective digital urbanism ... [and] a return to the continuously connected city, providing streets, parks, cafes, workplaces and public realm to be occupied by people in pursuit of social and economic exchange.[127]

Even Amin accepts that the unconscious "experience of public space remains one of sociability and social recognition and general acceptance of the codes of civic conduct and the benefits of access to collective public resources ... [which] still underpins cultures of sociability and civic sensibility."[128] Such collective impulses are not stimulated by all shared spaces, however, only by those "that are open, crowded, diverse, incomplete, improvised, and disorderly or lightly regulated."[129] Copenhagen appears to have achieved some of this potential. Jan Gehl reports survey evidence over three decades of efforts to design such spaces. The results were a four-fold increase in streetlife, shifting away from purely shopping activities to include many more cultural, spontaneous, and political uses, including "street music, street theater, vendors of all types, [and] people presenting ideological, political or religious messages to their fellow citizens."[130]

Laying aside the successes of political protest and revolutionary movements rooted in urban space, we argue that, even in "normal" times the shared urban commons can influence our political culture both reflexively *and* consciously, the one mechanism priming our receptiveness to the other. In this respect Amin is right to emphasize the subconscious effects. Clearly the subliminal effects of a public realm in which diverse cultures and subcultures are expressed artistically and socially will be much more positive than one dominated by commercial marketing. Moreover, we also argue that insofar as the virtues and character of well-shared urban spaces can be expressed in the virtual spaces that are now equally important to political activity, then the emergent politics is likely to be one of greater collective solidarity. In other words in the intersection between urban space and cyberspace lies the possibility of a genuinely cosmopolitan politics of justice and sustainability at a city scale, which can practically demonstrate autonomy from the centralized spaces of neoliberal globalization.

In different ways Las Indignadas, Occupy, and Podemos, as well as cosmopolitan citizen mobilization organizations MoveOn (in the US) and 38degrees (in the UK), both of which crowdsource their campaigns, and even President Obama's grassroots organizing campaigns, can be seen as early expressions of that potential. Benkler points out how the Obama campaigns married community organizers on the ground and bloggers on the web[131] (in a union of urban- and cyberspace) to promote progressive policies. Las Indignadas and Occupy—despite also unifying urban- and cyberspace—deliberately did not promote any particular political platform.[132] More ambitiously, they sought to challenge the practice of politics as a whole, and especially its corruption by the commercial and financial interests of neoliberalism. And they began to reinvent democracy in new practical, shared, collaborative, and participatory forms.

Yet there is a case for such movements to engage in conventional politics, too. They are not the only ones losing trust in apparently corrupt and opaque political systems. From the Tea Party in the US to the UK Independence Party, right-wing populist parties are riding a wave of declining public trust in politics, typically scapegoating immigrants and ethnic minorities for rising unemployment. Podemos—and to a lesser extent, the civic nationalists of Catalan and Scotland—buck that trend, suggesting that a more optimistic politics of solidarity could prevail, starting on the streets.

Streetlife: Sharing Streets

The street is the public space we interact on and with, every day. Success in sharing streets is vital to success in sharing across the public realm. The key modern challenge in remodeling streets into lively, open, and diverse spaces is how to share street space between car users and others. While Europe's narrow urban roads were historically shared between all users, growing dominance by motor vehicles has led several countries including Denmark, Germany, Sweden, and the Netherlands to act to support shared use. Such measures have resulted in *woonerven* (literally "living yards") in the Netherlands; "Home Zones" in the UK (residential streets, safe for children to play in); pedestrian-dominated shopping streets, such as Denmark's *gågader* (literally "walking streets"), where motorized traffic is limited to a speed of 15 kilometers (about 9 miles) per hour; as well as networks of cycle lanes as we saw in Copenhagen.[133] In the US, similar practices are beginning to take shape in Times Square and on Broadway in New York City, and through the promotion of "mixed-use" communities where land uses and services encourage walking and cycling, delivering a better balance between economic, social, and environmental outcomes.[134]

Since the early organization of modern humanity, roadways have been conduits for a mix of users, connecting people to their government, places of business, commerce, agriculture, leisure activities, and each other.[135] Some streets accommodated pedestrians on sidewalks as early as 2000 BCE, though many, as in Medieval Europe, lacked sidewalks so all users were mixed together.[136] Soon after the mass production of the automobile in the twentieth century, as consumerism took hold of identities, cars claimed a stranglehold over the street.

Government and industry embraced these revolutions. Even Copenhagen—in the early 1960s—wanted the car and American-style streets. Visionary urban designers such as Jan Gehl, and a succession of progressive mayors and management and leadership teams, steered the city down a different path (as we saw in the case study). Similarly, the Dutch love affair with the bicycle and designing streets for people is a relatively recent politically led development, spurred by the 1973 oil crisis and a series of child road deaths (*kindermoorden*). In the US by contrast, the regulation, maintenance, and oversight of the roadways have fallen to a patchwork of agencies, municipalities, and commercial interests that seek to consolidate their piece of the status quo.[137] Individuals' rights to the street as public space are sacrificed to automobility.

But change is afoot (as we like to pun). The US narratives of "Complete Streets," "transit justice," and "Livable Streets" are framing a message that streets are, ultimately, democratic public spaces, and that everyone in the community should have equal rights to space within them.

Complete and Incomplete Streets

The spread of these narratives is successfully reclaiming and reallocating space that has otherwise become appropriated almost exclusively by private cars. But it is less certain whether this enhances "spatial justice" and "democratizes" streets. Streets, we would argue, should not be thought of as merely physical spaces amenable to expert-led, urban design "improvements," but as co-produced, symbolic, and social spaces. When important voices and narratives are missing from the discourse and practice of Complete Streets, the result is Incomplete Streets. In other words, the ways Complete Streets are envisioned by elites and implemented by municipalities might be systematically reproducing some of the urban spatial and/or social inequalities and injustices that have characterized cities for the last century or more. This is the antithesis of the sharing paradigm, and can particularly undermine the possibilities of an inclusive shared politics.

Stephen Zavestoski of the University of San Francisco and Julian Agyeman challenge the ethnic and cultural "incompleteness" of the Complete Streets movement, noting how bike lanes have become "gentrification super-highways" in urban districts where they signal the entry of upwardly mobile white households into previously low-income and minority areas.[138] They cite residents' concerns that bike lanes somehow suddenly become necessary when it is rich white people on the bikes, rather than poor people of color. Mass transit expansions and upgrades, and even walkability improvements, have also become a factor in gentrification (to which we return at greater length in chapter 4). Realtors in the US frequently use the Walk Score as an index of the walkability of any given address and, according to Christopher Leinberger, a research professor at the George Washington University School of Business, "Walking isn't just good for you. It has become an indicator of your socioeconomic status."[139]

The built-environment researcher Anna Livia Brand also highlights how redevelopment strategies couched in the benevolent and nostalgic rhetoric of Complete Streets can fail to recognize the structural disadvantage of the incumbent community and their consequent vulnerability to displacement.[140] For example, the communities of New Orleans' North Claiborne Avenue were first devastated in the 1960s when a new interstate highway bisected them, and later by Hurricane Katrina in 2005. The latest redevelopment strategy, according to the planners, aims to develop "the most complete street in the world" by taking down the urban portion of the highway and promoting the neighborhood's historical roles in celebrated African American cultural traditions such as jazz and brass bands. But this glosses over the "brutalities of racial discrimination and segregation"[141] experienced in this community.

Similarly, the researcher Mark Vallianatos examines Los Angeles' long history of *loncheras* (taco trucks) and street food vendors. He asks:

> Can streets in a heavily immigrant metropolis, however multi-modal the distribution of lane space, be said to be "complete" if they fail to include the livelihoods and economic survival of vendors; the smells, sights and tastes of homelands; and places for people to pause, shop, and eat?[142]

And we would add: If whole communities are unrecognized even in the streets where they live, how will they be recognized as citizens of the wider city?

Intercultural Public Space

The sharing of streets, parks, buildings, and city squares has additional benefits apart from providing a political venue (as discussed earlier in this chapter). Done well—rather than carelessly, as we saw with Incomplete Streets—it offers a chance to promote inclusion, through interculturalism and empathy. Superkilen Park in Nørrebro, a diverse neighborhood in Copenhagen featured in the case study, illustrates the potential well. The Collingwood neighborhood in Vancouver is another case in which effective "intercultural community learning" helped "strangers become neighbors." The University of British Columbia professor Leonie Sandercock and the environmental engineer Giovanni Attili investigated a community that underwent rapid transformation: from 51 percent of the population being of English background, to 50 percent Chinese origin, 9 percent Filipino, and 8 percent South Asian. Focused on the Collingwood Neighborhood House, Sandercock and Attili's film shows how "we all benefit when we actively encourage connections between people from different cultures. It allows us to share our uniqueness, open up new ways of seeing and doing things and enables us to co-create something new."[143] Amin productively describes this mixing as *convivium*,

> a brush with multiplicity that is experienced, even momentarily, as a promise of plenitude. ... Is the shared experience of the well-stocked and safe, park or street and community center or library not such a brush, based on interest in the possibilities of serendipity and chance, the gains to be had from access to collective resources, the

knowledge that more does not become less through usage, the assurance of belonging to a larger fabric of urban life?[144]

In the 1990s, Amin and Sandercock, along with Peter Hall and other urban geography and planning scholars and practitioners, developed the concept of "interculturalism." The group highlighted how approaches framed in multiculturalism had supported cultural preservation, celebration, and tolerance, yet had produced culturally and spatially distinct communities. Multiculturalism as it was conceived did not require any fundamental change in thinking. James Tully, Distinguished Professor at the University of Victoria, Canada, argued instead that we need to be more sensitive to how cultures overlap, interact, and are negotiated in ways better described as "intercultural" and demanding of a "politics of recognition."[145]

The Council of Europe researcher Jude Bloomfield and Franco Bianchini, a professor of cultural policy and planning at Leeds Metropolitan University, argue for dynamic interculturalism, with important implications for politicians, planners, and sharing cities advocates:

> The interculturalism approach goes beyond opportunities and respect for existing cultural differences, to the pluralist transformation of public space, civic culture and institutions. So it does not recognize cultural boundaries as fixed but as in a state of flux and remaking. An interculturalist approach aims to facilitate dialogue, exchange and reciprocal understanding between people of different cultural backgrounds. Cities need to develop policies which prioritize funding for projects where different cultures intersect, "contaminate" each other and hybridize.[146]

Sandercock, however, demands an empathic response, appealing directly to our emotions:

> I dream of a city of bread and festivals, where those who don't have the bread aren't excluded from the carnival. I dream of a city in which action grows out of knowledge and understanding; where you haven't got it made until you can help others to get where you are or beyond; where social justice is more prized than a balanced budget; where I have a right to my surroundings, and so do all my fellow citizens; where we don't exist for the city but are seduced by it; where only after consultation with local folks could decisions be made about our neighborhoods; where scarcity does not build a barb-wire fence around carefully guarded inequalities; where no one flaunts their authority and no one is without authority; where I don't have to translate my "expertise" into jargon to impress officials and confuse citizens.[147]

Imagine, for a moment, a mayor or city leadership group who had the courage to move further in the creative and inclusive directions of Superkilen Park or the Collingwood Neighborhood House; to contaminate and hybridize across cultures; to feel seduced by the city; a mayor or leadership group that refused to go with the status quo, with what is probable, but instead focused on a vision of what is possible—the fully sharing city.

So is this more than a dream? First-hand experience of diversity in US cities, it has been suggested, contributes to a withdrawal from sociability with different neighbors.[148] But experience with desegregation in the US military suggests that where interaction and cooperation with "the other" is required it reduces prejudice and discrimination.[149] Shared space is a necessary but perhaps not sufficient factor for interculturalism. Gordon Allport's classic work on "contact theory" found both that cooperative interpersonal interactions reduce prejudice, and that interracial interactions in public spaces are more authentic than in workplaces.[150] In a meta-analysis of "contact" studies the social psychologists Thomas Pettigrew and Linda Tropp reviewed 515 studies, involving around a quarter of a million subjects in 38 countries who had engaged in face-to-face contact. They found that in 94 percent of cases, intergroup contact reduced prejudices and social divisions, with strong evidence for the effect between groups of different races and ethnicities, across the homosexual-heterosexual divide, and for people with and without physical disabilities.[151] Revisiting Allport's work, the psychologist Oliver Christ and his colleagues analyzed seven recent studies. These confirmed that positive contact reduces prejudice both among those "directly experiencing" interactions with members of other groups, and even among other members of the community without direct, face-to-face intergroup contact, through the establishment of new social norms.[152]

We believe that the collaborative and cooperative nature of sharing activities offers a strong base for promoting interculturalism, insofar as participation is open to, and enabled for, all cultural groups. While we do not expect intercultural participation, and transparency of those positive interactions, to arise automatically in sharing projects, there is encouraging evidence that, where mediated sharing highlights our interdependence with other groups, this would encourage a collective approach to politics. In psychological experiments, according to the Leiden University researcher Jojanneke van der Toorn and her colleagues: "Instilling a sense of intergroup interdependence can increase political liberalism and, in turn, foster concern for universal welfare."[153] Moreover they say, "To the extent that people do hold an appreciation for a diverse, multigroup society, the more they will lean toward political liberalism and concern for human rights."[154] So sharing, collective identities, and intercultural solidarity do run together.

Sharing a New Politics

In stressing the role of culture and identity in politics, we are aware that identity politics can be seen as a case of "false consciousness" in which people are encouraged to ignore their class interests in favor of more superficial group interests. For example it is true—as cogently stated by Nancy Fraser, a critical theorist and professor of political and social science at The New School in New York City—that aspects of feminism (and other forms of identity politics) have been co-opted by neoliberalism. Fraser argues:

> Feminist ideas that once formed part of a radical worldview are increasingly expressed in individualist terms. ... A movement that once prioritized social solidarity now celebrates female entrepreneurs. A perspective that once valorized "care" and interdependence now encourages individual advancement and meritocracy.[155]

But our focus in this chapter has been on our identity as citizens, not consumers, and on the common interests of the diverse cultural groups of the city, when acting as citizens. So we recognize that real human identities are complex and multilayered—combining genetics, culture, and ideological and political stances for example. But commercial consumerism is distorting everyone's identities. Progress demands a mature identity politics that challenges the pernicious role of marketing and consumerism in identity formation, recognizes the multiple identities we each occupy in modern society, and builds coalitions between different oppressed identities. Identity politics should be a broad church, in which the construction of political coalitions around the intersectionalities of race, gender, or sexuality can be genuinely liberating and powerful. Brittney Cooper, assistant professor of women's and gender studies and Africana studies at Rutgers University writes helpfully:

> I do think we have to remember that intersectionality was never put forth as an account of identity but rather an account of power. That we have taken up intersectionality as a way primarily to speak about ourselves and endless categories of identity is unfortunate, especially since it often means that we can't think productively about how racism, sexism, classism, heterosexism, ableism, and yes, neoliberalism, interact as social systems to disadvantage people multiply placed along these axes.[156]

Shifts in identity and identity formation that are enabled by the sharing paradigm can surely contribute to such a mature form of identity politics—they expose and challenge the taken-for-granted roles of consumerism and individualism, with sharing and mutuality, while simultaneously acknowledging difference and empowering diverse interests to work together. Moreover, by reclaiming the public realm for civic democracy—in both physical and virtual spaces—the sharing paradigm can stimulate its own virtuous cycle, as more participatory city authorities further enable the sharing paradigm to flourish.

We have argued that sharing can support democratic solidarity by promoting intercultural inclusion and challenging the commodification of urban life and urban spaces, at multiple levels. First, participation in sharing—especially forms of collaborative consumption that eschew ownership of goods for shared access to the services they provide—directly challenges identities rooted in consumer goods and brands. Second, the processes and skills of cooperation demanded by sharing challenge the individualistic model of the self promoted by contemporary advertising and marketing. Third, sociocultural sharing reinforces values of community and collaboration, which we then can expect to be reflected in the political domain. Fourth, co-production of public services and of the public realm as a whole, and redesign of public spaces for shared use, resists their domination by marketing. In these ways inclusive sharing promises to help put the "citizen" back into the city—a city richer in social capital and featuring a healthy intercultural public realm—both producing and reproduced by collective civic engagement and political activity.

Collective Governance: Cryptostates and Cryptocurrency

In chapter 2 we saw how public goods and public services are being delivered through co-production between citizens and state. And in this chapter we have seen how sharing in cities helps create a more open and collective political realm. Here we explore the possibilities for these to come together in citizen participation in the procedures of governance, and in the creation of parallel governance arrangements, as well as public involvement in regulation, and sharing practices in government (notably data sharing). We conclude the section with a brief discussion of land value and other questions of taxation in a sharing city.

Urban governance is a complex topic in its own right, with particular contestation over issues such as the powers of city mayors and councils, and the degree of autonomy cities should enjoy from national authorities. But cities do not simply need more autonomy from national governments so as to compete for inward investment, as advocates of charter cities suggest.[157] Instead they need a form of co-production which Harriet Bulkeley, a professor of geography at Durham University, UK, and her colleagues call "distributed autonomy."[158] Currently, local and individual autonomy in cities is fragmented, and—all too often—efforts to devolve responsibilities to local communities are not matched with commensurate resources. Distributed autonomy involves the networking of existing and new sources of autonomy within cities—including that in empowered communities and citizen networks. It devolves authority and resources in a clear framework, and draws on the existing capacities of citizen organizations to strengthen urban democracy. Yet it also remains clear about the need for appropriate universal standards and protections, contributions to global goals such as climate change mitigation, and the provision of resources through a shared system of redistributive taxation.

Bulkeley and her colleagues argue:

> To flourish and to contribute to global flourishing, cities and their people need both greater control over their own destinies, and a strong ethical compass. In moral terms, the concept of autonomy captures both these aspects: individual freedom, and self-control that respects the freedoms of others. ... For city leaders, [however,] frustrated at constraints on their freedom to implement their visions for change, enhanced autonomy may be found as much through collaboration with communities ... within the city ... as it is to be found in renegotiated relationships with national authorities.[159]

We see sharing as an expression of the communal values that underlie support for public services in many countries. But this does not mean that supporters of sharing naturally see a big role for the state. Many see collaborative provision of services and co-production as a way of rolling back and supplanting state bureaucracy. Some even harbor visions of alternative states.

Clearly betraying their cultural background in the US, Tapscott and Williams, for example, see microfinance (and Kiva in particular) as "democratizing" international development and an alternative to a caricature of a "bloated and inefficient UN type model."[160] They also see the possibility of co-production and open government "opening up free trade in public services," envisioning a future in which—rather like footloose multinational enterprises today—a citizen could choose "health-care from the Netherlands, business incorporation in Malaysia, marriage licensing from a municipality in the United States [and] education from a worldwide virtual network."[161] From the perspective of the sharing paradigm this would seem to completely miss the point. Shared consumption, production, and services are preferable to the conventional market-based alternatives because they guarantee basic needs, socialize costs within a community or nation, and deliver efficiencies that in turn protect our collective interests. Turning public services into some kind of globalized pick-and-mix private smorgasbord would likely undermine both justice and sustainability.

Some supporters of sharing choose to largely ignore the state. Rifkin makes a central claim that collaborative culture is gaining the power to displace capitalism as an organizing principle.[162] Yet, perhaps falling into the post-political trap of technological determinism, he does not consider what this might mean for politics nationally and internationally, nor for the states that are the current units of international politics. In the modern world states are as much the agents and creations of capitalism as firms. Presumably as Rifkin hints, the replacement of capitalism might also undermine the power of states.[163] We might well agree. In the sharing paradigm we would expect state economic power to decline insofar as the relationship between capitalist states and "global" enterprises is broken. But political power might also be transformed. Collaboration transcends national boundaries, yet is rooted in geographic focal nodes—the places we call cities. Like the economist Paul Romer,[164] the political theorist Benjamin Barber,[165] and Saskia Sassen,[166] we therefore conclude that cities may have a much larger part to play in global governance in the coming century than they did in the last—if for different reasons.

Even more radical visions also exist. In some quarters—notably among aficionados of the "cryptocurrency" bitcoin[167]—there is active discussion of "alternative states" in which people would consider themselves "voluntarily" citizens of an "online cryptostate" with its own governance. Of course this would not legally exempt them from accepting the rules of the spatial state(s) they live in, but like alternative currencies, could begin to create bubbles where some of the norms and practices associated with those real-world states are suspended, allowing experimentation.

Arguably, cryptocurrencies might help to distribute power more evenly, reducing the economic dominance of banks, and corporations—as do local alternative currencies—but without any particular link to a geographic community. (In chapter 4 we discuss local alternatives in "Complementary Currencies As a Foundation for Stronger Sharing.") Advocates of cryptostates argue that the same technologies and approaches could be applied to government, using encryption technologies to validate identities in these online states, and decentralizing power to citizens. Unfortunately the same discussions reveal active antidemocratic tendencies, justified in the name of meritocracy—another emanation of the "post-political" trap, perhaps.

Co-producing Governance

Fortunately other efforts to establish alternative economies and polities are more often than not explicitly democratic in their participation. From local cooperatives to the World Urban Forum, citizens are generating new ways to govern city spaces, often rooted in collective commons management or participatory democracy methods.

In Porto Alegre, Brazil, "participatory budgeting" has successfully enhanced social inclusion within a balanced budget.[168] In an elegant example of "distributed" autonomy, the central city authorities support local participation and deliberation in determining how a share of the city budget is spent, but also hold the local bodies accountable for operating in fair and effective ways.[169] The process has resulted in

> a reversal of priorities: primary health care was set up in the living areas of the poor, the number of schools and nursery schools was extended, and in the meantime the streets were asphalted and most of the households have access to water supply and waste water systems ... [while] districts with a deficient infrastructure receive more funds than areas with a high quality of life.[170]

Participation in Porto Alegre has been high, sustained, and fairly representative of all segments of the population.[171] It has also helped deliver wider social benefits: social capital has increased, corruption has declined, and tax compliance increased among the middle classes and affluent. Moreover, once people become involved in participatory budgeting, it appears that participation spreads more easily to other sectors, including education, health, infrastructural services, and sports facilities.[172]

Diether Beuermann and Maria Amelina, who are researchers at the Inter-American Development Bank, report similar findings from a controlled experimental trial in Russia, where participatory budgeting in administratively and politically decentralized local governments increased public participation in the process of public decision making, increased local tax revenues collection, channeled larger fractions of public budgets to services stated as top priorities by citizens, and increased satisfaction levels with public services.[173]

Participatory democracy is not always a top-down delegation of power. The Bolivian city of El Alto was the epicenter of anti-neoliberal uprisings—marked by resistance to the exploitation of natural resources long considered the common heritage of the indigenous population—that brought Evo Morales into power in 2005. Now the city has developed distinctive democratic practices blending communitarian neighborhood associations providing "collective local goods" with sectoral associations bringing together workers in the informal economy, such as street vendors, and in conventional labor unions. These "fuse culturally," promote "a collective sense of self" among the citizenry, and effectively co-govern the city autonomously.[174]

In Spain integral cooperatives, (a topic we introduced in "Cooperatives As a Catalyst for Co-production" in chapter 2), are using new technology to link consumption, production, community currencies (both online and local), and time-banking; they see their role as explicitly political in the sense of building a new polity.[175] This movement is urban in origin, adopting the methods and values of participatory governance created by Las Indignadas and Occupy, including the idea of "no leaders." These movements in turn adopted this norm from the Internet where, says Castells, "Horizontality is the norm, and there is little need for leadership because the coordination functions can be exercised through the network itself."[176] In the urban movements the rejection of political leadership also reflected a quest for authenticity in personal identity,[177] or could be seen as an extreme form of shared leadership, with all having the right to participate in general assemblies or as representatives in the "spokes" councils to which some decisions were delegated.

Many commentators argued that Occupy failed to agree any demands through its participatory governance mechanisms, but simply discussed issues endlessly. But this view is misguided. These movements were not seeking reformist changes in political programs. Rather they were refusing to recognize the legitimacy of the existing political order,[178] and rejecting the capitalist logic of production and its focus on material outcomes. They were inherently cultural, transforming minds toward a new politics that does not surrender "the meaning of life to economic rationality."[179] Moreover, argues Castells, "For a deep reflexive current in the movement, what matters is the process, more than the product. In fact, the process is the product."[180] Both the processes and the philosophy clearly echo those practiced in Christiania in Copenhagen (see case study preceding this chapter), which in many ways can be seen as a precursor of the Occupy movement. Some of the principles of Occupy have also been incorporated into the online discussion and decision-making platform Loomio. Loomio's developers at Enspiral, a collective of social enterprise consultants, collaborated with Occupy to develop and test the tool.[181]

The sense in which such groups and communities are seizing opportunities to exercise local autonomy, and weaving that together into forms of distributed autonomy, is palpable. Bulkeley and her colleagues argue that city authorities can in fact enable such processes, with initiatives like participatory budgeting, and supporting local alternative currencies.[182] In this way radical participative democracy linked to alternative, shared economic models could break through into the real world in our cities.

Even if its scope is less wide than in the integral cooperatives, shared community scale governance—applied to organizations, facilities, and services—is also being enabled by the same technological and cultural shifts that have simulated the emergence of the "sharing economy." "Commons governance"[183] and "Community governance"[184] models are touted as generic alternatives to both markets and states, with inherent advantages from their local scale.

More generally, the principles of commons governance offer one way to put the sharing paradigm into practice in the governance arena. Commons-based systems involve "mutually agreed on and mutually enforced community norms"[185] to govern any shared property, from parks to websites and from fisheries to squats. Elinor Ostrom showed that self-organized sustainable use can emerge from collective effort among relatively equal users of commons where they are enabled to play a central role in both setting the rules and devising accountable monitoring systems.[186] This also requires shared ethics, a high degree of collective autonomy, and the condition that resource is of similar importance to, and similarly well understood by, all participants.[187] Quilligan sees "a commons movement ... emerging as a potent counterforce to state capitalism ... a consciously organized third sector, including citizens as co-managers and co-producers in the shared management and preservation of their own resources."[188] He argues that such "co-governance" would be practical for a wide range of "resources" that might be managed as commons including a "new generation of cultural commons with unique forms of participation and social capital."[189]

An ambition to adopt commons governance forms for sharing activities is not excessive. Indeed, there are even historic examples of the extension of commons into previously marketized spaces. For instance, most turnpikes and private waterways were converted into public rights of way.[190] Although these are now mainly in public ownership rather than under commons governance, now, in the modern age—facilitated by online technologies for collaboration and real time monitoring—commons governance models would seem much more practical for a range of resources. Indeed the government of Ecuador recently launched a "strategic research project to 'fundamentally re-imagine Ecuador' based on the principles of open networks, peer production and commoning."[191] Such models could be especially valuable where the alternative to divisive liberalized markets is a corrupt public realm, as unfortunately remains the case in too many countries. Here the reputation and trust-building strengths of mediated sharing offer a lesson and model for public services as well as private.[192]

While the creation of alternative states and even commons governance may seem remote and theoretical, models of national sovereignty are clearly under threat from the growing autonomy and influence of cities and city networks in the contemporary economy. This suggests that the spaces for experiments in shared governance and shared economies are growing. Distributed autonomy could offer city authorities the scope "to negotiate new, less uniform and less constrained relationships with the global economy, rather than greater autonomy to try to be competitive within it as it is currently constituted."[193] However, if autonomy is seen instead as motivation for cities to compete for footloose inward investment, it would be a double-edged sword for the sharing paradigm. A flourishing cultural urban commons would be attractive (even while threatened by uniform redevelopment), but more generally, city authorities with their eyes on a competition with cities around the world can easily lose sight of the needs of disadvantaged groups and the co-produced services and cultural sharing approaches that can help meet those needs.

Benjamin Barber argues rather that cities can reconnect government with local participation (and civil society and voluntary community).[194] Nation-states, says Barber, are too big to nurture genuinely participatory democracy, yet too self-interested to collaborate effectively. But "regardless of size or political affiliation, cities deliver a nonpartisan and pragmatic style of governance that is lacking in national and international halls of power."[195] Building on what he sees as the growing effectiveness of networks such as the C40 climate cities group, Barber calls for a "global parliament" of mayors representing a cosmopolitan global system of networked urban centers. We note that the US Conference of Mayors, already in 2013, adopted a resolution to make their cities more shareable. Co-sponsored by 15 mayors including those from San Francisco and New York City, the resolution encourages better understanding of the sharing economy, regulations to enable participation in the sharing economy, and an active role for cities in sharing publicly owned assets.[196]

Barber's ideas might be seen as a more political version of the World Urban Forum run by the UN organization Habitat. The Forum is a biennial event, most recently held in Medellín, Colombia, in 2014, at which academics, mayors, ministers, NGOs, and more assemble to discuss common urban concerns. The World Urban Forums have sought to encourage wide participation. For example in 2005, Canada (as the sponsor of the Forum Secretariat, and in preparation for the forum in Vancouver in 2006), organized an online "Habitat Jam"—a 72-hour facilitated online discussion oriented around 6 forums framing critical urban issues, such as access to water and living in slums. More than 39,000 participants from 158 countries participated, with support and facilitation to break down barriers of language, literacy, disability, poverty, and even the digital divide. The discussions were recorded and 600 ideas captured, from which 70 "actionable" ideas were compiled into a workbook for the WUF meeting.[197]

Without adequate follow-up, such events can leave a legacy of disappointment and distrust. One might cynically suggest that actual jam making probably provides better social glue than such online jamborees, but the sharing paradigm makes space for both, and for shared mechanisms to follow up such set-piece events.

Such challenges to national sovereignty from cities and city networks are being supported by what we might describe as "cyberspace cosmopolitanism." As we explored in chapter 1, the Millennial generation is increasingly cosmopolitan in outlook. At the intersection of urban and online politics we also see the emergence of new "independence" movements. For instance, the political scientist Gerry Hassan describes a "third Scotland" supporting political independence from the UK, not on exclusionary nationalistic grounds, but rooted in a new form of cosmopolitan nationalism, echoing what we called elsewhere a "mature" form of identity politics. Hassan documents:

> The emergence of a self-organizing, self-determining Scotland [which] can be seen as a generational shift with the emergence of a whole swathe of articulate, passionate, thoughtful twenty-somethings. It signifies a shift in how authority and power is interpreted with people self-starting initiatives, campaigns and projects through social media and crowdfunding. Often dismissed as being middle class lefties and luvvies by detractors, the overwhelming social make-up of this group is drawn from what Guy Standing has labelled "the precariat": young, educated, insecure, portfolio workers. ... Its main groups include the arts and culture group National Collective.[198]

It is in many ways a Scottish expression of trends that are "evident across the Western world: the decline of deference, rise of individualism, the crisis of traditional authority, and an emergence of new ways of organizing and doing culture and politics."[199] These new ways draw heavily on online activism, participatory techniques, and artistic inspiration. In methods and participants, they actually match not only Western trends, but also those mapped across the Arab world by Castells,[200] reported in Brazil in 2013,[201] and apparently leaping across Asia, to Hong Kong in 2014.

There is also scope for the sharing paradigm to democratize the smaller scale mechanics of government and administration. The web has enabled a massive leap forward in sharing data about our cities, and doing so in close to real-time. Washington, DC, was one of the first cities to share data and encourage citizens to create apps. Vivek Kundra created a citywide data warehouse open to citizens and stakeholders, and to stimulate shared access he ran an innovation contest called Apps for Democracy. Entrants were "encouraged to tap into the city-wide data warehouse to invent new kinds of Web-based public services. The experiment yielded forty-seven Web, iPhone and Facebook apps in thirty days."[202] The apps ranged from help with reporting problems with parking meters to aggregating information about local services and planning routes.

Web-based feedback facilities for the public realm, such as the UK's fixmystreet.com, and seeclickfix.com in North America, are becoming more widespread, and provide simple opportunities for citizen participation in city services, enabling concerns about safety, vandalism, and maintenance to be more easily aggregated and addressed.[203] In part they rely upon the power of transparency to shame authorities into action. Similar tools have recently been deployed in India to aggregate reports of rape, so as to stimulate better and more sensitive policing.[204] In many areas, suggest Don Tapscott and Anthony Williams, "More disclosure and increased civic participation in regulatory systems could be a formidable complement to traditional command-and-control systems."[205] In the US, the Toxic Release Inventory, for instance, empowered community campaigns against polluting facilities, although in some locations and countries it was only with citizen co-regulation, in the form of "bucket brigades" taking environmental samples for themselves, that meaningful data became available.[206] In the modern world, Tapscott and Williams also suggest, such approaches could be extended to food quality as well as environmental quality, especially through equipping smart phones with simple sensors (as already prototyped by Nokia to measure air pollution, noise, and UV levels).[207]

Data openness and aggregation has been facilitated by the emergence of web-based geographic information systems (GIS) in which multiple "layers" of spatially distributed data can be overlain and interpreted. Google Earth provides a remarkable platform for simple GIS applications, which can overlay tax maps, property valuations, utility lines, emergency vehicle locations, resource maps, pollution levels, health data, and much more. These empower not only the staff of the administration, but also citizens groups and nongovernmental watchdogs (part of the process of enabling distributed autonomy). For example, Shareable has already coordinated two annual Map Jams for sharing cities, to bring activists together, online and in cities around the world, to map grassroots sharing projects, cooperatives, community resources, and the commons where they live.[208]

Tapscott and Williams are particularly keen on citizen participation in voluntary regulatory partnerships with business, monitoring compliance with codes of conduct and certification approaches at both firm and sector level.[209] But most such partnership approaches, which mushroomed after the Johannesburg World Summit on Sustainable Development in 2002, have succumbed to stakeholder fatigue, or never actually widened participation beyond the "usual suspects," leaving marginalized groups—such as indigenous peoples or women—framed out of the picture.[210] We argue, instead, that such approaches can only be a complement to public regulation, not the substitute the corporate sector has promoted them to be, and that genuine shared ownership by citizens is an essential factor.

We conclude this section on collective governance with brief discussion of two of the most tricky, and most fundamental, urban governance issues: tax and land.

Taxation in an Age of Collaborative Consumption

Clearly, any commitment to sharing politics has to deal with some fundamental issues related to taxation. Here we focus on general taxation, while below we also suggest that land value taxes may have a key role to play.

As sharing programs grow, and particularly as they become formalized, they can run into all sorts of barriers, such as tax treatment, rules, and a wide range of regulations. The tax treatment of payments for activities such as car sharing or renting rooms via Airbnb is delicate. Too heavy a tax and socially and economically valuable activity will be discouraged; too light, and successful sharing might dangerously erode the tax base—or even be abused as a deliberate tax avoidance strategy. In poor cities struggling to fund even basic public services, a cut in the tax base arising from individual sharing might do greater harm to the provision of the collective commons of public services.

However, rather than resisting sharing as a part of a "gray economy," tax and benefit rules need to recognize and permit it—with sensible thresholds for taxation of individual earnings from the sharing economy. Similarly, a better response to weak tax bases is tax sharing. In cities defined as "functional urban areas," including both urban cores and suburbs, tax sharing can ensure equitable functioning, even where the city is divided into different administrative areas.

While tax sharing in some form is normal in Europe, within the US it is unusual. In the fiscal disparities program of the Twin Cities of Minneapolis–St. Paul however, since 1975, local tax jurisdictions have contributed 40 percent of growth in commercial, industrial, and public utility property tax base into an area-wide shared pool, which is then distributed based on population and the relative market value of local property compared to that in the wider metropolitan area.[211] Seoul adopted tax-base sharing in 2008, in part learning from the Twin Cities, and has since successfully reduced fiscal disparities across its 25 autonomous districts.[212]

While we see some form of municipal tax-base sharing as essential in the move toward the sharing paradigm and sharing cities, we believe cities also need to develop tax rules and other regulations that facilitate sharing. Orsi and her colleagues set out many of the reforms needed at a city scale.[213] They argue for changes in rules on business registration, vacant land, public health, planning and zoning, insurance and liability, and in other areas as well.

In terms of car sharing, for example, they suggest "that cities more closely align taxes on car-sharing with the general sales tax for other goods and services."[214] Good practice examples include Chicago, Boston, and Portland (Oregon), which distinguish between car rental and sharing in municipal codes. As Orsi and colleagues argue further, "Cities [should] provide a tax credit to property owners who farm vacant or under-utilized lots, as such activities create food sources, economic opportunity, and civic engagement in otherwise blighted areas."[215] They highlight Maryland as having a tax credit for urban agriculture and the city of Philadelphia for levying "a yearly vacant lot registry fee, which is reduced if the land is cultivated and which may be eliminated altogether if the garden is registered under the new zoning code."[216]

Such proposals could be seen as a soft form of land value taxation, recognizing that ownership and rights over urban land bring power in city politics, and the better sharing of land is likely to be the most critical and contested dimension of sharing cities.

From Public Space to Public Land

Land is one of the dark horses of sharing. Sharing land is an essential feature of cities—whether in multistory buildings or shared streets, urban land is used intensively, and often by multiple users. Yet the mechanics of land markets are such that the capture of land value contributes massively to growing inequality, and access to land for basic needs is denied to many. High housing costs that exclude poorer groups and many young people from cities are largely the consequence of inflated land values. The extraction of land rent is a key driver of regeneration, gentrification, and redevelopment regardless of local social need.[217] Land ownership also brings political power and influence. This is about to change in Philadelphia where there are 40,000 vacant lots and buildings, many of them city owned. The Campaign to Take Back Vacant Land, a coalition of community, faith, and labor organizations, got together to break the insidious cycle of selling land to the highest bidder, irrespective of motive. They made the city pass a law in the form of a land bank in January 2014. This is a transparent, accountable public entity that, operating through community land trusts, converts vacant and abandoned lots and tax delinquent properties into productive use, thereby allowing communities to reclaim, reinvest in, and rebuild their neighborhoods by, for example, creating affordable and accessible housing, businesses that create jobs for residents, and spaces for community food production. The city is the largest in the US to do so and may encourage others to consider the same.[218]

There is a case therefore, for more than protecting and reviving public spaces, but for turning much more urban land into de facto commons or, in some way, socializing the value of that land for the community that (as we saw in chapter 2) co-creates its value in the first place. The innovative "Regulation for the Care and Regeneration of Urban Commons" adopted by Bologna, Italy, in 2014 offers one route. Within a few months, citizen groups had already initiated 30 neighborhood improvement projects in formal partnerships with the city, and other Italian cities are rapidly emulating the scheme.[219]

There are many sharing programs that share land and property in some way: from Airbnb to garden sharing to after-hours use of city facilities by community groups. There are relatively simple measures that could encourage such shared uses to proliferate. In US cities for example, as Orsi and her colleagues point out, zoning restrictions constrain sharing, and thoughtful revisions of zoning regulations could support community gardens, cohousing, shared parking, home business uses, and short-term rentals.[220]

Of course, if such measures were to become the norm, in most cities this would drive upward pressure on housing costs, with a particular risk that family accommodation would become less accessible to those on lower incomes. In university cities like Edinburgh in the UK, for example, the growth of a profitable student rental sector has led to reasonable calls for limitations on the conversion of dwellings to houses intended for multiple occupation.

It must be noted that we are not advocating any compulsion to share housing. Enabling the sharing of housing, whether short-term or long-term, is very different to the so-called bedroom tax introduced in the UK in 2013, which penalized residents who lived in social (i.e., public or subsidized) housing with "spare rooms." This policy was particularly iniquitous because there was (and remains) a significant shortage of smaller housing units in the UK, especially in the social-housing sector. Nor did such residents typically have an opportunity under their leases to sublet a room anyway, or offer it on a couch-surfing platform. As a policy to reduce under-occupation, and thus reduce the environmental burden of housing demand, encouraging sharing through positive incentives seems entirely more just, even recognizing the potential risks.

In particular higher development densities can be allowed in cohousing and car-sharing developments that naturally mitigate the potential negative impacts of high density, in turn reducing the relative costs and enhancing the potential inclusivity of such developments. Higher densities have long been advocated for environmental reasons, particularly their ability to reduce car dependence by supporting local facilities and public transport services.[221]

But in the face of high land values and the power of the development industry, planning rules can only achieve so much. Cities need to open serious debate about land taxation to support shared community use in affordable cohousing, urban agriculture, and community facilities. The political economist Henry George's idea of land-value taxation dates back to the nineteenth century, but has much to commend it today as a means of encouraging owners to put their land and buildings into uses valuable to the community, contributing to, rather than withdrawing from, the urban commons. Land value taxation aims to eliminate land speculation[222] and encourages landowners to generate income from the land, promoting density. The community effectively defines the land values.[223] Such reforms would not be an easy, short-term political opportunity, and while the debate is being conducted, cities might be well advised to put city-owned land into such uses, or to form urban land trusts to support housing cooperatives, for example. Cities should also recognize the value of squatting where sites or buildings are left vacant for prolonged periods, such as in the case of Christiania in Copenhagen.

Martinez celebrates "the outstanding qualities of most squats" as "utopian, heterotopian and liberated urban spaces"; his reasoning is worth considering at length, as he perceives political, communal, social justice, and environmental benefits:

> First, squats are built by squatters, active citizens who devote a great part of their lives to providing autonomous and low-cost solutions to … housing shortages, expensive rental rates, the bureaucratic machinery that discourages any grassroots proposal, [and] … the political corruption in the background of urban transformations. Second, squatters move but squats remain as a sort of "anomalous institution,"

neither private nor state-owned, but belonging to the "common goods" of citizenship, like many other public facilities. Third, since most squats have a non-commercial character, this entails easy access to their activities, services and venues for all who are excluded from mainstream circuits, which is a crucial contribution to social justice, equality and local democracy. Fourth the occupation of buildings is not an isolated practice but a collective intervention in the urban fabric that avoids further deterioration in decaying areas by recycling materials, greening the brown fields and the sad plots of urban void and, not least, by building up social networks and street life, which are palpable social benefits.[224]

So, among other benefits, squatting tends to imply more efficient sharing of resources in two dimensions: squatters typically target empty lots or buildings, thus increasing the intensity of use of the land area of the city. And squats are in turn typically shared buildings—an informal version of cohousing. Stigmatizing squatting would clearly be counterproductive in a sharing city. Instead cities should perhaps aim to build trust between regular citizens and squatter citizens.

Living together in shared accommodation has been the dominant mode of living throughout history, and remains so in much of the global South, where extended family groups share accommodation and groups of families share facilities. Modern cohousing in the global North—where a group of households share accommodation and facilities—reemerged as a distinct concept in the 1960s with a group of 50 families near Copenhagen who were inspired to raise their children in a closely connected community environment.[225] Elsewhere in Scandinavia, Swedish city living typically involves apartments with shared garden space, storage, and laundry facilities; formal cohousing in the country follows a similar model. There are around 50 developments, mainly in state-owned buildings, with shared facilities on the ground floor and basement of an apartment block (and sometimes also shared roof patios), and private spaces on the upper floors.[226] Many examples of the Swedish *kollektivhus* (collective house) date from the 1980s, but recent years have seen several new developments and—notably in Stockholm where housing is in short supply—high interest in cohousing.[227] In both countries new cohousing developments tend to offer larger shared spaces, and smaller private areas. In the UK cohousing is also a small, but growing niche, with just 18 built cohousing projects and more than 60 groups working up plans.[228]

Cohousing developments typically involve communal cooking, eating, cleaning and clothes-washing, garden, and leisure facilities. The benefits to the environment could be significant simply from the higher density, more efficient use of land, energy, and sharing of resources.[229] Cohousing projects also tend to have higher environmental aspirations than conventional developments—in part arising from the simple fact that the builders are going to live in the property, and thus naturally care more about running costs and issues such as the avoidance of toxic materials. Examples ranging from the LILAC Coop, which built straw-bale, eco-affordable cohousing (following Danish traditions) in Leeds, to the Lancaster cohousing development built to the extremely low-energy German Passivhaus standard, demonstrate the potential.[230]

This section has highlighted the ways in which principles of sharing and participation are making new polities, which are penetrating urban governance at all scales from the community to the city, in cohousing, apps for democracy, integrated co-ops and occupied spaces. Yet involvement in governance without resources is only half of the story. Participatory budgeting is a step in the right direction, but unless city politics engages with the tax base and the land values created by the urban commons, then the sharing of politics will fall short of its transformative potential.

Summary

In this chapter we shifted from looking at the consumptive (chapter 1) and productive practices (chapter 2) of cities, to exploring the political and cultural dimensions of the sharing paradigm in the city. We saw in Copenhagen, a city famed for its cycling infrastructure, how shared spaces contribute to urban civility and the sociocultural sharing of both city space and cohousing. We looked at issues of space and place, remembering how planning in Copenhagen pays particular attention to the areas where public and private spaces meet; on the sidewalks, for example, where encounter and eye contact can be fostered, where encounters with intercultural difference can be facilitated, and where the design of shared public space underpins democracy. As the Newcastle University professor emerita Patsy Healey puts it, urban planning is about "managing our co-existence in shared space."[231] We looked at the convivium of Collingwood Neighbourhood House in Vancouver, where encounters with strangers are productive and neighborly. We looked at how streets, the most used of our public spaces, can be seen as physical spaces waiting for expert-led, urban design "improvements" that often lead to gentrification, or as co-produced, shared, symbolic, and social spaces.

In particular, in this chapter we looked at the role of spaces and places in the public realm: *space as resistance* in political movements from Tiananmen Square to Tahrir Square to Slut Walks. We also focused on shared *space as possibility*: pop-up, insurgent, participatory, activist, countercultural movements with the potential to transform societies. We highlighted the impact of Occupy and similar movements, and the importance of the arts and leisure in this respect. We saw that both uses of the public realm now depend equally fundamentally on net-neutral public cyberspace. We suggested that sharing cities needed to consider ways to better turn not just public spaces, but land itself, into a shared public resource or commons. Whether city-led, pop-up, complete or incomplete, cryptostate or co-produced, this chapter makes the case for sharing cities as platforms for democracy in practice, for building social capital and reweaving Martin Luther King's web of mutuality, for supporting a flourishing public realm, and ultimately for decentering the link between rampant consumerism and human identities, so as to put people, as citizens, back into our cities.

4 Case Study: Medellín

Medellín has come a very long way in the past 20 years. Colombia's second city is now a thriving medical, business, and tourist center. The change strategy was driven by the philosophy (and department) of social urbanism of the Medellín Academy. In the mid-1990s these endeavors established a focus on empowering citizens, beginning in the poorest neighborhoods. The case of Medellín illustrates the vast potential for sharing in cities in the developing world, and particularly the importance of the shared public realm for increasing social equality.

Medellín has transformed itself from the former murder capital of the world to a model of urban social integration.[1] It was once home to the violent and powerful drug trafficking organization called the Medellín Cartel, headed by the infamous Pablo Escobar. By 1982, cocaine had surpassed coffee as Colombia's biggest export, as cartels transported billions of dollars' worth of the drug to the US.[2] In 1991, the murder rate climbed to 380 per 100,000 people, with over 6,000 killings.[3] Violence paralyzed the city with fear, which led to widespread abandonment of the public realm and most facets of civic participation. After Colombian special forces killed Escobar in 1993, city leaders, community activist groups, and residents alike dedicated their efforts to reclaiming the city through a fresh start.

Today Medellín is a city of about 2.4 million people, with a metro area population of 3.5 million.[4] The city sits in the Aburrá Valley at 5,000 feet above sea level and is bisected by the Medellín River. Its sprawling hillside neighborhoods, called *comunas*, are informal settlements created by displaced populations who had fled their homes in other parts of the country due to violence and conflict. The result was a highly segregated city with a strong disparity between the wealthy south and slums to the north.[5] As Laura Isaza, a consultant to Medellín City Hall, explains: "This displaced population didn't feel like they were part of the city. They used to say: "I live in this neighborhood and I don't live in Medellín." And that was one of our first steps: To gain their confidence and to make them feel that they are part of our city."[6]

While revitalization in Medellín first took root through national Colombian policy mandating architectural interventions to combat poverty and crime,[7] its regeneration is widely considered to have blossomed under former mayor Sergio Fajardo (in office 2003–2007), who made the bold declaration that the city's "most beautiful buildings must be in our poorest areas."[8] The economist Joseph Stiglitz highlights that indeed:

> Medellín constructed avant-garde public buildings in areas that were the most rundown, provided house paint to citizens living in poor districts, and cleaned up and improved the streets—all in the belief that if you treat people with dignity, they will value their surroundings and take pride in their communities.[9]

Alejandro Echeverri, an architect who worked with Fajardo on the city's transformation, highlights the role of the department of *urbanismo social* (which translates as social civic planning, or social urbanism). During the mid-1990s, Echeverri explains, "A small group began to think in terms not of top-down policy, but of one that would begin with the poorest neighborhoods. ... It was both a concept and a physical strategy."[10]

Social urbanism focuses on spatial justice: striving to bridge the city's socio-spatial divide and achieve inclusion through prioritizing historically neglected neighborhoods on the urban agenda. The new strategy revolved around reclaiming shared public spaces and connecting the isolated barrios, or neighborhoods, both to each other and to the rest of the city. For Medellín, this has meant a substantial shift in public investment in the form of infrastructure, and public services, buildings and spaces, all with high-quality architectural design.

Social urbanism explicitly prioritizes equity in its approach, meriting connections to the concept of "just sustainabilities" that highlights the need to integrate social needs and welfare in a more equity-focused approach to sustainable development.[11] As we mentioned in the introduction with respect to both Bogotá and Curitiba's BRT systems, many of Medellín's social urbanism projects are recognized for their environmental impact, even though they were initially planned and carried out for their social impact on access and equity.

The social urbanism strategy uses specific projects to inject investment into targeted areas in a way that cultivates civic pride, participation, and greater social impact. Medellín officials describe these projects as Proyecto Urbano Integral, or integral urban projects. They become catalysts for surrounding public space and infrastructure interventions to tackle poverty and violence, and are viewed holistically as part of a comprehensive plan for targeted neighborhoods.[12]

Integral urban projects that enhance accessibility and connectivity for residents of the *comunas* have come in the form of impressive upgrades to the city's public transit, which was named one of the top systems in the world by the Institute for Transportation and Development Policy in 2012. The Metro de Medellín is a network of efficient metro cars serving over half a million passengers each day.[13] One of its most impressive features is the Metrocable, a network of nine cable car or gondola systems that transport passengers up and down the steep mountainsides of the city, as many parts of the *comunas* have been built up on hillsides too steep to allow for buses or cars. Prior to completion of the cable lines in 2010, residents living in the steep city slums faced a long, dangerous commute down the mountainside, which could take hours on foot. Today, this commute looks very different. For less than one US dollar, a *comuna* resident can take a comfortable and scenic 25-minute ride down the steep slopes with direct transfer to the metro cars below.

Another visually impressive feature of Medellín's transportation system that is gaining global recognition is its seven-station outdoor escalator running through the neighborhood of Comuna 13. In the upper reaches of this neighborhood, the streets are so steep that they give way to staircases, amounting to the equivalent of climbing a 28-story building for some residents.[14] Opened in 2011 and financed through a public-private partnership, the escalator sweeps people up to the top within a matter of minutes free of charge. Similarly to the Metrocable stations, strategic placement of this escalator is perceived by many as a symbol of rebirth and investment in a neighborhood that has experienced extensive neglect and violence.[15]

There are some residents, however, who are concerned that these "flashy new projects" might be distracting people's attention from social issues that, despite great improvement, continue to plague the *comunas*, including gang violence and drug trafficking.[16] The design and placement of the cable lines and stations strategically disrupted drug trafficking routes while also providing affordable, quick access to amenities in the city center for residents.[17] Public facilities such as health centers, schools, and libraries have also been developed at the cable-car stations. In this sense, space is being used for both security, and possibility.[18] A similar cable car system was subsequently constructed in Rio's Complexo do Alemão favela, with local facilities at the stations also promised. But by 2014, three years after the system began operating, and despite a pressing need for community spaces, the only developments at the cable car stations in Rio were police stations, and residents appeared skeptical that community facilities would ever arrive.[19]

In Medellín, increasing accessibility and connectivity through the metro system has also funneled a new stream of commerce and investment into the areas surrounding these new nodes of transit hubs. Many of the new parks, schools, and other public buildings are integrated into the infrastructure of the metro system itself.[20] This is an example of how the integral urban projects are connected and spur greater revitalization in the neighborhoods in which they are located. So far, the Metro has created 3.4 million square feet of green space throughout the city, equivalent to more than 40 full size soccer fields.[21]

Parque Biblioteca España is one of a handful of library parks constructed in marginalized parts of the city, providing free access to computer and information technology, other educational classes, and space for cultural activities and recreation.[22] It was built in the neighborhood of Santo Domingo, just a short walk from a Metrocable station, and has become a potent emblem of the city's social urbanism projects and greater transformation. The structure's exterior looks futuristic: three massive, linked black boulder-like buildings perched 1,500 feet over the valley below.[23] The physical transformation of the public space holds a deeper meaning for many residents. Giancarlo Mazzanti, the Colombian architect who designed the library, calls it a "symbol that produces dignity."[24] Writing in the *Guardian* newspaper, Sibylla Brodzinsky recounts a visit to Medellín and a conversation that occurred on the grass of the Parque Biblioteca España with an 18-year-old resident, who explained that the most abandoned areas of the city are at the greatest risk for violence, and "today we are no longer abandoned here."[25] While once considered one of the most violent parts of the city, this neighborhood is now both a tourist destination and local gathering place. This library park and the many other public places constructed through social urbanism also provide a space for social engagement and cohesion between fragmented neighborhoods and socioeconomic classes. In this sense, such projects integrate physical and social infrastructures providing spaces of positive encounter.

Other efforts to extend social inclusion focus on housing. The Social Institute for Housing and Habitat of Medellin (ISVIMED) is responsible for managing social interest housing in the municipality. It builds new public housing and supports housing improvements, and oversees regularization of property and land ownership in some *comunas*. Some retitling initiatives provide "indivisible" collective ownership of plots allotted for multiple-occupation, but most focus on individual titling. The researcher Jota

Samper notes that individual titling has been abused by criminal gangs, suggesting instead that the city should consider collective titling of housing and adjacent open spaces, perhaps in forms where equity is shared with the municipality.[26]

The city's current mayor, Aníbal Gaviria, announced another ambitious project, the Cinturón Verde Metropolitiano, or Metropolitan Green Belt, a proposed 46-mile long (75 kilometer) park along the upper slopes of the valley surrounding the city.[27] The project, echoing the UK's post–World War II Green Belts, aims to contain urban expansion by curbing development of informal settlements along the upper ring of the city's hillsides, while also providing access to green space and recreation. Since many residents already live above the proposed Green Belt line, they will have to be relocated. This project is facing criticism from some who feel poorer residents are being displaced for the sake of another project in the name of "innovation" that will garner global media attention.[28]

Revenue from the city's public services company, Empresas Públicas de Medellín (EPM), has funded many of the integral urban projects. Some $450 million of the utility company's profits go directly to improving social welfare through funding these public projects. EPM has 20 new projects planned for completion in 2015, all of which are being designed and planned through a participatory process with the community to ensure the spaces meet public needs.[29] A local economic development consultant, Milford Bateman, reports that locals appear "genuinely proud of 'their' company's contributions to the city's economic development and culture."[30]

Medellín has also refocused its local economic development initiatives, away from conventional microfinance, establishing a network of municipal business development agencies called the Centers of Zonal Development of Companies (CEDEZOs), which support small businesses to develop and collaborate. Medellín also supports both marketing and workers cooperatives, such as the Coomsocial health center, which has 150 worker members across two sites and provides health care to tens of thousands of residents under contract to the city.[31] "With huge inequality still prevalent in Medellín," reports Bateman, "the hope is that the promotion of worker cooperatives will provide important examples of an enterprise structure in which it is perfectly possible to combine economic efficiency with high levels of equality, dignity and democracy."[32]

Medellín is one of the largest cities in the world to successfully practice participatory budgeting. Five percent of the city's budget is set aside for this form of economic democracy. Mayor Gaviria says civic participation in participatory budgeting and other city planning projects nurtures civic pride where citizens feel "they participate in the construction, design and approval of public works and government programs."[33] One beneficiary of participatory budgeting is Son Bata, an Afro-Colombian music group that has morphed into a community center and cultural initiative.[34] In this center, neighborhood children find refuge in the safe, colorful building and receive free music classes. With its participatory budgeting allocation, which covers 30 percent of its operating budget, Son Bata has been able to contract with music professors, purchase instruments, and construct a music studio. Other neighborhoods have gained schools and health clinics through the funding scheme.[35]

Son Bata is not the only arts-based cultural initiative that is pushing back against the gang violence that still exists in the *comunas*. Just as the city's investment in the public realm has been an effort to reclaim public spaces from violence and negligence, La Casa Morada, a shared studio space for musicians in Comuna 13, strives to keep culture alive and fill the city's spaces with life by holding concerts for the public in its front yard.[36] Similarly, the Moravia Center for Cultural Development, located in a neighborhood that developed on one of the city's former garbage dumps, focuses on engaging youth and their families in music, theater, and other art programs. Yeison Hendo of the Moravia Center describes their strategy as "using education and culture as a means to create conditions of peace and tolerance."[37] The *New York Times* architecture critic Michael Kimmelman, on a Medellín visit inspired by the city's use of architecture to combat crime, describes the Moravia Center as the most remarkable building of all, with a "dance studio and theater opening onto the outdoors, the library and courtyard, flanked by low ramps, providing a desperately needed safe and attractive public space."[38] The combination of participatory budgeting and grassroots efforts to provide at-risk youth with alternatives to joining a *combo,* or local street gang, provides an excellent example of the potential for co-creation to build empathy and transform the urban commons, even in the most challenging of circumstances.

All of these initiatives earned Medellín the US's Urban Land Institute's "Innovative City of the Year" title in 2013. The competition, developed in partnership with *Wall Street Journal* and Citi, recognizes the most innovative urban centers by public vote, and Medellín beat out finalists including New York City and Tel Aviv.[39]

To add to its rising fame, in April 2014 Medellín hosted the seventh session of the United Nations Habitat World Urban Forum (WUF) under the theme "Urban Equity in Development—Cities for Life." Medellín offered an ideal urban laboratory for exploring things like prioritization of vulnerable populations and the recovery of public space. The gathering convened over 20,000 participants representing more than 160 countries.[40] One product of the seventh WUF's discussions was the Medellín Declaration—a statement of intent to work toward Habitat III, the 2016 Third United Nations Conference on Housing and Sustainable Urban Development. Participants reaffirmed their commitment to integrate equity into the urban agenda. According to the Medellín Declaration this new urban agenda: "Requires new technologies, reliable urban data and integrated, participatory planning approaches to respond both to present challenges and emerging needs of cities of the future."[41]

Acknowledging that models of urbanization exist on diverse levels of social and cultural conditions, the Declaration provides a list of objectives to guide cities around the world in implementing this urban agenda. These aim to link current urban development with future needs through plans and policies grounded in the fundamental principles of equity, justice, and human rights; to promote inclusive participatory local governance that advances social cohesion and breaks down social divides by empowering all segments of society; and to support sustainable urban development that promotes equity, provision of basic services, affordable accessible transport, and access to safe public spaces for all.

The rhetoric of the WUF is strong, but as we noted in chapter 3, can lack commitment in practice, with its models of urban development overtaken by commercial interests. The presence of the Medellín meeting also stimulated an alternative even more inclusive People's Forum,[42] which issued a strongly worded declaration "denouncing"

> the current neoliberal model of urban development. ... From Medellín, a city heavily affected by violence and inequality, we call on everybody to fight for the urban area that we deserve, and to immediately start implementing a city project based on redistribution of wealth, human rights, the environment, common goods, and the responsibility of inhabitants to be recognized creators and leaders of territories, and not just consumers/users.[43]

It is an incredible feat for Medellín to have risen to "model city status" for urban equality planning strategies, but the fragility of the social and economic conditions implicated in this urban transformation should not be overlooked.[44] Like Bogotá in the late 1990s, while poverty, violence, and crime have been reduced, these challenges are still very much alive. Many residents feel overwhelmed by the city's sudden fame and are not fully convinced that the welfare of poorer residents is the true driving force of its social urbanism agenda.[45] Nonetheless, the case of Medellin highlights well the importance of design for equity; investment in a shared and intercultural public realm; government funded public services providing key shared infrastructures; the importance of the arts for growing empathy and shifting cultural norms; and the scope for international city networking to influence a wider stage. Overall, the progress achieved through the social urbanism strategy and grassroots community initiatives has made great strides in reclaiming the city for all its citizens.

Sharing Society: Reclaiming the City

Any city however small, is in fact divided into two, one the city of the poor, the other of the rich.
—Plato

Chapter Introduction and Outline

In this chapter we explore the scope for sharing to enhance equity and social justice in the city and beyond. The chapter splits into two broad parts. The first situates the sharing paradigm in contemporary theories of justice, with particular reference to just sustainabilities, the capabilities approach and recognition. The second part examines some of the emerging areas of conflict and tension between justice and sharing in practice, first illustrating the challenges with consideration of transport—and particularly carsharing. It goes on to address problems with exploitation of labor in sharing, the commodification of nonmarket aspects of life, and exclusion of the disadvantaged from sharing practices through the divergent processes of marketization, criminalization, and gentrification. We finish the chapter by looking at some key contributions to justice in sharing: building empathy, strengthening civil liberties online, and developing complementary currencies. This might seem an eclectic mix of examples, but they all illustrate how sharing can be inclusive and just, when such factors are considered from the outset.

Just Sustainabilities

In many ways, the sharing paradigm is a direct descendant of the concept of "just sustainabilities." As we argued in the introduction, equity and justice are too often ignored or assumed in initial "environmental" and business cases for sharing. Even though some sharing programs deliver greater benefits to poorer, "below-median income" groups[1], we see exclusion of low-income groups, for example where bike- and carsharing programs require a credit card, or up-front fees or deposits. Similarly, commodification in commercial sharing can lead to exclusion of low-income communities (and exploitation of those providing services, such as Lyft drivers and TaskRabbits), while co-production can be abused to divest responsibilities to the most vulnerable. Unsurprisingly, demographic data gathered by the sharing economy aggregators Compare and Share suggests that the users of commercial sharing platforms such as Airbnb are predominantly from the well-off middle class.[2] Thus, whether at its margins with markets, or with the public sector, sharing, like sustainability, has to be envisioned, designed, and managed to address questions of justice from the outset, with the participation of potential users.

Integrating social needs and welfare offers us a more just, rounded, equity-focused definition of sustainable development than the commonly cited definition from the 1987 Brundtland Commission ("development which meets the needs of current generations without compromising the ability of future generations to meet their own needs"[3]), while *not* negating the very real environmental threats. A "just" sustainability is therefore: "The need to ensure a better quality of life for all, now and into the future, in a just and equitable manner, whilst living within the limits of supporting ecosystems."[4] This definition focuses *equally* on four essential conditions for just and sustainable communities of any scale. These conditions are: improving our quality of life and well-being; meeting the needs of both present and future generations (intra- and inter- generational equity); justice and equity in terms of recognition, process, procedure, and outcome; and living within ecosystem limits (also called one-planet living).[5]

Theory: Justice, Capabilities, and Recognition

The sharing paradigm represents a huge opportunity to promote greater equity and social justice in cities. We have highlighted many ways in which sharing enables people to access resources and experiences from which they may otherwise be excluded on grounds of income. We have also emphasized how a shared public realm and services can support social inclusion of disadvantaged and marginalized groups. But we have seen, too, in the real world, where the proximity of wealth and squalor in cities exposes stark inequalities, how sharing might also be vulnerable to problems and politics of injustice. In other words, even if Internet-enabled sharing and co-production dramatically cut the costs of access to goods and services, and enhanced their availability by improving environmental efficiency—as Jeremy Rifkin argues, for example[6]—it would be falling into a post-political trap of technological determinism to assume that this will naturally reduce inequality. Here we explore the role of the sharing paradigm in delivering justice, and, in the process, exploring what we mean by justice, and some of its implications.

Justice is not a simple concept. Different ideological and cultural foundations can lead to distinct outcomes: for example, liberal conceptions of justice tend to focus on individual freedoms, and communitarian ones on shared values. But different approaches can also be helpful in highlighting commonalities in how different aspects of the sharing paradigm contribute to justice.

Seeing human skills and talents as something to be shared with our communities and wider society is actually fundamental in both liberal and communitarian approaches to justice. Similarly, most approaches to justice imply—or in some cases, explicitly demand—forms of redistribution. Sharing can clearly help redistribute access to goods and services in socially acceptable ways, and it can also help spread the capacities to flourish in society that are central to the "capabilities approach." The following sections explore how some of these different justice theories might help us understand sharing.

Liberal Justice: Sharing and Redistribution

The predominant theories of justice in modern Western society are liberal in origin, notably rooted in the work of the political philosopher John Rawls, who conceived of justice as egalitarian, reflected in a notional "social contract" from which two principles emerge: equal basic liberties for all citizens, and the "difference principle." The latter permits only those social and economic inequalities that work to better the absolute position of the least well-off members of society. This functions, says Rawls, as

> an agreement to regard the distribution of natural talents as in some respects a common asset and to share in the greater social and economic benefits made possible by the complementarities of this distribution whatever it turns out to be. Those who have been favored by nature, whoever they are, may gain from their good fortune only on terms that improve the situation of those who have lost out.[7]

This does not demand perfect equality of outcomes. Indeed it potentially legitimates a market society with wide inequalities if, as a result of tolerating those inequalities as incentives to wealth creation, the worst off in society are better off than they would have been otherwise. The political doctrine of "trickle down" is an extreme expression of this view. But the difference principle does acknowledge that wealth is based on "common assets," the benefits of which should be shared.

Unsurprisingly there is intense political debate over what level and mechanisms of redistribution might make liberal capitalist markets fair. Some of the most powerful critiques of Rawlsian approaches stress that relative inequality in terms of income and consumption can be so harmful to health and well-being that it is preferable to have lower, but more equally distributed, incomes and fewer material goods.[8] Such critiques have been given more weight by economic findings that suggest inequality is also a brake on economic growth and development.[9]

Progressive taxation to fund public services is the conventional form of redistribution in European liberal democracies, whereas philanthropy plays a greater role in the US. Sharing, especially in sociocultural flavors, can contribute directly, too. By redistributing access to resources rather than money or services, sharing would likely stimulate more productivity than taxation or philanthropy. This would be the case even if some of the gains were never traded in markets but directly consumed or informally shared, as would happen, for instance, as a result of the creation of more community gardens or allotments. But if we see sharing as redistribution, then exclusion from it is clearly a problem.

On the other hand not all sharing activity has to directly benefit the least well-off in society: neither tax breaks nor philanthropy exclusively target the poor either. And because something may help the wealthy or middle classes, too, is no reason to resist it: public services would be decimated if such illogical reasoning were accepted widely. Austerity has, however, added to pressures in many countries to "means-test" public services and benefits so as to exclude those above a certain threshold of income, rather than provide "universal" access. Paradoxically the costs of means testing often exceed the savings from its introduction, while discouraging many of the people most in need from applying, because of the stigma and shame of means testing. In sharing, too, we suggest a default position of universal access—along with targeted promotion—as the best way of maximizing uptake by, and benefits to, those in most need.

Rawls's difference principle also rejects the idea of "moral desert," rewarding virtue or other attributes as an underlying principle. Rawls recognizes that our underlying differences and capacities may be arbitrary or largely based on the efforts of others—parents, teachers, neighbors, and the like. Thus the formulation of the "rules of the game" should, in Rawls's view, be a matter for society as a whole.

Communitarianism: The Shared Society

Yet Rawlsian approaches to justice have also been criticized as too strongly focused on individuals. Communitarian concepts of justice are often found in Eastern cultures, in which more collective values flourish more strongly. (See our discussion in "International Variation in Sharing Cultures" in chapter 2). As the social psychologist Jonathan Haidt notes, many such cultures retain an "ethic of community" among members of overlapping collectives—families, tribes, religious groups, and nations—with strong claims to loyalty. In such societies the emphasis on personal autonomy and liberty in secular Western nations can look more like libertinism and hedonism.[10] In the West communitarianism was given a strong boost by the American sociologist Amitai Etzioni, arguing for a better balance between freedom and morality, autonomy and community, and rights and responsibilities. Etzioni advocates a "responsive" form of communitarianism that rejects an authoritarian state and promotes integration of individuals in multilayered modern communities, sharing core values through social norms yet having a choice of which groups to identify with within modern society.[11] He stresses that responsive communitarians "do not seek to return to traditional communities, with their authoritarian power structure, rigid stratification, and discriminatory practices against minorities and women. Responsive communitarians seek to build communities based on open participation, dialogue, and truly shared values."[12]

In such forms, communitarianism can be combined with more liberal concepts of rights. In Latin American cultures, as Arturo Escobar suggests, notions of community solidarity and collective rights (for example for indigenous peoples) can be blended constitutionally with individual rights,[13] thus delivering intercultural and plural forms of society, and inspiring both social movements and state action in various ways in cities like El Alto and Medellín. Escobar highlights however, how social movements in the region—such as Zapatistas and Oaxaca—offer visions of simultaneously post-statist and post-capitalist worlds. These can be seen as political expressions of the "third way" of sharing.

A similar nuanced view is helpful in considering how communitarian values might inform sharing. While we have frequently criticized commercial models of sharing, we do not advocate sharing purely in sociocultural modes within homogenous communities. On the contrary, our enthusiasm for sharing reflects its potential—in universally accessible mediated forms—to connect people from different backgrounds—helping build multilayered, intercultural communities and identities suited to a cosmopolitan connected world.

There are further useful lessons in communitarian theory. Michael Sandel—communitarianism's highest profile contemporary advocate—highlights that we can and do owe obligations as communities to communities: such as apologies or reparations for slavery. Moreover, for justice, we have to reason together about the good life in ways liberal theory resists: "Justice is not just about the right way to distribute things. It is also about the right way to value things," he says.[14] A just society would therefore: "Cultivate in citizens a concern for the whole, a dedication to the common good. … [It would] lean against purely privatized notions of the good life, and cultivate civic virtue."[15] And as the economist Amartya Sen argues, "A democracy cannot survive without civic virtue."[16] Sharing, of course, helps sustain a belief in society and civic virtue, which appears to have been largely misplaced in many modern wealthy "democracies."

Sandel therefore sees a public debate on the moral limits of markets as critical: as "marketizing social practices may corrupt or degrade the norms that define them."[17] But the blogger Max Holleran fears that this is exactly what is happening in the sharing economy. "Giving prices to acts that were previously priceless, in the literal sense, is a key innovation of the sharing economy," he says, referring to the way communal, sociocultural sharing behaviors like giving a lift to a friend can be converted into mediated commercial ones.[18] Gifting, in contrast, creates "unpayable bonds" that bring groups together in cycles of social interactions, preventing conflict and building cohesion. To Holleran then, the marketization of sharing does not just corrupt our individual relationships, but risks undermining the social solidarity on which welfare systems depend.

This is not a new concern, nor limited to sharing. As Sandel describes:

> The affluent secede from public places and services … [and] institutions such as schools, parks, playgrounds, and community centers cease to be places where citizens from different walks of life encounter one another. Institutions that once gathered people together and served as informal schools of civic virtue become few and far between. The hollowing out of the public realm makes it difficult to cultivate the solidarity and sense of community on which democratic citizenship depends.[19]

Sandel suggests this can be reversed through public institution building. We also see opportunities for reconstruction of the shared public realm (as described in chapter 3), and for sharing organizations and institutions to rebuild social capital, although recognizing the risks of over-commercialization of sharing, especially in societies with a strong sociocultural sharing tradition.

Capabilities: Shared Freedom?

Sen similarly argues for collective public reasoning and deliberation as a key tool in defining justice.[20] Rooted in liberal approaches, Sen champions justice as freedom, but takes a distinctive approach to how such freedom is achieved and defined. Freedom, for Sen, is impossible for any individual without a range of capabilities to function and flourish. And those capabilities—things like bodily health, affiliation, play, and control over one's environment, as highlighted by Sen's collaborator, a University of Chicago professor of law and ethics, Martha Nussbaum[21]—are not just individually generated. Such capabilities are clearly dependent on far more than markets, and more than our individual freedoms, critically also depending on our shared cultural and social commons.

In capabilities theory, justice is achieved not by equal outcomes in terms of money, or purely by equal treatment in terms of process, but by equal potential and equal capability to realize potential. In other words, capabilities level the playing field so that "equal opportunity" becomes genuinely fair. For example, if ethnic minorities lack access to good education, equal opportunities in the job market will not deliver fair representation of black people in high-earning careers.[22] What really matters is good public education and childhood support. The capability approach is firmly rooted in individualist, liberal philosophy, yet would be meaningless without the families and communities that enable education, psychological development, communication, and interaction. Both social capital and public goods can provide essential foundations. For instance, writes Sen, "Basic education, elementary health care, and secure employment are important not only on their own, but also for the role they can play in giving people the opportunity to approach the world with courage and freedom."[23]

It appears to us that like Sandel's, Sen and Nussbaum's views of justice reflect well the modern biological and evolutionary evidence of humanity as a naturally cooperative species, for whom culture enables and encourages communication and cooperation in increasingly large groups.

At the heart of our case for the sharing paradigm is an understanding of justice as universal access to the capacities and abilities we need to flourish. Sen and Nussbaum's work highlights how widespread injustice is, when understood as the denial of opportunity to realize one's capabilities. The emotional and cognitive capabilities of human beings are only realized intermittently even in so-called developed societies, even when more material needs may be met. And cooperation is essential for people to realize their capabilities. Yet, as Richard Sennett argues: "People's capacities for cooperation are far greater and more complex than institutions allow them to be."[24] Sharing is an opportunity to release that capacity, confined by competitive markets and bureaucratic states. Sennett also highlights how our cooperative capabilities can be damaged by inequalities in childhood, particularly through competitive schooling and the commercialization of status, but he sees hope for rebuilding cooperation in encounter, dialogue, and the practice of collaborative crafts and skills.

Sen argues that the practical challenges of increasing justice in the real world can be met in part by public interventions to promote more shared and equal outcomes. For instance, in the UK,

> during the First World War there were remarkable developments in social attitudes about "sharing" and public policies aimed at achieving that sharing. ... During the Second World War also, unusually supportive and shared social arrangements developed ... which made ... radical public arrangements for the distribution of food and health care acceptable and effective. Even the National Health Service was born during those war years.[25]

Moreover, the effects of sharing and greater equality in those war decades were substantial and immediate. They reduced undernourishment dramatically, and mortality rates (except for war mortality itself) declined sharply, delivering a rapid increase in life expectancy. The shared services introduced underpinned a basic enhancement of capabilities, especially for those previously disadvantaged.

Recognition, Identity, and Inclusion

David Schlosberg, a professor of environmental politics at the University of Sydney, concurs with Sen that justice is not only about securing a fair distribution of material goods or consumption. Indeed neither gives primacy to material well-being, but rather to social factors. David Schlosberg, drawing on the late political philosopher Iris Marion Young's concept of recognition,[26] argues that just treatment depends on recognizing people's membership in the moral and political community, as well as providing for the capabilities needed for their functioning and flourishing, and ensuring their inclusion in political decision making.[27] He sees distribution, recognition, capabilities, and participation as interrelated and interdependent.[28]

Recognition is particularly important in intercultural societies, where other dimensions of justice might be culturally distorted. Recognition of the rights of those with sexual and gender differences (including LGBT individuals), is an area where much progress can be identified in recent decades, but much also remains to be done. The importance of recognition is not just a reflection of the impacts of individual discrimination, but also rooted in the prevalence of both institutional and cultural misrecognition. As Nancy Fraser argues, institutions persistently devalue and misrecognize subordinate groups, on the basis of gender for example.[29] Not only are women under-represented in city halls and in boardrooms, but also procedures of recruitment, training, maternity leave, and support are all structured in ways that maintain those disadvantages. More generally, elites rely on hierarchies being "taken for granted" by subordinate groups. Such hierarchies are reproduced in cultural spaces and activities such as education and consumerism, so their true origins are misrecognized by those most disadvantaged by them.[30]

Sharing can help provide access to critical goods and services for the disadvantaged, and even help build capabilities while promoting social norms rooted in cooperation and fairness. But it can only do so for those that are recognized and thus included. If sharing carries social stigma, as much sociocultural sharing does—such as the wearing of hand-me-down clothes in the brand-sensitive marketing zones we call schools—then its power is diminished. Misrecognition extends stigma to entire groups based on race, gender, sexuality, or class—treated here as Pierre Bourdieu describes it—not a product of occupation, but of the visible signals of "taste and culture" constructed by groups. While gradations of class are constructed in consumption patterns, those who resist consumerism, or consume in different ways, risk being misrecognized. In building the sharing paradigm, cities and sharing entrepreneurs (both social and commercial) therefore need to work hard to make sharing inclusive, to recognize excluded groups and to erase stigma.

Justice, Property Rights, and the Commons

Our understanding of justice and equity does not exclude material outcomes, which in turn further determine capabilities. Material income and wealth provide very real capabilities to meet needs for shelter and security, and thus to avoid the stresses and insecurities of life without sound financial resources. We saw earlier that material inequality can harm our health and damage our economy. Sharing can of course help redistribute the resources we rely on, but the sharing paradigm also challenges the privileged position of property rights in modern society.

Highly inequitable distribution is typical of resources in which property rights are widely applied, such as land and both physical and intellectual property. Land rights can be seen as a legal fabrication to defend the past acquisition of land by force.[31] Even where ownership is demonstrably legal its social legitimacy might remain questionable, where, for example, the rules for inheritance follow the interests of the already wealthy. The persistence of primogeniture in many circumstances means the law leaves widows and younger children effectively disinherited—a clear example of institutional and cultural reinforcement of disadvantage and misrecognition.

Whether considering the land or other material or even intellectual property, there are dramatic implications for justice defined in terms of capabilities. Access to land, resources and technologies are basic capabilities for development and poverty alleviation, which can be denied to billions of people in the modern world insofar as it suits the financial interests of already wealthy elites and corporate interests. The sharing paradigm offers a tool to challenge such exclusion, and provide access, especially where it is backed by common ownership.

Directly contrary to conventional economic approaches to environmental problems that seek to privatize currently common property resources—for example by creating carbon markets—a just sustainabilities approach would look to create new forms of common property through sharing, land reform, and "open source" solutions that do not rely on proprietary technologies or intellectual property rights.

In this section we have seen how sharing fits into powerful and culturally diverse concepts of justice, both liberal and communitarian, especially if understood as underpinning both individual and collective capabilities in society. Yet we have highlighted that justice will not arise automatically in sharing: sharing initiatives will need design for justice and sustainability. To explore how that might be achieved we need to move from theory to practice.

From Theory to Practice

Our brief trip through some of the territory of justice theory suggests several strong reasons to support the sharing paradigm as a vehicle for justice. But justice theory and practice can remain very different, as our short discussion of property rights—typically defended as central to justice in liberal market societies—already revealed. In considering how best to develop and promote sharing, it therefore remains essential to ask at each step who is being empowered by and benefitting from sharing, and who may be experiencing disempowerment or exclusion. If already-wealthy elites in Silicon Valley take the profits, yet poor communities face displacement by gentrification, then that is not just. Such impacts are not an indictment of the sharing paradigm, but recognition that sharing needs to be actively envisioned, designed, and managed toward justice.

With good design we can expect significant equity and well-being benefits from sharing as the less well-off gain access to resources they could not otherwise afford (from borrowed library books to shared ownership of homes). Moreover in diverse sharing communities online, potential users are not limited to sharing only the resources that those within their existing social group own and are prepared to share. Of course, this implies that any "digital divide" becomes a more severe problem, and fair sharing cultures and cities must also work to eliminate that. At the same time, greater sharing at the whole society level would also benefit those who are better off. This is not just about avoiding tax liabilities, or the thrill of giving—although giving can have a significant positive impact on personal happiness.[32] It is also about benefits to health and well-being—arising from social capital and solidarity—that cannot be replaced by higher incomes and consumption.[33]

In practice, we are concerned not only with the way that poorly designed commercial, mediated sharing can be exclusive or exploitative, but also with the way in which the informal sharing arrangements of those on lower incomes are marginalized—or worse—in the current sharing discourse. Informal sharing describes sociocultural sharing activity such as carpooling and daycare systems undertaken collectively to reduce economic burdens and increase opportunity, rather than to meet any other higher environmental, or personal goals.[34] As Juliet Schor notes, such "practices remain more common in working-class, poor, and minority communities."[35] So, as we discuss later in the chapter, informal sharers and sharing need to be better recognized and their practices de-stigmatized. The growth of sharing in middle-class and bourgeois bohemian groups suggests that sharing in general is becoming less stigmatized, but this trend seems highly unlikely to spread to all forms and groups without specific support.[36]

In the remainder of this chapter we explore practical issues of achieving and developing equity and justice in sharing at various scales, highlighting threats to justice in sharing that arise from both commercial and government forces. We first look at the case of transport, then at some common problems in cities of both the rich and poor worlds, and finally at some generic solutions that can build recognition and inclusion.

Sharing as a Vehicle for Justice

Here we explore and contrast some of the various models emerging in transport (and particularly carsharing) to illustrate some key implications for justice. In many ways, carsharing is a flagship for sharing. After homes, cars are the most expensive assets individuals are likely to own. Cars are almost irrevocably wedded to identity concepts of individualism and freedom. Uber, the ride-service company, has the highest market valuation of all commercial sharing economy businesses; revenues in the sector, in North America alone, are projected to hit $3.3 billion by 2016[37] and continue growing at 23 percent annually through 2025.[38] Carsharing is also one of the most developed and diversified sharing activities, with models that range from public transport and community vehicles, through short-term rental, car clubs, ridesharing, and taxi-substitute ride services, to carpooling and hitchhiking.

Ridesharing models began with the insight that there were lots of empty seats in cars traveling to all sorts of places, and smart, mobile technology could match those seats with potential travelers. In the Uber approach, however, it has swiftly evolved into a substitute taxi service: the user "hails" an Uber car, using a mobile app, and the driver takes them to their destination for a fee. Uber takes a commission, and the driver looks for another fare. Many Uber drivers own their vehicles, and drive for Uber (or Lyft) on their own initiative, in a form of highly flexible, casualized labor. In some cases drivers rent cars from a third party, or are even employed by someone who owns a pool of vehicles. Uber tends to undercut regular taxi services but—at present—offers no employment benefits to the drivers, and it rarely meets regulatory standards for taxis with respect to insurance or access for people with disabilities, for example.[39] The service competes not only with private cars and taxis but with public transport, cycling, and walking, potentially crowding out these more socially inclusive models of mobility. Uber is disproportionately used in richer communities—maps of Uber services show a worryingly close correlation with the wealthiest districts of world cities.[40] As Uber has grown it has exercised market power, driving down fares to drive out competition to the point that some fear it obtaining both monopoly power over its users and monopsony power over its drivers.[41] It is also flexing its muscles politically, recently hiring David Plouffe—a White House senior adviser until 2012—as campaign manager in its battle for supportive regulation.[42]

With its commercial model it is questionable whether Uber takes any cars off the road, leaving air pollution, accidents, and congestion unaffected (or even exacerbated). We need to remember that these negative effects of traffic typically hit the poor hardest. Nor does Uber particularly seem to enable people in poverty to get around better or more cheaply. It is even questionable whether it provides useful quality work, under fair conditions, for those lacking other opportunities. And like all smartphone-connected uses of the Internet of things, it also raises concerns about data privacy and surveillance.

Are car clubs, like Zipcar, any better? Co-founded by Antje Danielson and Robin Chase in 2000, Zipcar had offices in more than 26 American cities and 860,000 members across the US, Canada, and Europe by 2014.[43] Each shared car is estimated to replace 9 to 13 typically more-polluting private vehicles.[44] Car clubs make the costs of car use transparent, loading all the overheads into the marginal cost paid—in Zipcar's case, literally by the minute. The cost ticker encourages efficient use, better route planning and combined trips, to the extent that "Zipsters" reduce their car miles traveled by 40 percent,[45] in part replacing them with public transport. Almost half of Zipsters are reported to increase their use of public transport. Access to carsharing both reduces average car use and makes enhanced transport services available for non–car owning households.[46] Analysis of Getaround, a large peer-to-peer car-rental platform, similarly suggests that below median income users gain more from participation – particularly where users can afford a car because they can also rent it out through Getaround.[47]

However, Zipcar focuses its efforts on building trust between users and the company, rather than between users. Late fees are levied as sanctions paid to the company, rather than paid to the subsequent user disadvantaged by the late return, an approach that might enhance community accountability between users.[48] A survey of users revealed they took less care of the vehicles than owners, exhibited little concern for subsequent users (in terms of lateness, cleanliness, fuel reserves, etc.) and were primarily motivated by factors of cost and convenience. Few expressed pride in the Zipcar brand, and most would only use badged cars on grounds of cost savings.[49]

Interestingly, Zipcar adjusted strategy in 2002 under a new CEO. Scott Griffith sought to introduce a business-performance culture and introduced more image-conscious vehicles such as BMWs. Griffith reportedly blames the "save-the-world, change-the-world culture" for the financial difficulties that led to his appointment.[50] But if Griffith's business culture has been reflected in relations with members, it would be unsurprising if they have responded by treating Zipcar as a rental business, rather than a sharing club.

Like Uber, commercial car clubs also typically concentrate their services in areas with most usage—which are rarely low-income, and increasingly face competition from carsharing operations run by car manufacturers such as Daimler's Car2Go, now available in 21 cities with a global fleet of 10,000 Smart ForTwos.[51] The concentration of enhanced mobility services in rich areas risks further downgrading of citywide public transport. These factors—along with the relatively high costs of commercial car clubs—have stimulated grant-funded nonprofits to provide carsharing services targeted at poorly served areas of cities such as Chicago, Minneapolis, Philadelphia, and San Francisco. In Chicago I-GO was established by the Center for Neighborhood Technology to serve both affluent and low-income neighborhoods such as South Shore and Bronzeville. In 2009, I-GO linked up with the Chicago Transit Authority to offer a joint smart card for public transportation and the I-GO cars.[52] In 2013, having expanded to 15,000 members in 40 of Chicago's 200 districts, I-GO was acquired by Enterprise in a deal that committed to further extend I-GOs coverage of Chicago's neighborhoods as well as to provide access to Enterprise cars in other cities for I-GO members.[53]

An alternative model achieves social and environmental benefits by raising occupancy rates on journeys that are already being undertaken. BlaBlaCar, a French ridesharing organization, has 6 million members in 12 European countries.[54] Users share intercity rides only[55] in a model that is effectively high-tech hitchhiking, helping the user find a ride going their way using a web app rather than by standing on the roadside. (With adequate transparency this practice has clear safety benefits, although the risks of soliciting help from strangers are likely to decline dramatically where a sharing culture is (re)instituted; see "Building Social Capital by Sharing" in chapter 3).

BlaBlaCar sets a price cap that means drivers cannot profit from sharing, they merely get a contribution to their costs. That also means they are not liable for tax, and are covered by their regular insurance. BlaBlaCar takes a small commission on the capped price.[56]

BlaBlaCar is one of several ridesharing web platforms that have overcome the key problems of trust and coordination that previously hampered the wider uptake of carsharing, despite the urging of public authorities concerned about oil scarcity. In Germany the leading ridesharing platform has expanded rapidly to 5 million registered users, facilitating over 1.3 million journeys monthly.[57] It offers users information about and links to public transport alternatives for each journey.[58] As with BlaBlaCar, drivers can recover costs but do not make any profit. Such approaches maximize the environmental and social benefits of sharing, without raising concerns about employment abuses, tax liabilities, or unfair competition with taxi services.

In most cities though, improved mass transport, especially BRT—as we saw in the Medellín case study—is a clear winner in the sharing stakes. "A BRT system with clean buses, exclusive lanes and state of the art service can provide 'metro-quality' service at a fraction of the cost," says Gunjan Parik, the director of the C40 Cities Transportation Initiative,[59] so cities can provide high-quality options to all citizens. Bus systems need not be managed and provided solely by the state to deliver benefits. For example the member-owned Egged Israel Transport Cooperative Society is the largest transit bus company in Israel. Egged operates around 3,000 buses providing about 55 percent of Israel's public transport services and employs more than 6,000 workers.[60]

City-led systems can be even more ambitious and innovative, linking the best features of carsharing and public transit. In Helsinki, Finland, plans are afoot to provide a comprehensive mobility solution as a public utility accessed and paid for by phone. The app will "function as both journey planner and universal payment platform" linking existing public transit with "everything from driverless cars and nimble little buses to shared bikes and ferries into a single, supple mesh of mobility."[61] In 2013 the regional transport authority launched a precursor to the service in the form of a shared minibus service called Kutsuplus, which aggregates journeys booked by smartphone or SMS (short message service) into flexible routes.

The commercial model of companies like Lyft and Uber might still help reduce the overall size of the car fleet, if its availability enables residents to forgo car ownership, and even take cars off the road; this may be plausible if Lyft Line and UberPool, the variant apps designed to coordinate multiple passenger pickups so that several people going in the same direction ride together in the same car—take off widely.[62] It might also provide some valuable income for people in poverty—those who can still afford a car suitable to operate in such a service. But a model that sweats the underused assets (the car, or the driver's time) to provide cheap taxi equivalents without licensing or provision for disabled people also has clear social disbenefits and might well draw journeys away from mass transit, cycling, and walking. And key questions remain about ownership and working conditions, which extend to issues like driver liability and driver safety. Revelations that Uber has coached its drivers to take rides in Lyft cars (in an attempt to poach their drivers) have fuelled concerns about its efforts to hamper competition.[63] Although this might be seen as a symptom of a scarcity of drivers, and a sign of competition that will allow drivers to claim a greater share of the profits from carsharing,[64] the jury is still out. Regardless, if driving for a rideshare company remains casual and un-unionized, it seems that drivers will remain at risk of exploitation.[65]

Our view is that as soon as a "sharing app" in fact draws a new vehicle onto the streets to service demand, then it probably should be understood as a taxi, and fairly regulated as part of that sector, and that includes treating drivers as employees, and enabling them to organize collectively. But if an app genuinely just facilitates sharing of trips already being made, then it deserves some exemption. The challenge of course is how to identify this line and distinction. Just like limiting the number of rental days permitted to ensure it remains unprofitable to buy an apartment simply to rent it out on Airbnb, if ridesharing drivers cannot recover more than their costs, as with BlaBlaCar, they won't make additional journeys. So restricting exemptions to that level of payment could be the target of smart regulation, perhaps policed through the tax system.

We need to be clear, however. We offer these views and suggestions not in an effort to eliminate commercial sharing businesses like Uber—there are still places and cultures where that model can help deliver important shifts in norms on sharing, consumption, and the environment—but to steer the sector more generally toward models that can fulfill sharing's promise of greater equality in an inclusive diverse society.

We have seen how poorly designed and commercially driven carsharing models can fail to recognize poor communities and their needs, and threaten to further casualize labor rather than build capabilities. In the following sections we explore unjust effects of and responses to sharing in a variety of places and sectors, starting with casualization. In each case we see arguments for inclusion by better recognition and empowerment by building capabilities.

Casualization: Exploiting Labor?

As we saw with ride sharing, sharing can be co-opted to create and casualize labor reserves. Evgeny Morozov pulls no punches in his critique:

> The erosion of full-time employment, the disappearance of healthcare and insurance benefits, the assault on unions and the transformation of workers into always-on self-employed entrepreneurs who must think like brands. The sharing economy amplifies the worst excesses of the dominant economic model: it is neoliberalism on steroids.[66]

Morozov calls out the resistance of some sharing platform operators to recognize "employees," citing media reports of an executive at Uber, commenting on a protest by Uber drivers concerned by recent firings, with the claim that "a 'driver contracting with Uber is not a bona fide employee' so that 'firing, in this case, amounts to deactivating a driver's account because he's received low ratings from passengers.'"[67]

The blogger Adam Pagnucco makes a similar point, if more temperately, raising concerns about the replacement of journalists by underpaid or unpaid bloggers:

> If bloggers fill their functions for free, the [*Washington*] *Post* will inevitably phase out [paid columnists]. In the labor movement, we have a term for workers who undercut

other workers and threaten their jobs: scabs. As a labor guy for sixteen years I have no intention of blogoscabbing.[68]

Online platforms for outsourcing simple tasks that can be completed online—such as Crowdflower and Amazon Turk—appear the most egregious examples of so-called sharing businesses threatening labor standards by paying below minimum wages with no employment contracts. These operations have all the disruptive potential of the "putting out" movement that broke the power of mediaeval guilds in Europe, isolating workers from any collaborative opportunities to improve their conditions.

In the UK casualization of labor through so-called zero-hours contracts, which place workers on call but don't guarantee any paid work, is stimulating opposition from unions and social campaigners. Some see the sharing economy contributing to this trend, and there is clearly a risk here. But there is also a potential difference between such contracts imposed by corporations on their workforces, and the provision of tasks on a negotiated peer-to-peer basis using a sharing platform. The key issue is power, which is why the use of platforms like TaskRabbit by companies is worrying, even though TaskRabbit now insists on a minimum hourly rate well above the legal minimum wage.

Schor helpfully encourages us to consider differences between models. She argues that, for example, compared to temp agencies, online platforms with a P2P structure enable

> low-paid workers to earn considerably more and have more autonomy over which jobs they accept. … The question is about how much value providers on these platforms can capture. This depends partly on whether they can organize themselves, [and partly] whether there is competition among platforms.[69]

Where platforms must compete for providers, providers will take a bigger share of the value created. We saw above how Uber is currently competing, but other Internet platform businesses such as Google and Amazon have already built virtual monopolies. Where scale is critical, sharing economy businesses might also build effective monopolies, but in many cases the models are relatively simple to replicate, so competition should survive.

We must recognize that the problems of casualization can also be exaggerated by incumbent economic interests. Moreover, opponents of the sharing economy risk romanticizing the conventional economy. Defending incumbents on the grounds that they support living wage and union jobs supported by norms, regulation, and collective bargaining overlooks both the oppressive conditions faced by many workers—however strong their unions—and the continued exclusion of others from the workforce. It overlooks as well the fact that these problems are worsening anyway with the spread of neoliberalism.

Yochai Benkler, echoing Sen, highlights the value in the "autonomy to choose to participate, to select opportunities for action, and to act when the participant wishes and in the fashion that she chooses" in sharing activities.[70] In this context, such autonomy depends on a high level of other capabilities, and alternative choices being available. April Rinne, an advisor to the World Economic Forum on sharable cities, argues that the majority of TaskRabbits and Lyft drivers, for example, are people with skills who don't necessarily want a fixed-hours job—including many mothers of young children.[71] But that desire for flexibility should not condemn people to a precarious and casual existence. Conventional business models mapped onto the sharing economy neither maximize the benefits to those working to provide their services, nor offer any guarantees that the platforms can resist incentives to exploit a largely unprotected de facto workforce. Constructive solutions are possible if we see in the sharing economy a possibility for a transformation of economic models toward co-ownership and solidarity. As Schor argues, "Achieving that potential [of sharing] will require democratizing the ownership and governance of the platforms."[72] Supporters of sharing such as Orsi and Neal Gorenflo also recognize the challenges raised by the commodification of sharing experiences, and advocate for new business models, themselves based in a sharing culture.[73]

Benita Matofska of Compare and Share—a sharing economy aggregator and comparison site—suggests a model drawing on the idea of the B-Corp or benefit corporation.[74] The sharing corporation, or S-Corp, would remain "for profit" but would have its social mission built into its legal structure, making the board directly and equally legally responsible for purpose as much as profit. Such a model could be used to target regulation and tax relief—but its primary purpose would be self-regulatory, locking in the social or community mission and purpose of the enterprise.

Sara Horowitz, founder of the Freelancers Union, shares the concern that venture capital–driven tech companies are not an effective foundation for what she calls the "new mutualism." She argues:

> New Mutualist organizations are driven by a social good and serve a true need in their community. ... [They] draw their power from the strength of community and a feeling of solidarity, those spiritual and economic connections that make a group more powerful than any individual. ... The people (the builders, the makers, the consumers) have to be in control. That could mean a worker-owned cooperative or maybe a membership organization. It's not about venture capitalists funding the next fancy app and receiving all the profits.[75]

Greater cooperative unionization of sharing economy workers could help reverse the tide of casualization. The Freelancers Union already brings together 170,000 American freelancers from lawyers to nannies, providing or collectively connecting them to benefits, resources, and community and political action. Yet so far, the union has only touched the tip of the iceberg: more than 40 million Americans work for themselves, the vast majority without access to such collective benefits, and as the sharing economy spreads, this number will only rise.

Some workers in the sharing economy are already pushing back against casualization. Some Uber drivers have formed their own union, the California App-Based Drivers Association.[76] Other Uber and Lyft drivers have won the first round of legal battles to obtain formal recognition and protection as employees, rather than as independent contractors.[77] The implications of such a reclassification could be broad and dramatic, cutting investment and constraining growth of commercial platforms across the sector.

Concerns for workers suggest a need for smart regulation of sharing platforms—not just relying on self-regulation. They also add weight to the case for cooperative models in which those providing labor share in the ownership of the platform. As SolidarityNYC suggests:

> The abuse of labor can exist in any enterprise or organization, no matter how progressive. Instead of signing up as an Uber driver, these drivers could form their own company as a taxi collective—like Union Cab in Madison, Wisconsin. Another way to do this that would improve upon the labor issues and the allocation and distribution of the surplus would be a consumer cooperative in which the vehicles were actually owned and shared by people who were the consumer-owners. You could even combine the two.[78]

Orsi agrees, using the example of Lyft. Orsi considers whether Lyft should be a cooperative owned by drivers, riders, or both.[79] As a user-owned cooperative, it would be clear that Lyft operated primarily to provide technology and payment processing to the users. Being a user-owned co-op would not immunize Lyft from employment-related lawsuits, but as a driver-owned co-op it would have neither obvious motive nor means to exploit its driver base. She argues that if Lyft's "highest priority is to create opportunities for drivers to make a living, then ... drivers should control the company."[80] But if the organization aims to "revolutionize transportation and reduce carbon emissions, then ... both drivers and riders should control the company."[81] Such an approach echoes the concept of a "solidarity co-op" in which the board includes representatives of all key stakeholders: workers, service beneficiaries, and the wider community.[82]

As a cooperative, an organization like Lyft would charge cost-based fees for the use of the sharing platform technology. Any surplus would be returned as a "patronage" dividend, proportionate to how much money each rider paid or each driver earned. Ultimately, says Orsi, "earnings [would] go back to the users, and not toward the purpose of making rich people richer."[83]

Orsi also maps out how an existing sharing company could voluntarily convert to a cooperative base. She suggests the company and its shareholders could come to a binding agreement to progressively sell the company to its customers as a cooperative over several years. Such a "buy-out could happen in one of at least two ways: users could form a cooperative corporation to slowly redeem the shares of the company, or the company could internally create a new class of shares for future co-op members."[84]

This would be a novel process. Mondragon has successfully restructured some conventional companies into cooperatives, but only after they had been wholly acquired by the cooperative group.[85] Neal Gorenflo argues that cooperative models may need to evolve if they are to flourish in the sharing sector. He says, "You typically see co-ops execute on well-understood business models like retail, distribution, and manufacturing. ... This can change, but it'll take a lot of time and hard work."[86] Don Tapscott and Anthony Williams suggest "hybrid models where participants both share and appropriate at the same time."[87] They envision a "digital-age co-op" for wikinomics-style operations, using peer-rating systems to apportion shares to contributors.

Schor agrees that existing platforms might become user-governed or cooperatively owned if their user or provider communities organize effectively to support such a change. Alternately, she suggests, organizations from the solidarity sector,

> such as unions, churches, civil society groups, and cooperatives ... could build alternatives to the for-profits, [that are] user governed and/or owned. ... Mounting a competitive challenge to business-as-usual should be easier when ... the platform is a broker, not a producer.[88]

Cooperatives or other social enterprise models for sharing platforms, or for provider organization or unionization, would both empower workers and build capabilities.

Commodification: Displacing Gift Economies and Informal Sharing

We saw earlier how commercial ridesharing threatens to crowd out more traditional informal approaches as well as compete with taxi services. In the rich world the threat of casualization might be the biggest concern, but in the poor world the addition of commercial sharing to the existing pressures for development and marketization through conventional policies is perhaps the greater threat. Such market development approaches, like microfinance, can bring desirable improvements in well-being and potentially enhance equity, but simultaneously threaten an existing infrastructure of gift and social economies.

Sociocultural sharing is of particular significance in developing countries. For example, what limited access residents have to power, water, and sanitation in informal settlements in cities such as Mumbai (India) and Cali de Santiago (Colombia) is typically shared, perhaps technically illegal, but often achieved through communal activity. In Sao Paulo, for example, cohousing in squatted buildings and land is offering new prospects for the urban poor in a city estimated to need half a million new dwellings.[89] In Rio's *favelas*, even landscaping to stabilize slopes prone to mudslides is being delivered through community skillsharing.[90] In such cities—as in Medellín—such sharing approaches are not alternatives to more conventional development, but rather they are important complements (and in some cases, the only option for many).

Julia Elyachar—the director of the Center for Global Peace and Conflict Studies at the University of California, Irvine—suggests helpfully that we need to transcend the dichotomy of gift and market economies.[91] Instead we should recognize and build on social infrastructures of communication and exchange in poor communities. This is not just local sharing of housing, food, and childcare duties in extended families. It also includes international networks of migrants—such as the millions of Egyptian men working in the Gulf states in the 1980s—sending "flows of affect, money, information, and faith" back to families in their home countries, "in forms as simple as a neighbor carrying an envelope and news, a friend carrying a cassette, or a fellow worshipper carrying cash."[92] Such social infrastructure should be treated and nurtured as a public good or perhaps a commons—says Elyachar, rather than valorized only as it is expropriated and exploited by multinational companies seeking to market their products to the poor. Ananya Roy, inaugural director of UCLA Luskin Institute on Inequality and Democracy, has described the same process as capitalizing the "shadow economies of the poor."[93] The lack of recognition of such "shadow economies" mirrors the lack of recognition received by the "environmentalism of the poor" described eloquently by the historian Ramachandra Guha and the economist Joan Martinez Alier.[94] The environmentalism of the poor is rooted in resistance to the unequal imposition of environmental burdens and unequal access to environmental resources, issues which rarely get much attention in global environmental debates and negotiations.

There are also strong arguments in favor of markets, based in the freedoms they enable rather than their potential to support economic growth.[95] We argue that sharing models might extend such freedoms to exchange beyond those with financial means to participate in markets. But we would need to be constantly alert to the ways that sharing platforms function to avoid establishing new obstacles to participation by particular groups. Many traditional social economies imposed effective serfdom on women or children or particular castes.[96] For sharing to be clearly preferable to marketization in such societies it must also tackle those unfreedoms (and not create new ones by creating economic compulsion to share). Recognition and valuation of women's work (for instance, by inclusion of domestic labor in national accounts) is just one step in this direction.

Elyachar's critique of development as marketization highlights how solutions such as microfinance were co-opted to help legitimate a neoliberal narrative in which development means only the extension of markets. At the same time poor states were forced to cut back or eliminate social development programs in the name of "structural adjustment" or other neoliberal reform, and the marketization of community level sharing economies left a cultural vacuum. In much of the Middle East, North Africa, and Asia this vacuum was occupied by Islamist movements that also worked on the ground to support social, health, and educational needs.[97] We see similar reactions to contemporary neoliberal austerity programs in Europe, which have created fertile ground for racist politics often legitimized in the name of "ordinary working people." An anonymous Spanish blogger supporting the efforts of Integral Cooperatives highlights the challenge well:

> Don't get me wrong, but we have to learn from Hezbollah, Hamas, and the Greek fascists. Why? Because they … know how to build community spirit. They offer food to those who are hungry. They offer shelter to those who are homeless. They offer help to those who are unemployed. … [But] they actively create divisions and encourage hatred by selectively serving one kind of people on the basis of their ethnic origin. … We have to do better. … We must create community spaces for the locals and the immigrants. … We must share what we have. … Direct democracy will not work if it doesn't stem from a society in which everyone feels represented.[98]

The integral co-op movement also emphasizes alternative currencies and freedom from formal monetary economics. The conventional development model on the other hand typically makes some far-reaching assumptions about the desirability and development of commodified economies.

Notably, the domination of money as a means of exchange is seen as an evolution from processes of barter, and thus as naturally superior to other modes of exchange such as gifting and sharing. Thus marketizing and monetizing exchange is taken for granted as progress. David Graeber challenges this assumption. Instead, he says, non-monetary cultures typically use "a very broad system of non-enumerated credits and debts," including social obligations and cultural as well as material exchanges.[99] Marketizing such systems destroys value, reducing complex contextual information and social capital to prices and quantities.

Sociocultural sharing is still prevalent even within so-called developed societies,[100] and potentially also at risk from commodification and marketization—in the form of mediated commercial sharing approaches. Sociocultural sharing is often particularly embedded in specific ethnic groups or communities of color. Nembhard has documented how African Americans, for example, have long used cooperative economic practices to help each other survive. Applying cultural traditions from Africa, African Americans, both slaves and free, pooled resources informally:

> Enslaved Blacks might share a small kitchen garden to provide more variety of food than what the master would give them. Those that had opportunity to earn money would pool those earnings to buy each other's freedom. ... The Underground Railroad was a collective interracial effort to provide a hidden and protected route North to safety for fugitives from enslavement. Gradually, enslaved as well as freed Blacks started mutual aid societies through religious and fraternal institutions.[101]

Richard Sennett highlights the role of institutions established to help revive cooperation among ex-slaves, recognizing that the psychological damage that slavery does to mutual trust among those oppressed required deliberate healing.[102] In the Hampton and Tuskegee Institutes, for example, ex-slaves trained in craft skills in partly self-governing workshops that were also gender equal. The daily practice of skills and collective reflection on the process helped to re-embed cooperative norms.

Such cultural sharing persisted in the face of continuing discrimination and became a part of community mobilization. In the 1960s, reports Nembhard:

> Members of the Black Panther Party used collective housing and promoted cooperative housing for the community; established cooperative bakeries, and free breakfast programs for children in the community. ... [And] after the uprising in South Central Los Angeles (following the police acquittals in the Rodney King case), Food from the 'Hood, a student-led co-op at Crenshaw High School in Los Angeles, started a school garden and gave the produce to their low-income neighbors. They also began

to sell their vegetables at a farmers' market. ... The student co-op owners mentored other students and the co-op continued even as the original students graduated. By 2003, seventy-seven members had graduated and gone on to college, using money earned from working in the co-op.[103]

Like the advocates of integral cooperatives in Spain, Nembhard now argues that empowerment movements—rather than emphasizing participation in commercial markets—should make greater use of cooperatives for their multiple benefits:

Cooperatives solve economic problems in different ways than conventional for-profit businesses. They operate on the values and principles of democratic participation, inclusion, solidarity, sharing, sharing, and "for need" rather than "for profit." Cooperative businesses stabilize communities because they are community-based business anchors, and they distribute, recycle, and multiply local expertise and capital within a community.[104]

In this respect cooperatives constitute a microcosm of what urbanists Ewald Engelen, Sukhdev Johal, Angelo Salento and Karel Williams call the "grounded city". Here justice is pursued not primarily through redistribution but through an emphasis on building basic services and production in those parts of the economy which are not exposed to competition.[105]

In a similar vein Peter Utting, the former deputy director of the UN Research Institute for Social Development, advocates for a focus upon the social and solidarity economy (SSE). SSE organizations, such as the integral cooperatives, says Utting, have two distinguishing features: "First, they have explicit economic AND social (and often environmental) objectives. Second, they involve varying forms of co-operative, associative and solidarity relations."[106] SSE effectively seeks to enhance equality through changing the productive model, rather than redistributing after the fact, and in this respect it is similar to both the sharing paradigm and the grounded city.

The researcher Golam Sarwar, however, fears that social enterprise faces three "paradoxes."[107] Insofar as social enterprise replaces public services, it risks *reducing accountability* and *increasing exclusion* (on grounds of inability to pay). Moreover, insofar as it relies on the broader economy, its resilience can be the least at just the point its services are needed the most (in times of economic crisis). Wright suggests further practical challenges for SSE organizations.[108] If they seek to compete in capitalist markets, they face poaching of key staff; predatory competition based in cherry picking of affluent customers, and ignoring of social externalities; and much greater difficulties accessing finance.

The challenge of finance is a pervasive theme in assessments of social enterprises and the SSE. Co-ops, for instance face credit constraints because worker-owners typically lack collateral and are seen as higher risk by banks than conventional firms.[109] A cooperative bank was at the heart of Mondragon's success. The Caja Laboral Popular functioned as both a savings bank and a credit union for its members, and invested in other cooperatives in the region.[110]

Our description of a sharing paradigm that spans production and reproduction, and includes both commercial and communal modes of activity includes (but extends beyond) the organizations of the social and solidarity economy. Yet if a sharing economy is seen as a substitute for public services—in the Big Society model advocated by Britain's conservative political parties, for example, or instead of the vital government-funded services we saw in Medellín—it would be subject to much the same risks Sarwar highlights, and especially so insofar as the business models deployed within the sector were conventional commercial models. But as yet, that does not seem to be the case. Commercial, venture, or equity funded models have flourished primarily where sharing businesses are competing with commercial incumbents—such as in ride sharing and short-term accommodation rentals—rather than as a replacement for existing public services; and they have done so even in a period of sluggish economic performance.

Nonetheless, where sharing needs to be, or is already, monetized, it can be important to remove barriers to participation by low-income households. But city authorities must also be alert to the risk that the downside of monetization may be larger than the benefit, even among marginalized groups. For instance, if removing barriers to short-term rentals makes rented accommodation scarcer and more expensive for the least secure in society, cities may need to prioritize other ways of sharing the housing resource—through measures such as land value taxation, or reducing the obstacles to squatting of un- or under-used property.

Marketization: Recognition in Cities of the Global South

Inequality is a serious concern in most cities, but particularly in the developing world. David Satterthwaite emphasizes the high costs of urban living in arguing for a stronger focus in development policy on the urban poor.[111] Not only is inequality more severe than in rural areas, but the scale of the problem goes unrecognized. Poor people in cities are often "invisible" and misrecognized compared with the rural poor. This is partly because city dwellers have higher financial needs measured against the same absolute "poverty line"[112] and partly because of poor definitions—for example "having a water tap within 100 metres is not the same in a rural settlement with 100 persons per tap and a squatter settlement with 5,000 people per tap."[113]

Higher urban land values mean higher housing costs. Life on the urban periphery means higher costs for public transport, a problem we saw being addressed in Medellín. Inadequate public provision raises the costs of schooling and healthcare. And urban dwellers rely on market provision for access to water, sanitation and garbage collection, energy and food, much of which is self-provisioned by poor people in rural areas. The urban poor are consequently more vulnerable to price rises or falls in income.[114]

In low- and middle-income nations, says Satterthwaite, urban dwellers face greater health risks where infrastructure, services and waste management are poor, but cities with competent governance tend to offer better health services than in rural areas, again as we saw in Medellín. Poor city dwellers however face a different and often more severe range of occupational health and safety risks (especially among groups such as waste pickers), and greater vulnerability to natural disasters because the only land they can access for housing is typically at high risk in floods, landslides, or earthquakes.[115]

The factors come together in the necessary reliance of the urban poor on informal solutions. In turn this exposes the urban poor to a further recognition injustice as—all too often—their survival strategies are criminalized, and they are treated not as citizens but obstacles to progress. Many city "households live on illegally occupied land, or illegal subdivisions, tapping piped water and electricity networks illegally,"[116] and live constantly with a threat of eviction from their homes. Yet in cities there is still a "greater scope for joint action, community mobilization, and negotiation with government for infrastructure and services, especially within democratic structures"[117]—even though the more diverse and transient populations in many city districts can weaken the basis for cooperation.

Insecure living is endemic in the global South. Mike Davis, the author of *Planet of Slums*, defines "slums" as areas of

> substandard housing with insecurity of tenure and the absence of one or more urban services and infrastructure—sewage treatment, plumbing, clean water, electricity, paved roads and so on. While only 6 percent of the city population of developed countries live in slum conditions, the slum population constitutes a staggering 78.2 percent of the urban population in less-developed countries.[118]

But most of these people don't appear in official poverty statistics. For instance, only "5–10 per cent of people in Cairo, Egypt, are [officially] poor, but up to two-thirds live in informal settlements."[119]

Davis continues, echoing our concerns about the marketization of socio-cultural sharing:

> This very large fraction of humanity ... is experimenting in a variety of ways how to survive. And they are doing it at a time when not only formal development strategies, including microcredit, are less effective, but also strategies of the poor themselves, such as squatting, ... [are hampered by] property titling that leads to ownership and thus increasing competition and rents the poorest of the poor can't afford.[120]

Yet, "the failure of the old strategies of development has a silver lining," says Davis, which is "the recognition that development from below is the better approach because the people directly affected are more efficient administering resources to themselves." But, he continues, their "resources are radically insufficient for addressing the scale of the problem" so ways of increasing the available resource base are essential.[121] One option, favored by Davis, is radical redistribution of wealth, especially by redirecting the wealth generated by exploiting natural resources. Redistribution may well prove essential, but the sharing paradigm also reminds us of the prospects for co-production to create and enhance the supportive underlying resource base in cities.

These features of cities in the global South are a reason both for concern, and for advocacy for the sharing paradigm to become a central element of support and practice for human development. While Satterthwaite wisely warns against generalization in the global South, the principles of the sharing paradigm and co-production seem well suited to the challenges of urban development in the global South.

In the Orangi squatter camps of Karachi, Pakistan, residents "successfully provided themselves with drainage and mains water faster and at a far lower cost than the more accepted top-down method."[122] Their co-production approach was—with modifications—applied effectively in Faisalabad, too, where a local NGO, the Anjuman Samaji Behbood (ASB) "demonstrated the capacity to support community-built and financed sewers and water supply distribution lines in the informal settlements in which most of Faisalabad's population lives."[123]

Implementation, however, was "difficult and time consuming. There were long negotiations with the Water and Sewerage Authority for permission to connect to its water supply network,"[124] and to overcome "the bureaucratic procedures necessary to obtain permission for the water connection to cross a road; ... the solution was to make the connection at night, without permission."[125]

The results though, were impressive. Hundreds of thousands of people were connected. Water- and sanitation-related diseases fell by over 60 percent. Repaying the loans cost the inhabitants less than they were previously spending on water, and saved "money that previously went on doctors' fees and medicines. The value of properties has gone up and quarrels over water and sanitation have disappeared."[126]

In cities of the global South the scope for sharing to build on their unique strengths and overcome their distinct challenges appears limited only by the imagination of the communities involved. Ushahadi, an Internet mapping facility, offers another very different example, levering the growing penetration of mobile communications in the global South.[127] Established by Ory Okolloh, a Kenyan lawyer, Ushahadi is based on open-source software whose users report incidents using their mobile phones. Initially used to compile reports of rioting, rape, and other violence in the aftermath of the controversial Kenyan elections of 2008, it has subsequently been deployed in other crisis situations—such as the Haitian earthquake of 2010—to help coordinate assistance. It is not fully automated, but relies on volunteers to interpret and plot the incoming data, but these volunteers can be anywhere in the world. Ushahadi is very flexible, and has also been used for election monitoring in India and Mexico. It offers a prime example of how sharing—in this case simply of information—can build (or rebuild) social capital, and how sharing in cyberspace enables the process. Like Ushahadi, and the facilities mapping exercises of Shack/Slum Dwellers International, the Missing Maps project recognizes and valorizes the information and knowledge of city dwellers themselves. Missing Maps is crowdsourcing data to compile open source maps of global South cities, often the first time maps have been produced in many informal settlements.[128] There can be tensions raised by such projects where the authorities still fail to recognize informal slum dwellers because "being on the map" can increase their vulnerability to clearance and eviction, as residents of Bengaluru have highlighted.[129]

For the vast majority of the world's people—in poorer developing economies—there are patent shortcomings in health and well-being. Some of these might be overcome through conventional economic growth and increased material consumption. But even in wealthy societies it appears that the majority of people are not able to experience a good quality of life, as a result of various sources of stress,[130] and there is growing consensus that a new model of development is needed. We believe that the sharing paradigm can offer valuable contributions to thinking about new development models. But that means challenging conventional market-based approaches, and normalizing, protecting, and enhancing existing sharing alternatives, rather than criminalizing them.

Criminalization: Sharing "on the Edge of Legitimacy"

Culturally appropriate forms of sharing might challenge our preconceptions of legality and legitimacy more often than might be commonly thought, and not just in the global South.

While commercial, mediated sharing opportunities are booming for the wealthy, much informal sharing by those living on the margins of society—where it isn't actively replaced by markets—is unrecognized, stigmatized, and even criminalized. We have already seen this with filesharing and with squatting. For justice it is important that society accept and normalize such behaviors, rather than criminalizing and resisting them.

Yet sharing systems, as they currently exist in the contemporary revival of sharing in Western countries, cater far more toward those who have means than those who don't. This is not entirely surprising. There are several factors in play. The digital divide plays a role. Those with cheap, fast web access on both mobile devices and personal computers can obviously access and exploit web-based sharing applications with relative ease. This is still substantial. Even in the richest countries substantial minorities lack smartphones and convenient web access. For example, a recent Pew Foundation survey found that one in five American adults does not use the Internet, with those with disabilities, senior citizens, Hispanics, and those with limited education or earning less than $30,000 per year disproportionately affected.[131] Existing wealth is also a factor. Homeowners with space, cars, and an existing stock of appliances can more easily benefit from sharing what they have, and are better placed to make use of what they can borrow. For example, those with the capital to buy their own solar panels (and a home to put them on) benefit more from grants and tariff support than those who are forced to rent their roof space to a commercial operator for a share of the energy generated.

Social capital is also critical, especially for those programs that are based on proximity (like Streetbank).[132] Stable neighborhoods, dominated by middle-class homeowners, can start with a much greater level of mutual trust, as well as a greater pool of resources to share. Neighborhoods marked instead by rapid turnover and high levels of crime offer more difficult terrain. But it would be misleading and simplistic to suggest that there is no social capital in poor neighborhoods, and that sharing approaches cannot work there. In both the global North and South, poor communities—such as the *comunas* of Medellín—often rely already on high levels of informal sharing, especially for critical services such as childcare. Failure to recognize and support these, instead attempting to introduce markets in such poor societies, exacerbates the divide and often degrades the existing social capital.

But the problems faced by poorer communities can be far worse than a failure to recognize and protect existing social capital. Rather we see official resistance to informal and "illegal" sharing, such as squatting and work or exchange in the so-called black economy. Such sharing on the edge of legitimacy is much more likely to be practiced by poor and marginalized groups—often mobilized by justice concerns in response to economic collapse. For example, the Plataforma de Afectados por la Hipoteca (Platform for those Affected by Mortgages—PAH) is a Spanish "network of over 260 groups of people directly affected by the mortgage crisis who defend one another against evictions using direct actions and organize in assemblies."[133] Through its campaigning, the PAH "is transforming the isolating stigma of eviction into a groundswell of popular outrage that is fuelling practical action,"[134] It has "taken over homes, buildings and in a few cases many adjacent buildings, creating homes for hundreds if not thousands of families." PAH activists also "organize in horizontal assemblies, where each affected person has an equal say in what is done and how."[135]

As Marina Sitrin, an activist and sociologist, says: "This is a clear case of creating commons—something that should be for all and literally breaking down the walls that are the enclosures."[136] In the global South, such informal—and officially illegal—sharing and activism often extends from housing to other basic services. In Caracas, Venezuela, an unfinished skyscraper, whose construction was halted by the global financial crash, was squatted for seven years before being cleared by the authorities. The *Quartz* writer Michael Silverberg reports, "The squatters organized their own electricity, running water, and plumbing, along with bodegas, a barbershop, and an orthodontist. The improvised community became known as … the Tower of David."[137] The writer, critic, and curator Justin McGuirk describes the Torre David as "a radical experiment in self-organized urban living," part of the "informal city."[138] He argues that accepting and improving "the informal city as an unavoidable feature of the urban condition, and not as a city-in-waiting, is the key lesson that this generation of Latin American architecture can offer the world."[139]

In some cities squatters have sometimes won assistance, and even formalization, from the authorities. The Saraí occupation in Porto Alegre persuaded the city to expropriate the property for public housing, in line with the Brazilian constitutional right to decent housing. The squatters transformed the building into a cultural space and used digital media to build a movement of supporters.[140] Many more cities are using land titling to formalize informal, self-built slum developments. This not only shows recognition, but in some cases—such as among the urban squatters in Argentina studied by the University of Maryland economist Sebastian Galiani, in collaboration with Ernesto Schargrodsky of the Universidad Torcuato Di Tella in Buenos Aires—it appears to help the newly titled slum dwellers access mortgage credit markets and improve their housing.[141] However, the researchers found no evidence of enhanced incomes, or wider improvements in access to credit among those given land titles, although these households subsequently appeared to restrain fertility and invest in more years of education for their children when compared with equivalent families that did not get titles.[142]

So, despite the gains in recognition and the importance of freedoms to participate in markets,[143] awarding private property rights appears at best a partial solution. It could even be counterproductive where slum dwellers lack the capabilities to participate in markets on equal terms. Even though those seen as illegal occupants typically face more severe impacts from slum clearance, squatters given land titles—but without secure incomes—remain vulnerable to market-driven clearance for redevelopment, gaining only meagre cash compensation.[144] In other words they would be "participating" in markets on profoundly unfair terms. Even without displacement, entangling squatters with such market rights risks destroying "collective and non-profit maximizing modes of social solidarity and mutual support,"[145] in the way microfinance tended to do once it became bound up with global finance.[146] Even the poorest communities have relationships of solidarity established through informal sharing. These provide some—however limited—social capital. Replacing solidarity with market relations—especially in the absences of basic skills, norms, and education—damages that social capital and increases insecurity and vulnerability. This is essentially the same problem as caused by the intrusion of commercial models into new sharing approaches in the rich world: thoughtless marketization increases the precariousness of the least well-off.

The problem does not necessarily lie with titling per se. However, *individual* titling clearly carries risks that inequalities will be recreated, or of other abuses by powerful interests—as we saw in Medellín. *Community* land titling might be preferable. In Mumbai's Slum Upgrading Program (SUP) "tenure was legalized on the basis of cooperatives. Mumbai's policy makers decided to use the cooperative structure because it was difficult to define individual land-holdings in the city's haphazardly laid-out slums."[147] Granting communal rights helped streamline the regularization process as well as strengthening communal governance, according to the MIT urban planner Jota Samper.[148]

In other words, similar problems merit similar responses. In the global South, as in the global North, to resist the intrusion of commercial flavors of sharing we should seek to strengthen communal, sociocultural, and unmediated models of sharing first and foremost, while investing in building capabilities in the broadest sense. This means supporting the acquisition not just of skills that help disadvantaged people join markets, but also of all the skills and resources needed to participate in society, politically, culturally, and economically. Pedagogical models of popular education, for example the thinking set forth by the Brazilian philosopher and educator Paulo Friere,[149] may offer a way forward, as Eurig Scandrett explores:

> [Paulo] Freire's methodology starts by challenging the assumption that illiterate people simply lack a skill—reading and writing—which can be imparted to them decontextualised from their socio-political reality of oppression. Rather, landless peasants are illiterate because they are oppressed, but they also inhabit a social context and are a source of knowledge about that context which is necessary for challenging oppression. Literacy education therefore must become a dialogue between the knowledge and skills of the educator that the peasants desire, and the knowledge and skills of the peasants, that must be shared with the educator if together a liberating education is to be achieved.[150]

In such ways the necessary skills and capabilities might be stimulated, so that communities with strong informal sectors can use mediated sharing as a development tool within those sectors, rather than marketizing them, which risks destroying social capital that must later be rebuilt.

Another misrecognized form of sharing on the edge of legitimacy is found in the food sector, and here it illustrates how sharing can challenge and begin to shift mainstream culture toward inclusion and sustainability. Scavenging lifestyles include "skipping" or "dumpster diving" and "freeganism"—the recovery and consumption of edible but discarded food from the garbage skips (dumpsters) of shops, together with "foraging," collecting edible wild herbs in urban parks. In many countries shops still routinely dispose of food that has reached its "sell-by" date, regardless of its actual condition. While some companies make a genuine effort to recycle surplus food (for instance, the UK-headquartered sandwich chain Pret-a-Manger has arrangements to donate it to relevant charities[151]), others go to extremes to prevent it from being consumed, even to the extent of actively discoloring or contaminating it with chemicals. And others have sought to use the law to treat the recovery of their waste by freegans or skippers as theft.[152]

In the rich world, behaviors such as skipping and squatting are also spreading as part of a counterculture of anticonsumerism. Here they are not necessarily the product of need, but represent a deliberate rejection of corporate-led consumerism. Skipping has spread in part as deliberate protest over the waste and environmental damage implied by throwing away perfectly edible food. The Real Junk Food Project in Leeds, UK, symbolically combines the idea of the right to food with freeganism, running a "pay as you feel" café using recovered and donated waste food only.[153] Illegal filesharing also can be motivated in such ways, as an act of protest,[154] but it has become far more widespread and normalized with almost half of all Internet users worldwide admitting to illegal downloading in one survey.[155] The power of such countercultural approaches to help shift norms in mainstream culture can be seen in the arrival of web-based intermediates for practices like couchsurfing and meal sharing, as we noted in chapter 1. Cities can actively help broaden and de-stigmatize sharing behaviors—for instance by giving them an "official stamp of approval" through public provision—or as a formal "standard," as used in Seoul. They can also help tackle other forms of exclusion from sharing.

Gentrification, Displacement, Enclosure, and Social Exclusion

Many city dwellers are faced with displacement and exclusion as a result of inequality. We have already seen how whole districts can be redeveloped or gentrified in ways that displace existing—often diverse—communities and replace them with the dominant cultural group. David Harvey puts it starkly: "The results of … increasing polarization in the distribution of wealth and power are indelibly etched into the spatial forms of our cities, which increasingly become cities of fortified fragments, of gated communities and privatized public spaces kept under constant surveillance."[156]

This works to exclude minorities and people on lower incomes from the shared facilities and shared spaces of those districts and developments—even as those facilities and spaces improve. A pernicious contemporary expression of such exclusion is found in the phenomenon of "poor doors." As housing costs have spiraled, many cities have insisted that new developments include a proportion of affordable housing units. To maintain exclusivity developers have taken to installing separate entrances to the affordable portion of the development. These entrances—which do not give access to the shared services enjoyed by those in the rest of the building, such as concierges and leisure facilities, have been dubbed "poor doors."[157]

We have also seen how privatization of public services can create divisions and inequalities eroding social capital; and how enclosure of public spaces and streets in the form of privatized commercial or leisure facilities, and domination of the urban commons as a whole by commerce and consumerism, leaves less physical space for communal activity, public art, and collective politics.

The importance of the urban commons for leisure, arts, and politics can be partly substituted by the burgeoning sharing economy, which extends a semipublic realm into new facilities and even into private dwellings. By contrast, the processes of privatization of public spaces often involve deliberate efforts to exclude those without the means to participate in the consumer economy, those seen as disruptive to consumerism—such as young black men, those who might seek to oppose corporate interests, and so on. Moreover, they also tend to incidentally exclude equally large groups from marginalized cultures who would choose to use those spaces differently.

To those who see cities as competing for inward investment in global markets, and facing a constant struggle to balance budgets, gentrification can appear positive. It raises property values and tax revenues. Less frequently stated openly, but equally valued by some administrations, gentrification often displaces populations that place higher demands on services, thus allowing authorities to cut service expenditure. Leicester geography professor Loretta Lees argues that gentrification

> is increasingly promoted in policy circles both in Europe and North America on the assumption that it will lead to less segregated and more sustainable communities. Yet ... despite the new middle classes' desire for diversity and difference they tend to self-segregate and, far from being tolerant, gentrification is part of an aggressive, revanchist ideology designed to retake the inner city for the middle classes.[158]

This is not to deny the potential for improvements in the quality of the built environment associated with processes of gentrification—nor that such processes are less environmentally harmful than demolition and redevelopment.[159] But it is clear to us that the social costs of gentrification may far outweigh any economic or environmental benefits. Moreover, in processes of urban redesign that focus on desirable features such as public spaces, parks, and "complete streets" it is possible to contribute to the displacement and exclusion of culturally marginalized or disadvantaged groups even while espousing ethics of sharing or inclusion.

For example, the Internet entrepreneur Tony Hseih's Las Vegas Downtown Project and rhetoric of "startup urbanism" proposed redevelopment of 20 city blocks into a sustainable, walkable, spontaneously creative space deliberately invoking Jane Jacobs's "ballet of the sidewalk."[160] The redevelopment aimed to stimulate innovation and growth by supporting co-working and prospects for serendipitous meetings. The plans were ambitious: co-working spaces in refurbished warehouses, a cultural center, private charter schools, Wifi everywhere. But, reports Leo Hollis—author of *Cities Are Good for You,*

> to build the new center, the old community had to be "disappeared." ... Local opposition to changes has mostly been ignored, [despite] rising real estate prices and the closure of local stores. ... And for those that remain: Does everyone have to buy their daily food from the Downtown 3rd Farmers' Market? ... Despite the marketing, this is the opposite of a diverse neighborhood in the making.[161]

Similarly, even clearly progressive local developments can be co-opted to serve an agenda of gentrification and displacement. For example, the anthropologist Andrew Newman explores an apparently successful political struggle by low-income immigrant communities for a new park in northeast Paris, to provide clean air, "gathering spaces for parents and children outside of deteriorated, overcrowded apartments, ... a space for political associations to meet, ... and a space of cultural production for residents of diverse ethnic, national, and cultural origins" on a contested brownfield site.[162]

The completed park design embodied "collective action into the operation of the space,"[163] involved a new residents' committee in its management, and has produced a vibrant urban space, popular with low-income immigrant communities. The park is ungated and open through the night, unique features for a park in Paris. But the campaign was also co-opted by political interests and urban planners into an "urban vision based on ideals of urban sustainability and green design,"[164] with features that emphasized sustainability through water management and recycling of materials taking precedence over the demands of locals for culturally specific elements, such as plants and trees from their cultural heritage. This in turn reflected a political vision of "reconquest" of these Parisian districts as part of "long-term strategies geared towards global interurban competition" which, through processes of gentrification and displacement "has led to a deepening housing crisis for [the same] working-class and low-income residents, and in particular, those of Maghrebi and West African origin."[165] For instance, the park development was accompanied by "the demolition of two full residential city blocks adjacent to the park to 'remedy' northeast Paris of 'insalubrious' conditions."[166]

Newman's study suggests ways in which powerful interests co-opt immigrant and low-income discourses and needs around open space by using urban sustainability measures as a veil to maintain hegemony. Contrast this with the antigentrification measures taken by a low-income community in Greenpoint, a neighborhood in the Brooklyn borough of New York

City. In a debate over options for Newtown Creek, an industrial waterfront with fabulous views of Manhattan (and therefore extremely high development pressure) the locals argued for "environmental remediation without environmental gentrification." As a result the "cleanup of Newtown Creek will be *just green enough* to improve the health and quality of life of existing residents, but not so literally green as to attract upscale sustainable ... residential developments that drive out working-class residents and industrial businesses."[167]

Sarah Dooling of the University of Texas, Austin, identifies in Seattle a direct disregard of a vulnerable minority: in this case homeless people living in public parks, under constant threat of expulsion or arrest. She accuses the city of *ecological gentrification*: "The implementation of an environmental planning agenda related to public green spaces that leads to the displacement or exclusion of the most economically vulnerable human population—homeless people—while espousing an environmental ethic."[168]

Dooling stresses how the norms and regulations of public spaces reflect particular conceptions of the "legitimate citizen," specifically in this case, "housed individuals."[169] We would argue that the boundary of "legitimate citizen" is actually often drawn much more tightly—if less deliberately—around moneyed and consuming adult members of the dominant culture, excluding youth, minorities, and countercultures. Thus, in contrast with practice in Copenhagen (noted in the case study that follows chapter 2), "defensive design" is widely used to exclude such "non-legitimate citizens" and the uses they might make of urban spaces.[170]

Similar misplaced conceptions also underlie the dominant discourse on squatting and gentrification—that squats (particularly artistic ones) can be desirable temporary uses of land and buildings because they enable gentrification. But as we saw in the work of Miguel Martinez, cited in chapter 2, this is an exclusionary way of understanding squatting.[171] Most practical interventions cities might take to assist homeless people involve enhanced sharing of land and buildings; such as the Philadelphia Land Bank we saw in chapter 3, or access to permanent squats, housing cooperatives, and rent control legislation to "expand the pool of affordable housing."[172] However, Dooling also stresses the importance of politically empowering homeless people to challenge the binary of "ideological constructions of home and homelessness."[173] Much the same might be suggested for squatters.

These examples show how citizenship can be devalued. But it need not be so. In several Latin American countries we have seen indigenous peoples and informal settlement dwellers campaigning to enshrine recognition and citizenship into national constitutions.[174] In our case study of Medellin, we saw how the recognition of the needs and interests of poor communities is being combined with practical measures such as land titling and public transport provision, helping deliver effective capabilities and genuine recognition.

In a remarkable study of New York City that exposes major failings of recognition, Sarah Schulman links physical gentrification with the impact of AIDS. As reviewer Olivia Laing summarizes:

> In New York … [gentrification] was facilitated by tax incentives for developers and moratoriums on city-sponsored low-income housing. The role of AIDS in all this was both coincidental and expedient. Because of rent control, properties couldn't be moved to market rate unless the leaseholder either moved out or died. AIDS accelerated turnover, changing the constitution and character of neighborhoods far more rapidly than … would otherwise have been permitted. … The new residents, for the most part the clean-cut citizenry of corporate America, were almost wholly ignorant of the people they'd displaced. In short order, an entire community of "risk-taking individuals living in oppositional subcultures, creating new ideas about sexuality, art and social justice," had almost disappeared from record.[175]

Schulman describes this as accompanied by a "gentrification of the mind" that is eliminating these people and their struggles from accepted history—which we might describe as de-recognition—an active reverse of recognition. Again Laing summarizes elegantly, how the

> undigested, unacknowledged trauma of AIDS … brought about a kind of cultural gentrification, a return to conservatism and conformity evident in everything from the decline of small presses to the shift of focus in the gay rights movement towards marriage equality. The sorry thing about this is that the true message of the AIDS years should have been that a small group of people at the very margins of society succeeded in forcing their nation to change its treatment of them.[176]

In the UK active gentrification has been encouraged by sales or transfers of housing stock from the public sector into individual ownership. Under Margaret Thatcher's government in the 1980s the innocuously named "Right to Buy" for council tenants shifted hundreds of thousands of (better quality) council properties into the private sector, facilitating further refurbishment and renovation, and subsequent resale, and at the same time reducing the diversity of housing tenure and limiting opportunities for many less well-off people to access housing in popular areas.

In their various ways these cases in Paris, Seattle, the UK, and New York City, like the example of Claiborne Avenue in New Orleans (see "Complete and Incomplete Streets" in chapter 3) all confirm Harvey's diagnosis of gentrification as an "insidious and cancerous process of transformation."[177] They also show how the construction of the cultural meaning of places is part of a politically contested process that we ignore at our peril.

To disregard the risk of gentrification of urban space through abuse of the opportunities of sharing platforms such as Airbnb would therefore be entirely inappropriate. If people purchase homes simply to rent them short term, or transform conventional rental property into short-term rentals the risk is very real. This also contributes to house price inflation, making it increasingly difficult for those reliant on the rental market to access housing, and indeed, more generally preventing young people from getting on the "property ladder." In chapter 5, we discuss some of the ways these risks can be countered in sharing design (for example with limitations on rental frequency) or by providing new shared access through provision of cohousing or enhanced land sharing through application of land-value taxation, for example. In these ways the sharing paradigm, not only offers potential to change the cultural context toward empathy and inclusion, but also offers practical approaches that might directly prevent or mitigate the impacts of gentrification.

Building Empathy, for Recognition and Justice

Here we turn from common justice "problems" to some generic "solutions," first examining empathy as a route to recognition.

That commercial sharing platforms such as Airbnb and Uber have been reasonably criticized for their poor provision for people with disabilities[178] may not reflect deliberate discrimination, but definitely implies a failure of recognition, which is endemic where commercial motivations rule. Both exclusion from sharing activities and misrecognition of informal sharing suggest that someone lacks empathy. In terms of justice, empathy appears to be a natural companion to recognition.[179]

For justice, we need to recognize and empathize with the different other citizens of our shared cities, whether they are able-bodied or not, happily take a Lyft to their Airbnb, time-bank their hours and freecycle their surplus stuff, or eat from trash bins and live in squats.

Or indeed even if they are too proud to accept what feels like charity. Some "self-exclusion" from sharing may be an expression of cultural disempowerment. Poorer families may feel forced to buy new clothes and shoes for their children to avoid such an obvious expression of their poverty, while middle class children, comfortable in their families' relative wealth and power, happily sport hand-me-downs and second hand clothes from bring-and-buy sales organized by middle-class institutions such as the UK's

National Childbirth Trust (NCT). Such cultural obstacles might be mitigated by greater empathy between the classes. However, direct peer-to-peer sociocultural sharing behaviors, at least within families and neighborhood communities, appear more prevalent and sustained in poorer communities. This may be because those on lower incomes are typically more generous, charitable, trusting, and helpful when compared to those on higher incomes, exhibiting greater compassion and commitment to egalitarian values.[180] Yet as we have seen, their sociocultural sharing practices are less well recognized by elites and decision makers than the mediated approaches of the contemporary commercial "sharing economy."

Still, sharing—like other forms of altruism—can always be seen in part as an expression of empathy. Our ability to identify with others emotionally, and to recognize their needs and concerns as similar to our own (even if not identical), both enables and is enhanced by sharing activity (especially face-to-face).

Empathy is not an opposite of individualism, but a complement to it. To empathize with someone else we must first have a clear sense of our own individual identity and recognize that other person as an individual like us—in other words we must practice a "theory of mind." In mentally healthy humans this is normal, as is the next step, of recognizing that the other person will have similar experiences, feelings and needs as we do. Empathy is the first fundamental sharing—a sharing of identity and individuality. Where it breaks down we see individualist culture and economic relationships dominate our empathic social instincts and solidarity. Empathy can therefore complement a culture of individuation—which "emphasizes the projects of the individual as the paramount principle orienting her/his behaviour."[181] Manuel Castells emphasizes that individuation is distinct from individualism, as "the projects of the individual" can "be geared towards collective action and shared ideals."[182] This helps us understand "autonomy" not as a self-interested goal for an individual, group or city, but as a goal of justice: the capability to "become a social actor" with projects constructed "independently of the institutions of society."[183] Or as we saw in good co-production, to be an agent, rather than a patient—and, in Sen's terms, to have the capabilities to be, and to do, what we value.

Empathy is not just expressed between individuals within groups or tribes, even if its evolutionary roots are to be found there. As we have already seen, cities have been critical to its spread. (See "Naturally Adapted to Share" in chapter 2.) Cities permit us to express our multiple identities more fully. And more secure in our own identities we can more easily empathize with others.

Yet at least two factors are hampering the spread of empathy in modern cities. First, a decline in social capital symptomatic of what the sociologist Zygmunt Bauman calls "liquid life": a situation in which our very identities are constantly shaken and distorted by the rapid turnover of our economic and cultural surroundings.[184] In other words, our empathic impulse is being undermined by the extreme instability of identity defined increasingly through consumerism. The growth of empathy across group boundaries is also harmed by "self-sorting" into segregated communities on grounds of race and income. As the *New York Times* columnist Charles Blow argues:

> We need to see people other than ourselves in order to empathize. If we don't live around others we do ourselves and our society damage because our ability to relate becomes impaired. … It's easy to demonize, or simply dismiss, people you don't know or see. … It's nearly impossible to commiserate with the unseen and unknown.[185]

We believe that the rediscovery of sharing in cities could also herald a revival of empathy, which could reach out beyond the city, but be rooted in growing empathy in the cosmopolitan spaces of the modern city. Similarly, we strongly believe that more could be done by city authorities to stimulate the "intercultural extension" of empathy. Roman Krznaric advocates human libraries "designed to promote dialogue, reduce prejudices and encourage understanding"[186] or empathy museums and other educational empathy programs.[187] Rifkin emphasizes the role of mutual support groups.[188] As long as these are not seen as substitutes for public services, they can serve to expand empathy, and to increase the inclusiveness of sharing activities within the city. Other examples include the intercultural learning programs at Collingwood Neighbourhood House in Vancouver mentioned in chapter 3, Amsterdam's cultural integration policies (see the Amsterdam case study below), and the Restorative Listening Project in Portland, Oregon, whose aim is "to have white people better understand the effect gentrification can have on the city's longtime black and other-minority neighborhoods by having minority residents tell what it is like to be on the receiving end."[189] Medellín's arts and music-based projects also provide spaces for empathy building, particularly for and between those young men—studied by Adam Baird[190]—who might otherwise be locked into divisive and violent gang cultures.

We are in no way suggesting that empathy is a substitute for rights and more other institutional approaches to justice. Rather as Krznaric suggests, extended empathy is one foundation for political activism to extend rights. As a direct basis for policy, empathy would be risky. People more easily empathize with those closer to them or more like them, and with visible victims. But this is exactly why sharing's potential to *extend* empathy and reduce the influence of such bias and discrimination is so important.

Civil Liberties: Privacy, Anonymity, and Justice

The digital age is helping extend empathy, but holds dangers for justice too. Not just in the impacts of the digital divide on equal access, but also in the ways privacy is being eroded. "Smart cities" are rolling out sensors everywhere from street lights and waste bins; homes are being wired up with smart meters and smart appliances that feed real-time data back to utilities; and buses, cars, and city bikes all have GPS or RFID identifiers. Data is also being harvested from smartphones. These features all help with efficient sharing of urban infrastructure and resources, yet exacerbate risks for privacy that could overcome the benefits of the anonymity in city life. The data gathered is increasingly retained, shared, and analyzed for a host of (often initially unforeseen) purposes. When data protection slips, personal data might be revealed, and even in ostensibly anonymized data sets, information on individuals is often easy to extract through comparative analysis. This can easily expose things we might reasonably wish to keep private. For example, in the London bikesharing program, a breach allowed one analyst to reveal the habit of an individual user to regularly sleep somewhere other than his home address.[191] In a case relating to unintentional release of filmsharing data, the sexuality of a user was exposed.[192]

Analysis of so-called big data allows much more detailed profiling and prediction of behavior. This could easily exacerbate the risks of discrimination according to "propensities" revealed in the data, yet with a high risk that people will be labeled incorrectly. Viktor Mayer-Schönberger, a professor of Internet governance and regulation at the University of Oxford, writing with the *Economist*'s Ken Cukier, cite several cases where data analysis successfully raised the targeting rate for interventions (such as identifying buildings in multiple occupation with high fire risks) to as much as 75 percent. This is clearly a massive gain for service managers and infrastructure maintenance, but applied to crime or health, 25 percent misidentification rates would mean very large numbers of people wrongly accused or excluded from cover. Mayer-Schönberger and Cukier are particularly exercised by the risk that the authorities might be encouraged to use such analysis to "prevent crime," but in doing so, will prejudge, and thus breach the individual liberties of those who are labeled as potential criminals.[193] For instance, in London in 2014, the Metropolitan Police Service was already trialing software from Accenture to track and predict gang violence.[194] Similarly, companies might use such approaches to screen out people from insurance coverage for certain activities or health risks.

To tackle these problems Mayer-Schönberger and Cukier argue for three safeguards: transparency of process, independent certification of the analytical methods, and "disprovability" (specified ways in which individuals can challenge their treatment).[195] But we believe a broader and stronger approach to civil liberties online, potentially enshrined in an Internet charter of rights, will be needed too.

The Obama administration is considering the case for new privacy laws to combat potential discrimination by data, following a major review which recognized the "potential for big data analytics to lead to discriminatory outcomes and to circumvent longstanding civil rights protections in housing, employment, credit, and the consumer marketplace,"[196] or "what some are already calling "digital redlining.'"[197]

Among other protections the review advocated enhanced consumer privacy legislation and measures to ensure that data collected on students in school is only for educational purposes. Wim Elfrink who heads up the "smart cities" team at Cisco, has also called for citizens' rights to opt in or opt out from monitoring by smart devices. He argues that in the absence of measures to protect privacy, citizens will actively resist the deployment and use of such devices.[198] Mayer-Schönberger and Cukier argue that, in the era of "big data," anonymity is impossible, even with opt-outs, while prior consent appears impractical.[199] They argue therefore for stronger accountability of data holders and users for abuse, supported by risk appraisal methods and data use auditors and ombudsmen, and firm limits on the periods for which particular types of data can be held (reflecting the "right to be forgotten" principle and the arguments for "deletion" we discuss in the "Cybertrust and Identity Online" section in chapter 5). Zittrain suggests it might help to require web companies entrusted with personal data and preferences to act as "information fiduciaries."[200] Like financial advisors and doctors, for example, who also obtain sensitive information about their clients and patients, information fiduciaries would be expected to use the information they get only in the interests of the individuals concerned.

The protection of our civil liberties online might appear to reflect uniquely liberal and culturally specific concepts of justice. But as Brazil illustrates, breaking new ground with its legal charter of Internet rights,[201] in the real world such approaches can be applied alongside more responsive communitarian approaches in intercultural societies.

Complementary Currencies for Stronger Sharing

Empathy for recognition and greater civil rights are important ways to protect and enhance justice. But where problems of injustice arise in the marketization and monetization of sharing, the sharing paradigm suggests a third way to make sharing fairer and more inclusive. Complementary and local currencies—whose value derives from community capabilities rather than financial institution lending—could support most activities in a sharing economy, not just those that might rely on gifting or barter.

There is a range of models of complementary currencies from the Bristol pound to bitcoin.[202] Virtual currencies could theoretically be designed to support the social and solidarity economy in ways that target particular sectors or segments,[203] as local real-world complementary currencies do. The Bristol pound is both a physical and electronic currency—based on the UK's pound sterling—which can be used to make purchases in participating local enterprises.

Bitcoin is a decentralized digital currency, which allows peer-to-peer online transactions, which are recorded and verified in a digitally distributed and encrypted "blockchain" ledger or database. As a "cryptocurrency" (discussed briefly in Chapter 3: "Collective Governance") bitcoin is issued and backed, and its supply controlled, by the security of the underlying software algorithm, not by banks or governments. Bitcoin thus enables a high degree of financial privacy from the authorities. The motivation behind bitcoin is not therefore—unlike most complementary currencies—to reduce economic leakage and maximize local recirculation to support the local economy. But this is a very real need. Sennett reports, for example, "In 2000, only about 5 cents of every dollar spent in retail commerce in Harlem remained in Harlem."[204] In Bristol, UK, the local government participates in the currency program, so citizens can pay taxes in Bristol Pounds, and to promote the program the mayor takes his salary in Bristol pounds. Many more complementary currencies are purely virtual and mutual in form, in that they are not convertible to conventional currency, but their value rests in the promise of real work that underlies them.

Jem Bendall—a professor of sustainability leadership at the University of Cumbria, UK—points out that the transformative value of alternative currencies arises in the way money is created out of credit between peers, rather than a central authority issuing currency into circulation.[205] Where peers can always create more "money" by doing tasks for one another it becomes impossible for the already wealthy to dominate the money supply. Bitcoin's algorithmic approach moves in this direction, but remains relatively undemocratic, as the software limits the overall supply of bitcoins, so it becomes vulnerable to domination by an elite. As Bendall argues, with collaborative credit, "Outside agencies cannot limit the amount ... available, as collaborative credit simply requires members of a network to trust each other rather than a bank."[206]

In both Local Exchange Trading Schemes (LETS) and time-banks (mentioned in chapter 2), participants create credits by undertaking useful work for others, on an equal basis. It is easy to see that these are not only forms of sharing in themselves, but ones which offer massive potential synergies if mediated sharing models were to accept complementary currencies as payment, rather than, as is too often the case, insisting on credit cards or PayPal.

Time-banking has spread widely, from Edgar Cahn's Timebanks US to Estonia's "Bank of Happiness." Time-banks are great levelers: everyone's time is worth the same in a time-bank—the janitor's as much as the corporate lawyer's—and no one, regardless of conventional financial worth, can command more hours from others than they give themselves (over the long term). Research suggests that 72 percent of people using time-banks feel a strengthened sense of community and 86 percent say they have learned a new skill through their participation.[207] Across the UK, 28,000 people use 300 local time-banks, sharing their time and skills with those who need help with anything from childcare to job applications.[208] Time-banking builds social and human capital, and the main constraints to its expansion appear to be psychological.[209] Participants reportedly find it easier to offer help than to request it, and struggle to grasp the concept of having a time-bank debt incurred by making use of a service before contributing something themselves.

In Schor's US-based study, however, she also highlights that the flat value of time in time-banking appeared to discourage highly educated people from offering their most valuable skills (like programming or web design) and thus reduced the scope of time-banking.[210] More innovative approaches might help. In the UK, for example, Spice time-bank encourages "volunteering in public services in exchange for time credits which can be redeemed from local partners who accept the 'Spice Network Credits'—such as a local theatre or babysitting service."[211]

In most countries it appears that LETS and time-banking have continued to grow, although not as rapidly as the booming commercial sharing economy, despite the shared economic driver of recession. Countries like Greece and Spain seem to have experienced particularly high interest in such models as austerity politics have taken hold,[212] but the US has not experienced particularly rapid growth.[213] Generally, it does not appear that participants have generally taken up opportunities to share and exchange in cyberspace as an *alternative* to doing so on the ground in their local communities, but as a *complement*. However, commercial sharing intermediaries do not generally appear to have recognized the opportunity to use complementary currencies to enhance their inclusivity and ensure that people are not excluded from sharing either because they don't have the financial means, or because they lack a particular means of payment (e.g., a credit card). Nor do P2P lending and even crowdfunding models create space for complementary currencies.

Some mediated forms of sharing have bypassed existing complementary currencies, and effectively created their own (perhaps as a means of locking users into a particular platform). Bartercard, for example, is one of a series of massive online barter exchanges whose business users barter their goods and services, facilitated by a dedicated points-based "currency." Such approaches could be fairly socially inclusive (the digital divide permitting), but miss out on the benefits of local community building and reducing local economic leakage that motivate time-banking and LETS.

We see much greater opportunities for the integration of complementary currencies both with sharing activity generally and specifically with P2P lending and crowdfunding, to begin to build an urban-scale financial commons. Imagine the best aspects of Kiva, Kickstarter, and credit unions, all denominated in a complementary local currency that could be used also to pay for services on local commercial sharing platforms. Such a strategy could generate big benefits for the sharing city for two reasons. First, the value of local currencies is directly related to real investments in social capital and the shared urban commons. Second, it would challenge the power of bankers and the financial industry, enabling a more citizen-oriented approach to city politics.

All this actually suggests a wide scope for a sharing approach to straddle longstanding ideological differences regarding markets. Yet if it is to meet its full potential, this needs to be a considered process. In some cases, sharing cities might actively raise barriers to sharing in the commercial economy to enable more sharing in sociocultural forms which might otherwise be crowded out by commercial forms.

But we must be clear. Our vision of a sharing paradigm is not a one-size-fits-all prescription for all countries and economies. It includes sociocultural, direct, peer-to-peer, and nonmonetized sharing in families, kin groups, and wider communities, as well as mediated commercial sharing. Understanding the paradigm cannot be separated from an understanding of who benefits and who is excluded from sharing innovations. With such understanding it can be deployed to target support for sharing that builds social capital and cosmopolitan values of inclusion, using models and approaches that are appropriate to the specific cultural and economic context, as we will see in chapter 5.

Summary

In this chapter we began by arguing that the sharing paradigm is a direct descendant of the concept of "just sustainabilities": improving our quality of life and well-being; inter- and intra- generational equity; justice and equity in terms of recognition, process, procedure, and outcome; and the need to live within environmental limits. We asked the question: What is the role of the sharing paradigm in building capabilities and delivering justice?; and in the process we explored what we mean by justice and explained some of its implications. We argued that sharing can be a vehicle for justice, with some of the most developed and equity-focused sharing programs being around public transit. Medellín approached this through the strategy of social urbanism: striving to bridge the city's socio-spatial divide and achieve access and inclusion, not simply mobility. The city's public transit system, a central plank in its strategy, was voted one of the best in the world.

In the second part of the chapter we moved from theory to practice, examining how sharing enhances or endangers justice. Some common themes arose repeatedly. We saw sharing practices and programs enabling people to participate in society regardless of ability to pay. We saw the importance of recognition to avoid exclusion through ecological gentrification, in valuing the people and products of existing gift and social economies, in protecting privacy and anonymity online, and in the need for fair and inclusive treatment of squatters and other sharers on the edge of legitimacy. We also saw, however, how easy it is for investments in the "sharing commons" by commercial sharing platforms to fail the tests of inclusion: carsharing programs that don't cover poor neighborhoods, P2P finance programs that ignore complementary currencies, and sharing models that demand a credit card as the minimum price of access.

We saw opportunities to build capabilities by supporting and enabling informal co-production in cities of the global South, in resisting the casualization of labor in commercial sharing platforms, and in overcoming the stigmatization of "illegal" and poor sharers alike. We saw the potential value of alternative ownership and management forms such as cooperatives; institutions for empathy building and cooperation including unions or associations for sharing economy providers; and of complementary currencies as a fair medium of exchange. And we saw a constant need for meaningful participation on citizens' own terms, especially for marginal and vulnerable groups, to help define the scope and nature of the sharing city.

In various ways we saw contrasts with the experiences highlighted in the Medellín case study. In that city, where sharing practices are primarily communal and sociocultural in origin, the marginal population of the *comunas* were recognized as full citizens; they were given new access to the city center to overcome spatial injustice, as well as support to build skills and capabilities with library parks and web access. In Medellín, efforts at political inclusion extended to participatory budgeting. Despite the challenges Medellín still faces, these contrasts make an eloquent case for why sharing cities must build in equity and justice from the beginning, and throughout all dimensions of their sharing programs.

Yet to ensure recognition and inclusion of all groups, to minimize surveillance and loss of privacy, and to aim for well-being and flourishing, cities must value sociocultural sharing traditions, and seek ways to recognize and normalize them, rather than simply marketizing and commercializing society. In turn this implies new models of development that focus on building capabilities with a focus on the collaborative and collective mechanisms of the sharing paradigm. Charles Eisenstein, a leading advocate of the gift economy, challenges us to reconsider the

> remote village ... where everybody occupies that oft-lamented condition of "living off less than two dollars a day." Imagining ourselves with such an income, we see a life of relentless hunger and deprivation. The truth may be quite different. Consider that the people there grow most of their own food within extended families that may number over a hundred people, so they don't need money to buy food. Similarly, everyone knows how to build a house out of freely available materials, so they don't need money for housing. If land is owned in common by the extended family, no one needs money for rent either. Entertainment, drama and play are functions of village life that don't require money as well. There is no need for insurance, as people take care of each other. There is no need to pay police, as informal social pressure and perhaps village councils enforce social norms. Of course, in the extended family there is no need to pay for cooking, cleaning or child care.[214]

In conventional "development" these sociocultural sharing and gifting practices are marketized. Resources are commodified. Commons are replaced with property. This creates business opportunities but locks communities into the market economy, often without access to the safety nets of public services and support. But we should not inappropriately glorify preindustrial society. We cannot overlook the lack of choice it embodies, or the authoritarian moral community that might imply. But the sharing paradigm can perhaps offer a modern intercultural version of such collective culture, realizing the strengths of mediated as well as sociocultural sharing.

Imagine then, if you will, a city which enables citizens to self-build sustainable low-energy cohousing, that restores mutual aid with universal co-produced care and health services; reinvigorates commons and community land rights, and uses land-value taxation to share the value of private land. Picture a city where gifting is common (and if people need money to pay for things they use alternative currencies), and where they invest in new initiatives through crowdfunding and credit unions. In such a city shared meals, pop-up restaurants, and community kitchens might replace fast-food restaurants; and shared streets for walking, shared bikes, and mass transit could displace private cars. Public art would be more common in public spaces than commercial billboards, while live performances in shared urban spaces might outnumber commercialized industrial entertainment outlets. And what if the city also refocused education to build capabilities in collaborative, student-led fashion for both children and adults, and involved citizens in direct democracy, growing participatory budgeting to encompass the majority of its public functions, and encouraged citizens and workers to associate collectively to defend economic and civil rights? Could not such a city produce more of its food in shared gardens and on community allotments and city farms, make and repair tools and products in shared local cooperative-owned fab-labs, and generate its electricity using community energy systems and rooftop power renewables—reducing its reliance on environmentally damaging imports of commodity crops, manufactured materials, and fossil-fuelled electricity? Would not such a city make the identity derived from brands redundant, instead recognizing multiple and different identities rooted in collaborative production and creativity?

That is what we think the sharing paradigm might look like in practice. Although the people of such a city would generate much less financial wealth by conventional measures, they would also need much less money to sustain a high quality of life. They would also enjoy health, capabilities, access and opportunity massively above that of pre-industrial societies. They might even finally transcend Plato's division of the city into rich and poor. In the next chapter we attempt to sketch out how we could get from the challenges of cities today to such a vision of cities of the future.

5 Case Study: Amsterdam

Europe's first official "Sharing City," Amsterdam is the capital city of the Netherlands although not the seat of the Dutch Parliament. In 2012, the city was home to just over 790,000 people, with a metro area population of around 2.2 million.[1] Amsterdam boasts a network of eighteenth-century canals that intersect the city center with some 1,500 connecting bridges. The canals shaped Dutch culture in a way that cultivates strong identities of both individuality and collectivity. The Dutch had to work together, district by district, in a shared challenge to maintain the canal system and prevent their land from being reclaimed by the sea.[2]

Is Amsterdam the "ideal city"? The geographer John Gilderbloom and his colleagues pose that question and highlight that: "People live longer because of Amsterdam's walkability and bike usage and access to parks."[3] As a shared city, Amsterdam exemplifies positive tolerance to immigrants and Amsterdammers have the capacity "to put up with another's fully recognized differences from self ... with a mild appreciation for, or enjoyment of, those differences."[4] The result, as Gilderbloom and his colleagues note, is that: "Unlike the USA, ghettos and/or highly segregated places, which are nearly all poor and made up of one race, do not exist in the Netherlands, because of ... the integration of immigrants."[5]

Ethnic minorities make up more than 45 percent of the city's residents, representing at least 175 different countries.[6] As an anti-discriminatory measure, the city's civil service is required to reflect the diverse population of the city.[7] Amsterdam plays a critical role in modeling cultural tolerance through its integration of immigrants and newcomers.[8] An international comparative study of Muslim integration found that 66 percent of people identifying with a Moroccan ethnicity in the Netherlands strongly identify as "Dutch," compared to 43 percent in France who identify as "French."[9]

The Netherlands, and Amsterdam in particular, constitutes a favorable, open context for minority political participation.[10] In 1985, the Netherlands began allowing non-nationals to vote in local elections after five years of living in the country legally. The local municipality even automatically mails voter registration cards to all residents who are able to vote. The Dutch concept of citizenship, the Netherlands' electoral system, and this right to vote at the local level have "helped bring about one of the Western world's highest levels of minority representation at the national level as well as—and especially—at the local level."[11]

However, national laws toward immigrants have become less tolerant in recent years. For example, before a foreigner can stay in the Netherlands, he or she must take a challenging civic integration test, which some believe is a primary reason that citizenship applications have dropped in recent years. Additionally, contrary to the earlier findings of Gilderbloom and his colleagues, some spatial segregation is now emerging. Harro Hoogerwerf, Amsterdam's manager of education and civic integration, notes that the city is seeing "people tending to live in certain quarters and going to certain schools and certain cultural events."[12] Nonetheless, Amsterdam still emphasizes tolerance and a high quality life for its diverse residents.

Urban practitioners and policymakers have been guided by the idea of social mixing, in their cultural integration efforts (see "Intercultural Public Space" in chapter 3). This is based on the premise that spatial concentration of certain populations, by race or class for example, perpetuates social and economic problems. Mixing of people from different backgrounds on the other hand improves social interaction and appears to improve educational attainment and job prospects and enhance a sense of community.[13] Due in part to policies and planning efforts deliberately mixing socioeconomic groups, in response to emerging trends of segregation, notably in the 1980s, Amsterdam has remained significantly less socially segregated than many European cities. For example in the neighborhood of Slotervaart, where many Turks and Moroccans have settled, just as many native Dutch residents have remained.[14]

Social mixing has helped maintain social capital and neighborhood trust. A longitudinal study from 2001 to 2009 found that greater housing and income diversity in neighborhoods had an independent positive effect on levels of trust in Amsterdam's historically working-class districts.[15] The social mixing approach is not without its critiques, notably that its focus on socioeconomic mixing has ignored a degree of ethnic segration.[16] This has been described as Amsterdam's own brand of "semi-mild" gentrification.[17] As we saw in chapter 4, other scholars, such as Loretta Lees and David Harvey, question the feasibility of such strategies. And indeed, even with government regulation, it appears the effects of "natural gentrification" or "incumbent upgrading" are finding their way into Amsterdam's neighborhoods.[18]

For many years upgrading was also an indirect result of progressive squatting laws, which legalized squatting in 1994, encouraging "landlords to fix up abandoned housing units rather than face losing these unused structures."[19] But in 2011, the policy was overturned, generating controversy and protests. However, at the same time it was made easier for local authorities to take over unused buildings and return them to beneficial use.

Policies in the social housing sector have been used to help the city maintain its level of social mixing. In Amsterdam only around 30 percent of householders own their homes,[20] and the stigma that is associated with social (public) housing in the US and some parts of Europe does not exist.[21] Dutch policymakers actively seek to prevent segregation in the housing stock. Generally speaking, the quality of social housing is high, and rent subsidies are available for low-income renters.[22]

However, budgetary and political pressures have led Amsterdam to adjust its social housing allocations. The policy focus has shifted from the universal provision of housing across the social spectrum to targeting those who have greater difficulty accessing market housing.[23] As a result of this shift, in 2011 at least 90 percent of housing association dwellings went to people with an annual household income of less than €33,614. To prevent this worsening segregation, the city is considering promoting targeted shared ownership.[24]

This could further support cohousing. A *centraal wonen*, meaning "central living" in Dutch, is a community of people in which each household has its own private home as well as a shared community space to allow for enhanced social contact and sharing of amenities and management of resources.[25] There are more than 100 such cohousing projects in the Netherlands, many of which consist of rented apartments or houses that are owned by a housing cooperative or public authorities,[26] allowing low-income households to be part of cohousing communities.[27] There are also projects with a mix of owned and rented homes.

Newer forms of home sharing are growing in Amsterdam's sharing economy. In 2014 Amsterdam became the first city to officially legalize short-term rentals of personal property through home sharing platforms like Airbnb.[28] City Councillor Freek Ossel said: "Occasional rental of privately-owned property as an additional form of accommodation dovetails with a hospitable Amsterdam."[29] Prior to the new law, renters were required to obtain a permit in order to list on Airbnb, which acted as a barrier to participation. The policy creates a new "private rental" category, under which hosts may rent out their homes, though they are still required to pay income- and tourist taxes.[30] A subsequent deal between the city and Airbnb has clarified Airbnb's responsibility to collect and remit tourist tax on behalf of hosts.[31]

The law limits hosts to 60 days a year of renting out their property in an effort to preserve the existing dynamics of each neighborhood.[32] The arrival of Airbnb does not appear to have increased numbers of tourists—Amsterdam has long been a popular destination. But participating hosts suggest that the Airbnb exchanges have allowed for a new kind of understanding between locals and travelers.[33]

In February 2015, following strong advocacy from the Dutch national sharing economy platform, ShareNL, city officials and businesses announced their support for sharing policies and experimentation, making Amsterdam Europe's first Sharing City.[34] ShareNL has worked with the Amsterdam Economic Board to promote this goal. Many sharing platforms and initiatives have already popped up around the city.[35] Amsterdam was the home of the first "repair café" in 2009, and now has 15 as well as hosting the Repair Café Foundation.[36] It is also where Peerby, an online platform that links individual lenders and borrowers of virtually any item or service, began; and the home of Floow2, a B2B sharing platform for underutilized skills and equipment, from printers to excavators.[37] Amsterdam also houses Konnektid, a skillsharing platform which facilitates users forming groups around shared interests, or in particular localities. According to Michel Visser, Konnektid's founder, within a few months, most of the dozens of languages spoken in multicultural Amsterdam were already being offered on the platform.[38] In the spirit of co-working, the new library in Almere (one of Amsterdam's satellite towns) offers "a Seats2meet location where patrons are empowered to help one another in exchange for free, permanent, co-working space" as well as a café, gaming facility, and reading garden.[39] The library "surpassed all expectation about usage with over 100,000 visitors in the first two months."[40]

There is substantial appetite for more sharing among Amsterdam citizens. A recent study found more than 80 percent thought it likely they would take part in at least one form of collaborative consumption. Respondents reported slightly greater enthusiasm for exchange modes that do not involve money (such as swapping). The study identified strong social and environmental intrinsic motives, such as "meeting people" or "helping out." Older and higher income groups exhibited slightly less willingness to share. But, strikingly, "the 54 non-western immigrants in the survey demonstrated the highest willingness of all ethnic groups."[41]

Amsterdam is beginning to integrate sharing into its Smart-City program. This focuses on cutting carbon emissions in energy and transport systems, but also engages citizens in participatory service evaluation and design.[42] The city also has an active Open Data program, which will enable public access to data from the city government. Open Data aims to provide "Amsterdammers with new insights and the chance to make decisions based upon actual facts and figures."[43] Among the "Apps for Amsterdam" supported by the city is BuurtMeter Amsterdam, which utilizes open data to provide scores for neighborhood participation, pollution, and safety based on a user's current location.[44] Among other smart-city projects is "Amsterdam Free Wifi." First provided around the harbor of IJburg in 2013, the aim is to extend this to other public places in Amsterdam.[45] Another is to establish co-working, with a pilot "smart-work center" in IJburg, to relieve traffic congestion. A simple e-payment system is being developed, and high-quality video conferencing will be provided.[46]

Amsterdam also rivals Copenhagen for cycling. Sixty-seven per cent of all trips in the city are by foot or cycle, and many people commute by combining bike use with the excellent suburban and regional trains.[47] Amsterdam began the world's first community bikesharing project in 1965. The White Bicycle Plan—a small-scale program providing free bicycles for temporary use—failed because it proved vulnerable to abuse, but was a precursor of the modern wave of technology-enabled city bikesharing. Amsterdam also boasts a modern carsharing program: MyWheels allows participants to rent each others' cars using a simple electronic swipe card—the same card that is used for access to public transport.[48]

Amsterdam's openness welcomes immigrants and innovation alike. The city is demonstrating many of the possibilities for sharing to help meet the challenges of delivering inclusion in a smart city, with co-production linking the city authorities, its business sector, and citizens.

The Sharing City: Understanding and Acting on the Sharing Paradigm

A city is a state—of mind, of taste, of opportunity ... where ideas are traded, opinions clash and eternal conflict may produce eternal truths.
—Herb Caen

Chapter Introduction and Outline

In this chapter we aim to bring together the concepts and challenges raised by our book so far in order to demonstrate how the various flavors of the sharing paradigm (mediated, sociocultural, communal, and commercial) and its domains (economic, environmental, social, cultural, and political) could reinforce one another at the city scale, and to outline the crucial ways city administrations should act to deliver such a virtuous cycle of sharing activity. The chapter divides roughly into four parts. First, we revisit the scope and territory of the sharing paradigm, rooting it conceptually in well-being and capabilities. Second, we bring together and explore possible reasons for inertia, rebutting some common objections and highlighting some genuine obstacles and challenges to the development of sharing. In doing so we emphasize the importance of emergent collective governance and positive social norms for stimulating and enabling sharing practice. Third, we explore the opportunity for broader sociocultural transformation through the sharing paradigm, focusing particularly on the potential for it to underpin new identities and thus challenge the cultural hegemony of consumerism and associated growth-fixated economic policy. We also examine the self-reinforcing synergies between the unconscious practices of sharing in everyday life, new sharing habits and trust building. Fourth and finally, in the light of this understanding of how sharing might spread, we examine the prospects for implementing and scaling up the sharing paradigm through active policy, planning, and practice at the city scale—where the realization of the sharing paradigm in practice, across the various dimensions of the urban commons, offers the possibility of a genuine alternative to neoliberalism.

The Sharing Spectrum

Earlier we explored a range of definitions of sharing and the sharing economy. By now it should be clear that most conventional definitions and categorizations, however helpful, are rather narrow. Even our two-by-two matrix of the flavors of sharing—highlighting variation in motivations and modalities of sharing—does not entirely capture the richness of the concept. What is critical to the paradigm approach is not the categorization, or any privileging of one category over another, but the new ways of thinking it offers in all flavors and categories: sharing resources fairly, rather than by ability to pay; nurturing the collective commons of culture and society; and treating those resources, commons, and the natural world as common heritage and common property.

Once we understand the sharing paradigm as offering new ways to create and use collective commons of physical and virtual resources, spaces, infrastructures and services, a focus on sharing simply as a way of allocating access to conventional goods and services is obviously too limited. We need to not only distinguish things that are produced collectively and collaboratively in P2P networks from those produced *individually* (or by organizations), but also to recognize the processes of *communal, collective production* that characterize the collective commons. Figure 5.1 presents our "sharing spectrum," which incorporates collaborative consumption, collaborative production, and collective political participation; and the shared processes and services that it produces into a single framework. But figure 5.1 stretches even further. It also highlights that a focus on goods and services can miss opportunities to share both *inputs* to the economy such as materials and water, and the *outputs* that people really value—the well-being obtained from our activities, and the capabilities (or real freedoms) to participate in society that we all seek. We deliberately use Amartya Sen's term "capability" here to describe the fundamental things we value as humans.

It is in this consideration of outputs that we break most distinctively from previous analyses of sharing. Most scholars of sharing agree that a central benefit of sharing is that it allows us utility without ownership. Recognizing and understanding this allows us to begin rethinking what we mean by needs, flourishing, well-being, and "the good life." People are interested in *goods* only for the *services* they provide—recognizing that in some cases one of those "services" might be the well-being or satisfaction

Sharing domain (what is being shared)	Concepts	Examples	Arena(s) where this may change norms
Material *Tangible*	Industrial ecology	Circular economy, recovery and recycling, glass and paper banks and collection, scrapyards	Relations to nature; forms of production, exchange, and consumption
Production facility	Collaborative production	Fab-labs, community energy, job sharing, open sourcing, credit unions, and crowdfunding	Forms of production, exchange, and consumption; labor processes
Product	Redistribution markets	Flea markets, charity shops, Freecycle, swapping and gifting platforms	Forms of production, exchange, and consumption
Service	Product service systems	Ridesharing, media streaming, fashion and toy rental, libraries	Forms of production, exchange, and consumption; labor processes; conduct of daily life; social relations between people
Experience	Collaborative lifestyles	Errand networks, peer to peer travel, couchsurfing, skillsharing	Conduct of daily life; conceptions of the world; social relations between people
Capability *Intangible*	Collective commons	The Internet, safe streets, participative politics, SOLEs, citizen's incomes	Conceptions of the world; social relations between people; institutional, legal and governmental arrangements

Figure 5.1
The Sharing Spectrum

derived from possession or ownership of something specific—a unique work of art for example—in the same way as we are interested in *material inputs* to the economy only because they can be transformed into *useful products*. As long as the *products* are as useful, delivering the *services* we want from them—recycling the concrete in our buildings, or the fiber in our magazines—is just as good as extracting new lime or timber.

Our thinking on transformation, however, goes further. Humans transform materials into products, and products into services. But in practice we believe that most people are only a little more interested in those services than they were in the raw materials. Most of us are much more interested in how those services transform into human-experienced well-being or happiness. And overarching all these transformations are our capabilities to live our lives in ways we have reason to value. Without the capabilities to transform them, neither materials nor goods nor services will necessarily deliver well-being or meet our needs. So our thinking about sharing should begin from the question of how it contributes to those capabilities, and in that light, sharing approaches and shared resources that more directly enhance capabilities for all are the most important to encourage.

Moreover, as we have shown in the preceding chapters, an exclusively economic, transactional focus on sharing is limiting not only in scope, but also in nature. If one assumes the exchanges of shared goods and services are purely economic, market transactions, then much of the potential value they bring to citizens and society is marginalized. The possibilities of sharing motivated by altruism, the pleasures of giving and helping, are missed, and equally, many potential gains or losses to social inclusion might be overlooked. And, as our spectrum illustrates, such intangible benefits are potentially of much greater significance than the tangible ones. Like an iceberg, we see the material impacts above the waterline, but without a change of perspective we miss the mass below. Critically, it is primarily in the sharing of experiences and capability that we can rediscover political solidarity across society, rather than individual or group interests.

The final column in figure 5.1 highlights arenas in which the adoption of sharing practices could be expected to challenge existing norms, or stimulate the emergence of new norms or values. Again we have drawn on David Harvey's description of the arenas in which the norms and practices of neoliberal capitalism shape our lives,[1] to point out how wide-ranging the impacts of the sharing paradigm could be, and the importance of extending our thinking to all the different spaces set out here.

We have explained the scope of our sharing paradigm and how, fundamentally, a sharing perspective could build our individual and collective capabilities—the fundamental things that by sharing, we can best ensure all humans can enjoy—so why isn't sharing already changing the world in this way? In the next section we identify objections to the sharing paradigm that create resistance and act as barriers to its development.

Objections to Sharing and Sources of Opposition

There are many possible objections to the sharing paradigm. Here we engage with five of the more common and far reaching. Our aim is partly to explain where we think they are mistaken, but mainly to identify legitimate concerns about sharing, and the obstacles they represent.

Objections to sharing can arise both from supporters of the current neoliberal system and its opponents. They can arise in beliefs about human nature, consumerism, and freedom of choice, about the power of vested interests and global competition, or about the scope of sharing to transform culture and society. While we highlight five objections or obstacles, these are not entirely discrete, with each bleeding into the others.

First we consider objections that suggest sharing cannot work as a paradigm because of human nature and the competitive capitalist systems we live in. We note how some practical concerns about the management of sharing may actually be expressions of such deeper, ideological objections.

Second, several commentators see neoliberal vested interests and incumbents beginning to co-opt and control sharing, making it exclusive, unfair, and meaningless. Such objectors see sharing as practical, but open to abuse; this distinguishes them from those in the first group that see similar challenges as reason to expect that sharing will not be widely adopted.

Third, there are some who might see sharing as both desirable and possible in the abstract, but in practice subject to grave dangers for personal liberties because of its implications for privacy and surveillance.

Fourth, others have posited that consumer culture is so dominant, and our identities so vested in consumer habits that there is no prospect of sharing alternatives reaching most people.

Fifth and finally, we look at the obstacles created by our related obsession—in policy and planning—with economic growth, and the idea that because many of the benefits of the sharing paradigm will not appear in standard economic accounts, it will not gain traction with decision makers.

Human Nature, and the Nature of Capitalism

Some opponents of sharing recognize the growing scale of the sharing economy but see it as something temporary and incapable of ever becoming transformative. Some even believe it quite literally goes against human nature. Such "sharing skeptics" believe that humans are not collaborative, but competitive and naturally distrustful of strangers—so sharing cities couldn't work. This is not just a position for radical libertarians. One doesn't need to see sharing as a communist plot to believe that economic models based on treating people as self-interested, rational beings have delivered well enough for there to be something in their underlying assumptions; or to hold that there are sound political reasons to support markets and individual rights as a basis for political freedoms.

We can respond to such objections on three levels. First, as we outlined back in chapter 2, biological and psychological science is coming to a consensus that humans are naturally and socioculturally evolved to share and collaborate *as well as* to compete; and that we can and do exhibit our capacity to collaborate across large groups not just with kin or local communities. Second, evidence suggests that we are natural reciprocators and will treat others in the ways they treat us (more on this later in this chapter). This means that even though our sharing nature might not always be expressed, we can be confident that in the right circumstances it could be. Third, sharing initiatives can be designed to build reputation, trust, and confidence, especially for the owners of shared products, making participation convenient and secure (also elaborated later in this chapter).

Survey evidence confirms that distrust of strangers may be a deterrent to sharing, although across the surveys of sharing we found[2] only Latitude[3] and NESTA[4] explicitly investigated barriers to sharing. In research for Cooperatives UK, Rachel Griffiths found that "residents understand the value of sharing but remain a little uncomfortable with sharing personal things with people outside the family."[5] Chris Diplock's Vancouver survey participants reported that they were more likely to share their own things with peers if they knew the borrower well (86 percent), and could trust that the lent item would be kept safe and looked after (85 percent).[6] Similarly, Latitude noted: "The most popularly cited barrier to sharing was having concerns about theft or damage to personal property."[7] It went on to highlight that, however, "88% of respondents claimed that they treat borrowed possessions well," and that sharing "communities that offer transparency (such as through open ratings and reviews) encourage good behavior and trust amongst members."[8] For NESTA, Kathleen Stokes and colleagues report findings from a UK focus group that also highlighted trust, security, and privacy as the main barriers to participation, alongside a lack of time to explore and experiment. They note: "To instil more confidence in such activities, people suggested a combination of government regulation and organisational transparency would be needed."[9]

In practice, whether using online reputational checks based in social networks, or evolved norms of behavior,[10] sharers have found ways to build trust and confidence to share. Arguably, modern technology is massively extending the reach of those tools, so sharing networks can scale up manyfold as strangers turn into real people with online identities and a verifiable reputation. Our desire to be able to choose who to share with, and how much, however, is not just about practically being able to avoid those we do not trust to play by the rules.[11] It is also an expression of our psychological desire to participate in the governance of the systems we use, so realizing the full potential of sharing will take more than transparency and reputational mechanisms.

It is possible to be concerned by the negative impacts of capitalist consumer markets *and* to see sharing as infeasible, or even a distraction. If all countries are trapped on a treadmill of debt—forcing us to pursue perpetual growth—and constrained by globalization to minimize constraints on markets, joining a "race to the bottom" in environmental and social standards, then measures to promote sharing might well seem infeasible or counterproductive. Worse, if such a system is controlled by anyone, it must be by global elites who have a vested interest in maintaining control, through any means possible.

Again this is a deliberate caricature, but aspects of this argument are certainly credible. To the extent that capitalism functions by maintaining labor reserves to keep labor costs down,[12] it would face an existential threat from a well-functioning informal or sharing economy allowing people to withdraw their labor from the system. Those benefiting from the existing circumstances—banking and corporate interests—would have strong incentives to resist it, using their financial and associated political leverage. Alternatively the same vested interests might—even unknowingly—use their cultural power to maintain demand for consumption habits and patterns that reproduce class inequality, alienation, and high levels of conspicuous consumption, leaving sharing as inevitably only a niche practice. Other vested interests, namely incumbents, clearly intervene in ways that marginalize sharing competition through legal challenges, or through co-option or purchase.

But such objections rely on an assumption that the conventional system can continue, with no constraints to it from environmental or other factors. Even setting aside the underlying challenge of environmental sustainability, the capacity of the economy to generate continued growth has been cast into question by the crises of recent years, which were not random, but predictable—and predicted.[13]

In addition the sharing paradigm includes many practices that can be adopted—and deliver positive financial returns—within the current economic model, but which would simultaneously shift norms and values away from those that the model relies upon. Such approaches—like the self-organized learning environments we saw in chapter 2—fit into the model of social change described by Erik Olin Wright as "symbiotic,"[14] and by one of us as "subversive":[15] they are less confrontational than revolutionary or "ruptural" approaches that aim to "smash the system," and engage more directly with the mainstream than "re-invention" or "interstitial" approaches that "ignore the system." Later in the chapter we sketch out ways in which the sharing paradigm as a whole can be seen as such a transformative strategy—socially emancipatory, yet able to win support from powerful cultural, political, and economic interests.[16]

Paradise Lost (Sharing Co-opted)

If "sharing skeptics" argue that it can't work, sharing "pessimists" fear that it might work all too well, but in forms controlled and co-opted to further the interests of elites.

Typically, these concerns are raised by those on the political left fearful of the influence of venture capital and libertarian technologists—directing sharing processes so as to exploit collective reserves of labor, knowledge, and resources to swell their profits. The blogger and author Tom Slee argues that as sharing companies go mainstream they are extending casual employment, deepening the so-called precariat and letting Silicon Valley "billionaires" override city-level democracy for the sake of return on investment.[17] He sees TaskRabbit, for example, becoming a temp agency driving insecure jobs—lacking social protections, standards, and insurance—into all sorts of activities. And he argues that larger enterprises are using sharing as a smokescreen to push for further deregulation.

There are reasons to agree that all might not be exactly what it appears to be in the sharing economy, and in particular that a specific interpretation of collaborative consumption could benefit existing elites. Pierre Bourdieu's term "habitus" describes the socialized norms or tendencies that guide behavior and thinking, so as to exert and legitimize power. Bourdieu argued that cultural capital—now largely dominated by consumerism—plays a central role in societal power relations in the modern world. It permits "a non-economic form of domination and hierarchy, as classes distinguish themselves through taste."[18] The shift from material to cultural and symbolic forms of capital hides the causes of inequality—as the elite define themselves through tasteful consumption and positional goods (which can include the experiences of the sharing economy as well as conventional "bling").

So although "pioneer" groups in modern society appear to be adopting less materialistic values and ethics that embrace some forms of sharing, we should be alert to the likelihood that this shift may materialize in ways that maintain distinction and privilege, rather than in ways that support justice and democracy. In this sense, collaborative consumption within middle-class clubs, and co-production as a hobby for wealthy downshifters, are warning signs.

Another expression of this objection arises from a concern that sharing—as an alternative form of consumption—reflects and engages too narrowly with conventional white, middle-class consumption patterns in the rich world. Instead, consumption can form part of a deliberate and understandable effort to overcome discrimination. For instance, the sociologists Michèle Lamont and Virág Molnár stress the use of consumption practices to define positive ethnic and racial identities: "Consumption is uniquely important for blacks in gaining social membership. Their experience with racism makes the issue of membership particularly salient, and consuming is a democratically available way of affirming insertion in mainstream society."[19]

In this light, an emphasis on sharing as a behavioral counter to the ills of consumerism could be argued to discriminate against those whose consumption practices or motives differ from the cultural norm. But this too can be better interpreted from the perspective of habitus. By defining identity and status through consumption practices, elites distract attention from the political task of identifying discrimination and misrecognition, and of challenging the power relations that maintain them.

This suggests that we should not see the exclusion of disadvantaged groups from sharing as simply an oversight, or a failure to target behavior change at those groups. Indeed we should not treat sharing as simply a matter of behavior change—which would leave structures of power unaltered. We return to these challenges later in this chapter as we discuss the opportunities to use sharing to deliver wider cultural and political transformation.

Breaching the Castle Walls: Sharing as a Threat to Privacy and Freedom

An alternative form of sharing pessimism shifts focus from the commercial benefits of co-opting sharing to the implications of high levels of sharing for our privacy and civil liberties.

At the personal level people might object to the loss of privacy implied by sharing their homes or cars—especially if they feel obliged to share by economic pressures or social norms. Sharing places us in circumstances where we may be vulnerable to criticism of our private behaviors or attitudes. This can be a serious deterrent especially where those behaviors or attitudes are important factors in identity formation. For instance willingness to share one's home with couchsurfers—exposing much about oneself, would obviously be influenced. Even more extremely, to participate in cohousing might expose to scrutiny and criticism not only one's taste in furnishing, but also one's approaches to childcare.

The explosion of sharing facilitated by smart technologies and online applications also vastly widens the scope of concerns about breaches of privacy. Legitimate concerns can arise incidentally—if data is hacked, for example—or as a result of deliberate sale of sharing-related data for marketing, or through its use in state surveillance activities.

From a liberal perspective the implications of some sharing for Internet freedoms and civil liberties can be particularly disturbing. The idea that sharing—notably of vehicles—might open one up to greater surveillance and tracking of one's movements will be repugnant to many (if not necessarily a great change in practice). Maintaining net neutrality, while using the potential of new technologies to minimize impacts on civil liberties and to maximize data privacy, is likely to be critical to delivering sharing in a just manner, and realizing the synergies between the sharing economy and sharing politics. On the other hand, if governments and corporations collaborate to enable intrusive control over mobile and web technologies for security, and surveillance purposes, this would dramatically undermine the value of these platforms for sharing—whether economic or political.

The potential for breaches of privacy—from ridesharing or bikesharing records for example—are real and legitimate concerns. Yet they can also be transformed into ideological objections in defense of individualism as a political doctrine. In a provocative and polemical piece, Milo Yiannopoulos—the founder of a technology gossip website called the Kernel—recognizes and resists shifting norms of privacy in the sharing economy:

> It's a terrifying development: just as we were getting over having our private lives public and permanently online, suddenly the demands made against our privacy are encroaching on our physical space and our possessions too. This is more intrusive and more oppressive than any snooping government.[20]

Yiannopoulos goes on to complain of feeling forced to share, objecting to:

> The social pressure [sharing] generates ... being made to feel selfish or wicked, just because you don't feel like throwing open your home or giving away everything

you've worked hard to achieve. Call me selfish, but I work hard to provide nice things for my family and no, I don't want strangers touching them.[21]

It is slightly ironic that other libertarians complain rather of being prevented from "doing what they like" with their possessions—such as renting their apartments out on Airbnb—because of restrictive regulations and "unfair" zoning laws.[22] But the central question here is whether sharing can be governed in ways that protect market freedoms—like being able to choose whether to share or not—and civil liberties. Along with new forms of privacy protection online, we see participative ownership and governance—revisited later in the chapter—as critical in these respects, especially if sharing politics is also to be further enabled and stimulated.

Consumer Cornucopia: Sharing as a Threat to Identity

It might appear that human behavior in markets demonstrates that—whatever we might say—in reality people love their stuff, and capitalist, consumer markets deliver happiness. Fundamentalist economists even argue that there is no other way to understand human needs than as expressed by what we choose in markets. From this perspective, sharing has been forced on people by recession, and will decline again once growth resumes (and among those holding this position there is no doubt that growth will resume).

It is true that in contemporary society—at least in the global North—much of our sense of social standing and belonging comes from what we consume, even though those same consumption habits and patterns are also central to the reproduction of class inequality, alienation, and power.[23] Contemporary consumerism—with its emphasis on "bling" characterized by luxury, expensiveness, exclusivity, rarity, uniqueness and distinction[24]—has become a comparative or competitive process in which individuals try to keep up with the consumptive practices of the eponymous Joneses. Preference for access and utility over ownership may lead to some favoring sharing economy models, but this might even be because such models allow for more rapid turnover of identity-affirming consumption goods in a fast-moving postmodern world.[25]

Arguably, for many, while "ownership allows sharing, feelings of possessiveness and attachment towards the things we own or possess discourage sharing."[26] Yiannopoulos explains this from the perspective of one trapped within a consumerist identity or "extended self,"[27] clearly feeling his identity threatened by perceived exhortation to change:

The sharing economy is … emasculating, dispiriting and demotivating, because property rights are the basis of our society and the sharing economy fails to respect how much of our identity we invest into the things we buy. When we make big purchases, we are engaging in a process of consumer choice that advertises to the world who we think we are, what our aspirations are and how we would like to be considered by others.

Sharing the results of those purchases devalues our personal investment to the point of irrelevance and robs brands and consumer products of the ability to reflect something about their owners' identities.[28]

Psychologically speaking, we should not be surprised to see sharing appearing more difficult where people are insecure in their identities. Unpacking the implications of Yiannopoulos's tirade though, we see something much more encouraging. Put in context, his fears seem to be that if sharing takes hold, wealth and the trappings of wealth and "good taste" might no longer be used to express status and exert power in society. In other words, in countries that have come to resemble wealthy oligarchies,[29] power might be returned from wealth to the people. Democracy might actually mean something again.

We rebut such objections not only on moral grounds but for two connected empirical reasons. First, there is abundant cross-cultural evidence that higher consumption does not lead to greater happiness and it may be connected with certain increasingly prevalent forms of mental and physical ill-health, together with levels of environmental degradation that threaten both the long-term stability of the economy and well-being all around the world.[30] Second, as we discuss later in this chapter, there are good reasons to see our desires for "stuff" as merely a culturally specific expression of more fundamental needs, and to believe that well-designed sharing systems could not only meet those needs just as effectively, but also fulfill a much broader set of needs for identity and belonging.

Growth Is All: Challenging Growth-Dependent Economic Ideologies

The opposite side of the coin of consumerism is growth. Neoliberal economists tend to see the future success of cities as depending on continued growth achieved by attracting footloose capital to invest in new development and infrastructure, and on attracting skilled and entrepreneurial people by offering masses of consumer and cultural choice. The sharing paradigm challenges both of these assumptions. The University College London professor Yvonne Rydin skillfully critiques the current growth-oriented British planning system, and its failure to deliver well-being, tackle inequalities, and promote just sustainabilities.[31] Instead she highlights the potential for approaches that embody aspects of the sharing paradigm, such as community land trusts "that vest ownership more directly in local communities and enable them to build control and a stewardship role," self-build "using community labor or 'sweat equity,'" use vacant property and land, and value community ownership and management of assets.[32]

One further major political obstacle to sharing is that the benefits do not appear in standard economic accounts, especially where no money changes hands. In other words sharing might support human development, but not "economic growth" (although, as we have argued elsewhere, a healthier conventional economy may be an indirect consequence of more sharing). When politicians regularly measure the success of policy in terms of growth rates and formal employment figures, activities that generate human well-being through informal labor and non-monetized exchanges are unlikely to attract much political capital. And within the sharing sector, commercial models are likely to get more attention than sociocultural ones, regardless of their relative overall value.

Such fixation on conventional growth is seen on both the left and right wings of politics, as exemplified by Moira Herbst, writing in the *Guardian*:

> One big problem with claims that the "sharing economy" can lead the way out of our economic morass is that proponents often advocate less consumption. How can that be a solution for an economy that—for better or worse—is fueled by consumer spending?[33]

Sen demonstrates that market activities, however, are a poor proxy for well-being: "Important matters such as mortality, morbidity, education, liberties and recognized rights … get—implicitly—… zero direct weight."[34] African Americans might enjoy higher incomes than the Chinese, or than Keralan Indians, but they have shorter lives. Relative disadvantage within the US society does more harm than is gained from higher incomes. Discrimination and misrecognition cause both physical and psychological damage. In the sharing paradigm, however, we recognize and include factors—such as social capital and interpersonal relations—which get zero weighting in the conventional economy. This means that by supporting and managing the sharing paradigm, cities are likely to do a better job in delivering the well-being citizens want. Moreover, the sharing paradigm helps enable the political and social processes of public discussion and evaluation that are central to increasing justice.

Much of the value of sharing is indeed missing from conventional measures of economic activity. Social production is only accounted for when marketized.[35] And even in commercially flavored sharing, only a fraction of the overall value appears in economic transactions (even assuming any payments involved are declared for tax). This lends weight to the case for new indicators of well-being[36] at both national and city scales.

The Contested Politics of Sharing

The objections and obstacles discussed above demonstrate that the territory of sharing is inherently political. And it is getting more so. Over the past decade mediated sharing has grown rapidly to the point where it might be perceived as a threat to incumbents in both business and politics. The ways in which that engagement plays out—between the disruptive innovation of sharing and the conventional, incumbent approaches of organizing markets and public services—will be critical.

A bleak analysis might suggest that collective consumption is already being co-opted, commercialized, and bought-up by the business mainstream, even while it is being resisted through regulatory and legal means. Similarly the politics of sharing might be dominated by disowning responsibility for public services on the one hand and facilitating exclusive sharing strategies within wealthy communities on the other. The proposal to construct a shared "citadel" around an arms factory in Idaho[37] is perhaps a mad-cap extreme, but gated communities, within which private facilities and resources are shared, are becoming more common everywhere. For example, in Buenos Aires in 2008, there were more than 500 gated communities sharing services and facilities—up by 500 percent in two decades. The spread of gated communities however increases both "inequality in the city" and "feelings of insecurity" for those inside the gates.[38] At an even larger scale, many plans for smart cities built from scratch exhibit the same tendencies, feeding a neoliberal utopian fantasy of entire exclusive cities built on a shared high-tech platform.[39]

Co-option of sharing approaches is seen in other areas too, reflecting a wider political battle. The political economists Jathan Sadowski and Paul Manson argue that, while the technologies of fab-labs and decentralized co-production may offer democratizing potential: "The maker movement is born out of, and contributes to, the individualistic, market-based society."[40]

Firmly caught in a post-political trap, the maker movement portrays "a kind of naively apolitical, techno-economic, capitalist utopia that thrives on individualistic values and discounts the very public contributions to science, infrastructure and society that enable them to do what they do."[41] Its refrain of "Love the 'makers,' deride the 'takers,'" echoes the divisive politics of "strivers" and "scroungers" in the UK, and underpins a disdain for welfare and other collective solutions, despite the "aura of grassroots community building and self-empowerment" that surrounds "hacker-spaces," and "makerspaces."[42] "No" say Sadowski and Manson:

> The maker movement is not ushering in a decentralized, noncorporate, democratized world. Rather ... [it] is serving as a convenient veneer, which hides the gears of corporate capitalism that have been turning all along. Instead of manufacturing jobs, we get manufacturing as a hobby.[43]

In these ways it is a microcosm of the politics of Silicon Valley, and a very different expression of the same widespread distrust of increasingly corrupt and financialized politics highlighted by Occupy and other protests. The political scientist David Runciman describes the problem: "Technology has the power to make politics seem obsolete. The speed of change leaves government looking slow, cumbersome, unwieldy and often irrelevant. ... Technology isn't seen as a way of doing politics better. It's seen as a way of bypassing politics altogether."[44] And, Runciman says, despite

> countless local experiments around the world in how to use the internet to promote more accountable and efficient government ... [such as] online town hall meetings, interactive consultation exercises, [and] micro-referendums, ... government ... fails to pick up on what works in time to take advantage of it. ... These failures help breed contempt for politicians not only among citizens but from the tech industry, which often assumes that government is simply an obstacle to be overcome, an analogue annoyance in a digital world.[45]

Other critics are similarly disturbed by high-tech sharing startups that present themselves as innovators and problem solvers, but identify and address only those problems that can be solved through technology.[46] And such startups typically ignore the ways in which state investments in shared infrastructure underpin those technological developments anyway.[47]

Like many critiques of the sharing economy, Sadowski and Manson's challenges raise very real concerns but are framed as a false dichotomy. It is not necessarily a question of embracing sharing *or* challenging the exclusionary and commodifying traits of conventional economic models. We can do both, as long as we also resist the recruitment of sharing to individualist, behavior change–oriented, post-political ideologies. Instead we must actively design sharing to be inclusive and just, and to stimulate collective values, while also promoting sharing in political spaces and as a political strategy in the form of genuinely participatory democracy.

Otherwise sharing could become mired in anti-politics, pursued where it furthers individual freedoms, but set in opposition to government and collective support; turned into a new opportunity for those with the skills, resources, and privilege, just another new dimension of exclusion for those without. This would be sharing without justice. The seeds for such a perversion of sharing are all around us. Hobby manufacturing; personalized fashion bags from Etsy; car- and bikeshares with fewer vehicles stationed in poor areas; white hosts on Airbnb getting higher returns; sharing programs that exclude anyone without a credit card; community-building websites used to target consumerist marketing; Incomplete Streets and farmers markets that support only the dominant food culture; and self-provisioning for the wealthy elite who have paid off their mortgages and downshifted. The list goes on.

Such outcomes are a travesty of the true possibilities of sharing. Many better opportunities exist to enhance sharing and ensure fairness. Our complaint is not about the fundamentals of sharing models, but with those emanations that suppress the values of fairness—the inclusion and collaboration that underpin the sharing paradigm as we have characterized it.

In this section we have explored five types of objection to sharing: that it (1) goes against human nature; (2) threatens our personal privacy; (3) undermines our ability to express ourselves fully; or alternatively that it will (4) harm the economy; or (5) be so completely co-opted by commercial interests that it will lead to social exploitation. In responding to these we have highlighted the importance of broadening participation and ownership; rebuilding trust to enable cooperation to flourish; going beyond simple behavior change to redistribute power, and to establish new ways of measuring well-being and meeting needs (including our need for identity); and finding ways of promoting change that do not simply divide existing social interests in conventional ways, but offer benefits that cut across established interests to lock in more just and sustainable approaches.

In the remainder of this chapter we develop our sharing paradigm as a combined cultural and political project. Taking all these objections, obstacles, and responses together suggests that to deliver transformation, sharing must be able to challenge the hegemony of consumerist identities, rebuild and extend trust across the barriers of difference, and spread and lock in prosocial behavior and values. It must also do this in ways that enable sociocultural and political change at the city scale. The following sections focus on these themes, before turning to the smart policy and regulatory measures that could help deliver sharing in such ways.

The Sharing Paradigm as Cultural Transformation

In this section we explore how identities, norms, and culture change through the interplay of practice, dialogue, and challenge; we apply that understanding to the sharing paradigm, with particular attention to the findings of "reciprocity theory." Reciprocity theory suggests that humans naturally respond to cues in language, culture, and practice by reciprocating with behavior that expresses similar values. This is just one of several potential positive feedbacks that can accelerate cultural change at, and across, different scales: others exist between empathy and identity; between norms and expectations; between trust and social capital; between imitability and network growth; and between shifts in cultural norms and political change. We will briefly examine all of these in the pages ahead.

Identity is central to cultural change, both in respect of the psychological processes we use to construct our individual identities, and the cultural groups with whom we identify. As we have seen, modern identities in wealthy societies are strongly bound up with consumerism. We use our purchases and our consumption practices to signal group membership and identity, to demonstrate taste, and to reassure ourselves (and others) of our status; in other words, they address deep-rooted psychological needs—albeit in culturally specific ways. Yet this has made us vulnerable to the darker sides of consumerism. The writer and campaigner Alastair McIntosh describes how motivational manipulation of advertising and marketing has engendered an addictive consumer mentality.[48] As with other addictions, we continue to consume even when it degrades our personal or collective well-being, by damaging our mental health or our natural environment, for example. Insofar as consumerism and ownership of consumer goods address deep-rooted psychological and cultural needs, people might resist sharing, even where it would enhance well-being conveniently and cheaply (as we saw from Yiannopoulos above).

We believe there is a potential for virtuous feedback, however, if we can trigger a shift to sharing. The more we associate our identities with the relationships embodied in sharing behavior, rather than with material goods, so the more our psychological needs are instead fulfilled by sharing (and in ways that enhance well-being), encouraging us to share even more. Social capital derived from sharing grows with use. We shall see below, that as our behavior changes so do our opinions and the values we express and promote as sociocultural norms. This reinforces both the prevalence and practicality of sharing practice as its participants proliferate.

Below we first explain reciprocity theory, and then outline the possibility that sharing can challenge the hegemonic power of consumer culture in a bottom-up redefinition of consumption that is rooted in collaborative co-production of services and products supplying basic needs. We subsequently explore how such changes in behavior and norms relate to changing values (or, more precisely, changing expression and emphasis of values) with particular reference to trust; we then explain how such a cultural meme could spread rapidly, even across commercially motivated networks, especially if supported through participatory politics.

Paying It Forward: Reciprocity Theory, Values, and Social Behavior

Both models and evidence show that "as kind, sharing, and reciprocal behavior increases in society, so does the tendency to trust others, reciprocate, and behave pro-socially."[49] In other words, "By practicing sharing people come to value it more, or come to learn to trust other participants."[50]

In the conventional economic model, we are treated as if we were all selfish, individualist welfare maximizers (*homo economicus*), and moreover, we are told that we are legally no different from corporations built around the same economic theory. It is unsurprising that we increasingly act in the same way. In sharing activities, we are treated as humans, and as collaborators with one another. Every time we share, we build trust, every time we give of ourselves we learn to expect that others will reciprocate—if not to us, then to someone else—and gain confidence that "what goes around, comes around"; in this way social capital is accumulated.

The ways in which sharing contributes to social capital vary between flavors, and with the social and cultural context. Most importantly, mediated sharing for financial or commercial ends can undermine functioning social capital in cultures where it remains strong, and yet begin to rebuild it in cultures where it is weak or lacking. Enhanced sociocultural sharing can strengthen social capital anywhere, but its uptake in individualistic market cultures may prove very limited.

Why is this? As we explained in chapter 2 (in "The Social and Evolutionary Roots of Sharing"), most people are natural reciprocators. However, if "free-riders" (following the creed of *homo economicus*) remain in the population, the critical issue is which behaviors are most salient. In social interactions and economic transactions reciprocators can reproduce either collaboration and trust, or distrust and selfishness. In such social systems there are at least two potentially stable states, typically separated by a "tipping point": below the tipping point distrust dominates, and reciprocating behavior runs in a vicious circle, until only a handful of hardcore altruists remain active. Above the tipping point, trust dominates, and reciprocating behavior drives a virtuous cycle, until only the most recidivist individualists fail to cooperate.

If the system is in a "vicious mode," the risks of sharing—that shared goods might be damaged or simply not returned, or that voluntary efforts or investments will not be reciprocated but exploited—appear high. Cultural norms then militate against sociocultural sharing beyond families and close friends. On the other hand, mediated sharing, in which people can see the reputation of potential collaborators in advance, can work well in such circumstances, even if it incorporates financial incentives for participation.

However, if the social system is already in a "virtuous mode," providing checks on reputation and financial incentives to share would be a negative cue, indicating to people that there is reason to distrust, and indeed that they themselves are not trusted and should disengage. So the intrusion of commercial and mediated sharing models into already strong sharing cultures is a threat, but in atomized, commercial cultures the same models can start to rebuild social capital.

Extrinsic motivations—such as monetary payment—can displace intrinsic motivations for participation because they impair both self-determination and self-esteem. As Yochai Benkler says:

> They cause an individual to feel that his internal motivation is rejected, not valued. … Being offered money to do something you know you "ought" to do, and that self-respecting members of society do, implies that the offeror believes that you are not a well-adjusted human being or an equally respectable member of society.[51]

This crowding-out effect is confirmed in a review of 41 experiments in the field by the economist Samuel Bowles.[52] The large majority of them showed evidence of displacement, and the remaining handful suggested complementarity between incentives and moral motivations. Bowles concludes that these behavioral experiments "suggest that economic incentives may be counterproductive when they signal that selfishness is an appropriate response; [and] constitute a learning environment through which over time people come to adopt more self-interested motivations."[53]

The effects spill over into politics and political behaviors, with particularly serious risks in monetization where public services or other collective public goods or commons are concerned.[54] Moreover, tax evasion becomes more likely when people believe others are cheating, and falls when people perceive a social norm of compliance, and fear exposure and moral condemnation. Dan Kahan notes:

> These are exactly the factors one would expect to influence tax compliance were individuals behaving like moral and emotional reciprocators. A strong reciprocator wants to understand herself and be understood by others as fair, but she loathes being taken advantage of.[55]

Kahan suggests that tax regulators should "carefully select cases to nourish the perception that evaders are deviants, not normal citizens."[56]

High-profile prosecutions of celebrity and wealthy tax avoiders, and indeed of corporations too (and closing corporate tax loopholes) are likely to encourage ordinary citizens to pay up, by emphasizing justice in the system and strengthening the legitimacy of the institutions involved.

This implies that willingness to support higher rates of taxes to enhance social equality and environmental protection should rise where there is inclusive politics and visible evidence of cooperation and sharing—as we saw in Amsterdam. This is what we might call the "granddaddy norm" of society, as Benkler puts it: "When we believe that the systems we inhabit treat us fairly, we are willing to cooperate more effectively."[57] Put more broadly, unless—unlike Margaret Thatcher—we share a belief in *society*, we are unlikely to indulge in *prosocial behavior* at all.

Michael Sandel is similarly concerned about the effects of commercial markets on morals. "It is likely," he writes, "that a decline in the spirit of altruism in one sphere of human activities will be accompanied by similar changes in attitudes, motives and relationships in other spheres."[58] Astonishingly, some economists challenge such findings by further extending economic and market norms: to the very idea of altruism itself. For example, Lawrence Summers claimed:

> We all have only so much altruism in us. Economists like me think of altruism as a valuable and rare good that needs conserving. Far better to conserve it by designing a system in which people's wants will be satisfied by individuals being selfish, and saving that altruism for our families, our friends and the many social problems in this world that markets cannot solve.[59]

Our reading of the evolutionary literature convinces us that it is much more likely that "our capacity for love and benevolence is not depleted with use, but enlarged with practice"[60] and as the philosopher Jean-Jacques Rousseau's argued, "civic virtue" is built up by the "strenuous practice of citizenship."[61]

Trust-building mechanisms used by sharing platforms might therefore be expected to encourage reciprocal cooperation and raise public involvement in other spheres. In other words, private sharing through web platforms can be expected to stimulate more prosocial behavior in the public realm, helping to co-create a thriving urban commons. But because of the contextual and cultural factors set out above we can also conclude that in most circumstances, cooperatively owned commercial platforms will better stimulate "collective norms" than equity financed or privately owned commercial platforms. In Amsterdam, The Circle Economy is a workers cooperative helping businesses and other organizations apply circular economy solutions (including things like product-sharing systems or co-production approaches) in their activities; and to help businesses share such solutions with others.[62] It naturally spreads cooperative norms alongside its advice on sharing practice, but much more could be done with strategic support from city authorities. Targeted support from cities for cooperative models or even an active conversion strategy would be sound policy in this respect.

So we see the spread of mediated sharing as often valuable, even where it is commercialized. The introduction of sharing alternatives in conventional markets acts to convert some existing market spaces into places where cooperation is essential, and thus where norms of reciprocity and mutuality are reinforced. This is—ironically—partly driven by a competitive mechanism, as the value of the platforms both to users and owners increases with the number and density of participants. So the competition between even commercial mediated platforms and conventional markets can spread positive norms. However it also—as we discuss elsewhere—creates a risk that commercial sharing approaches might exploit their contributors in an effort to compete with conventional market alternatives. This further strengthens the case for cooperative models that ensure the interests of those contributing labor are properly protected.

Reciprocity, Corporations, and *Homo Economicus*

We must, however, offer one cautionary note about reciprocity theory. Its consequences depend on the fact that it applies to interactions between *people*. It clearly applies in sharing activities that are based on direct interactions between users (even if mediated), but will have more limited impacts in commercial sharing economy interactions, especially those that focus on the relationship between the platform and the user (think Zipcar's B2P model).

The same holds for the potential for trust building as a result of sharing. Sharing interactions build trust in interpersonal relationships between two equals who can empathize with each other. Trust between two individuals is therefore reciprocal and reciprocated, but trust between an individual and an organization does not take the same form. This is partly because the average individual can be harmed by a corporate breach of trust much more than the average individual can harm a company. In this respect we should be acutely aware that when institutions or corporations actively seek to build trust they are also building power in society.

The potential of reciprocity to change behavior and even influence values therefore does not reduce one jot the case for an explicit framework of human rights, and interventions to guarantee capabilities and accountability, especially where the threats to those arise largely in the actions of institutions and corporations. And it would be a grave mistake—in our view—to attempt to apply reciprocity theory to relations with corporate persons, as they are not subject to the same cognitive and psychological processes as human individuals, but rather to legal or institutional duties to profit maximization.

Furthermore, we would argue that the problematic tendency of policy and markets to treat real humans as selfish, individual welfare maximizers (*homo economicus*) is exacerbated by the oxymoron of "corporate personhood." Where the public see corporations getting away with environmental pollution, paying below living wages and bribing politicians with campaign contributions, is it any surprise that they believe such behaviors to be normal and widespread among people, or that moral appeals by politicians to "do your bit" by saving energy, giving to charity, and so forth are so often ignored? Multiple aspects of reciprocity theory are clearly in play here: the behavior of politicians and companies both suggests that abuse is widespread, and reduces the legitimacy of key institutions in society to offer incentives—whether regulatory, financial, or exhortational—for better behavior.

We would expect many of the problems that reciprocity theory wrestles with to be insoluble until Western society comes to terms with the fact that core economic institutions and policies consistently assume *homo economicus* (rather than *homo sapiens*), and tend to value only what is done and measured in markets. In this respect, the sharing paradigm offers a critical tool to begin redefining this fundamental false caricature of humanity, and instead treat people as responsible members of communities. This is not just a cognitive framing effect that temporarily shifts opinions, but a sustained psychological and cultural impact, changing behavior and values, with serious implications for the design of policies and institutions, for the media and marketing and much, much more.

Challenging Consumerism's Cultural Hegemony

Consumerism too treats us as *homo economicus*, even though the tools that marketing uses to create and promote our consuming passions are far more subtle and complex than the simple theory might suggest. The cultural hegemony of consumerism is a problem, not only for the environment, but also for justice and democracy. In sharing activities we saw how particular cultural constructions of taste can exclude ethnic minorities or people on low incomes—especially in more sociocultural and communal flavors of sharing. And despite the Occupy movement and books such as Richard

Wilkinson and Kate Pickett's *Spirit Level*[63] and Thomas Piketty's *Capital in the 20th Century*,[64] an awareness of the dangers of the increasing gap between rich and poor has not been matched by public or political understanding of the role of consumerism. As we saw earlier, in the section "Paradise Lost: Sharing Co-opted," consumerist culture sustains itself in ways that simultaneously maintain and obscure the causes of inequality.

Moreover, consumerism reduces empathy and desensitizes people to their social responsibilities:

> Studies in positive-psychology, theories of human needs and well-being economics demonstrate that the "perfect consumer," motivated by extrinsic consumerist values, has a lower well-being than others, lowers the well-being of those around them, has a higher than average environmental footprint and is more closed than others to pro-social and environmental behavior change messaging.[65]

Tackling this isn't just a matter of changing consumption ethics. Lindsey Carfagna and colleagues at Boston College and New York University highlight how aspects of sharing feature in new consumption ethics, including "minimalist" living, describing the potential for what they call a new "eco-habitus."[66] But if sharing becomes part of a new habitus of the cultural elites, then we might expect it to be "out of bounds" to poorer groups in society, and instead serve to celebrate minimalist style, international mobility throughout the "global city," and artistic and cultural discrimination. Its celebrants will be already wealthy downshifters, who, having paid off their student loans and mortgages can afford to indulge in slow food, self-provisioning, and creative hobbies. The place of sharing in such an emerging eco-habitus is contradictory—it can obviously influence values and norms within the groups adopting it, yet at the same time might help sustain, rather than delegitimize, the power of existing elites.

There may not even be an instrumental argument for supporting the emergence of "sustainable consumption" in such groups. Elite groups cannot be expected to lead the widespread adoption of similar practices by "fashion."[67] Rather, the mainstream needs interventions that directly target needs for basic services and economic stability through shared services and resources in the face of economic challenges. Moreover, the quest to recognize sociocultural sharing, co-production in the core economy, and even sharing on the edge of legitimacy all suggest ways in which existing habitus as cultural power might be confronted, exposed, and even dismantled in genuine sharing cities. We are not arguing for trickle-down from a newly fashionable niche of "sustainable consumption" but instead for a bottom-up—simultaneously *countercultural* and *intercultural*—redefinition of consumption that is a collaborative, shared, co-production of services and products supplying fundamental needs. In turn this will challenge identities rooted in consumerism, not with different products or services, but with different activities and—fundamentally—different relationships and capabilities.

Empathy, Trust, and Sharing

Consumerism discourages trust and empathy, encouraging us instead to rely on external markers of identity and distinguish ourselves from others through our consumption choices. Even consumerist markers of group identity need not imply trust. Commercial capitalism and state socialism alike tend to replace person-to-person trust with trust in institutions—of the market and state respectively. And growing inequality damages all forms of trust, in institutions as well as in other people.[68] But interpersonal trust is both the social glue that enables sharing to function in practice and a lubricant for the spread of collective values through sharing. When trust and empathy extend, so does the potential for collaborative action and the reconstruction of society.

The practice of sharing—even the unconscious habits of sharing in public spaces and facilities—builds trust between members of a sharing community, even if in mediated spaces, reputational markers and guarantees might be needed to facilitate that process. On web-based platforms, user reviews can provide measures of good reputation and allow trusting, altruistic behavior to dominate in a community. Trust in strangers, say Rachel Botsman and Roo Rogers, is not an abstract unfounded naivety, but a product of the right circumstances and transparency.[69] Rating systems online provide the equivalent of personal recommendations—and in a world where only 14 percent of us trust advertisers, but 78 percent trust peer recommendation[70]—this is incredibly powerful.

The philosopher Onora O'Neill emphasizes that to build trust, it is not trust per se we need to see, but *evidence of trustworthiness*.[71] This can be achieved in part through ratings systems, but also demands accountability and openness. In business, too, trust is "the expectation that another party will be honest, considerate, accountable and open."[72] O'Neill similarly highlights that trustworthiness is rooted in open transactions: "If you make yourself vulnerable to the other party, then that is very good evidence that you are trustworthy and you have confidence in what you are saying."[73]

As the sharing economy has boomed in the US, it appears to be growing trust. Jason Tanz, a writer for *Wired*, argues—perhaps a little effusively:

Many of these companies have us engaging in behaviors that would have seemed unthinkably foolhardy as recently as five years ago. We are hopping into strangers' cars (Lyft, Sidecar, Uber), welcoming them into our spare rooms (Airbnb), dropping our dogs off at their houses (DogVacay, Rover), and eating food in their dining rooms (Feastly). ... We are entrusting complete strangers with our most valuable possessions, our personal experiences—and our very lives. In the process, we are entering a new era of Internet-enabled intimacy.[74]

Some sharing platforms—Freecycle and Craigslist for example—seem to have generated trust and effective community-level governance in an emergent fashion, not dissimilar to that described for commons management by Elinor Ostrom.[75] The members and groups involved moderate and self-police the community. In these "light-touch" systems it seems that it is the "very lack of helpful features that signals us to activate our own methods of reassurance."[76] And they work, even though the conventional, physical reputational signals we rely upon in real-world interactions are absent.

Botsman and Rogers suggest that this is because we can empathize with our online partners and so behave in ways we consider to be fair to them too, trusting that "we are part of a system of durable relationships that could benefit us in the future." This "shadow of the future" establishes "clear incentives for honesty and trust."[77] Trust does not rely only on rational reasoning about the future, however. As we saw in chapter 2, our evolved tendencies for cooperation—which we might call "the shadow of our past"—stimulate exactly the same type of behavior through social and cultural norms.

Yet it is easy to see how over-commercialization of sharing might undermine these effects by instead simulating the anonymity of market interactions. Markets work because we trust the institutions that embody and surround them, not because we trust the specific individuals involved. Sharing requires trust in the individuals concerned, and web-based reputation tools enable this. But a push for complete security, with powerful reputational tools, insurance and so forth reduces the need for such personal trust. Research on the CouchSurfing website has shown that the result is also to reduce the strength and durability of the social relations stimulated among couchsurfers and their hosts.[78]

Yet many sharing sites, particularly those of a commercial mediated flavor, have invested heavily in highly developed systems of identity confirmation and reputation management to reassure and safeguard users more familiar with conventional commercial exchanges. For instance:

Lyft riders must link their account to their Facebook profile; their photo pops up ... when they request a ride. Every rider has been rated by their previous Lyft drivers. ...

And they have to register with a credit card, so the ride is guaranteed to be paid for before they even get into [the] car.[79]

Similarly, eBay guarantees every purchase. Airbnb provides substantial insurance for its hosts, as well as providing a platform for hosts and guests to communicate with one another. Airbnb says the system is highly successful; of 6 million guests in 2013, the company paid out only 700 host claims for loss or damage.[80] Airbnb's systems also track behaviors so as to sniff out efforts to build up false reviews.

The more commercial the platform, it seems, the more elaborate the procedures. But trust is not just a matter of procedures and institutions. In 2012 the car-sharing platform RelayRides switched—for financial reasons—from using on-board technology to face-to-face handovers of keys. The results were remarkable:

The face-to-face meeting caused renters to take better care of the cars—and it made the experience better for both parties. Owners made significantly fewer damage claims under the new approach, and both renters and owners reported much higher satisfaction rates after meeting in person.[81]

Reciprocity theory suggests that we should be careful in designing incentives, whether legal or financial. Even insurance or guarantees might seem double edged. By offering insurance, a sharing platform cues awareness that there could be a problem (your car might be damaged, your drill lost, etc.), and creates a moral hazard that might make the user more careless. But it also offers reassurance that the process is legitimate and accountable, as well as the comfort that the "worst that can go wrong" is not so great. In a predominantly commercial culture, on balance, insurance probably sends positive signals. Elsewhere insurance may not be so effective to help sharing grow, although it may provide a business model for "social" entrepreneurs who want to mediate free exchanges (such as Peerby) and therefore cannot charge a commission.

RelayRides found that interpersonal connections reduced their insurance payouts. Unsurprisingly, many sharing platforms actively seek to maximize interpersonal connection. It is central to their business model. Lyft uses the slogan, "Your friend with a car," and encourages riders to sit in the front seat—like a friend—rather than in the backseat like a fare.[82] Kiva's model also emphasizes personal connections. Kiva's founder Premal Shah says, "Connecting people creates empathy. Empathy creates generosity. In a world where so many people are disengaged with the problems we face, it's essential that we "human-scale" the problem and create pathways for everyday people to help."[83]

More than Brand Communities

In the sharing paradigm our reputation and relationships—our "social capital"—matter more than our financial capital, enabling sharing to be more inclusive, regardless of our financial means. Traditional communal and sociocultural forms of sharing have always brought reputation to the fore.[84] Now we see commercial sharing platforms actively investing to replicate these benefits of communal sharing.

We also see conventional businesses trying, too. But despite decades in which marketing theory has promoted "relationship marketing," even the best conventional companies have struggled to establish the sort of relationship with their customers that comes almost naturally in a sharing organization. Ultimately we all know that our relationship with a brand is not the same as a relationship with a real person. Although the employees of conventional companies are real people, we know that "when we interact with them, they are operating as agents of a commercial enterprise. In the sharing economy, the commerce feels almost secondary, an afterthought to the human connection that undergirds the entire experience."[85]

Cameron Tonkinwise also sees this difference as critical, although interpreting it less optimistically than Jason Tanz. He sees the "friction" of real interpersonal relationships as reintroducing humanity into commerce, even as the sharing economy pushes commerce into new spaces. Clever platform design and reputational tools might ease the friction in these encounters—between Lyft driver and rider, Airbnb host and guest—but "in these kinds of 'sharing systems' people cannot hide behind alienated service employment roles. ... This same "encountering otherness beyond economic utility" happens in any of the face-to-face, peer-to-peer economies," says Tonkinwise.[86]

Such human connection is facilitated by what Botsman and Rogers call "anchors of commonality" enabled by digital "architecture for participation."[87] In the same way as a common friend triggers a sense of trust between people who have never previously met, mutual membership in a flourishing virtual community with a common purpose or ideal gives "diverse people a sense that they fit in and ... permission to collaborate [and] form new social bonds."[88] It can also "break down the emotional barriers and stigmas we often have around sharing or asking for help."[89] Overall, sharing fuels our sense of belonging.

The deliberate efforts of commercial brand managers to establish brand communities and to insert brands into our identities are clearly targeting the same psychological and cultural responses. But the personal interactions that stand at the center of the sharing paradigm establish an authenticity to the relationship that the even the best-funded brand advertising campaign cannot match, however many brand evangelists and celebrity ambassadors it might recruit.

And many brands are turning to ordinary people, rather than celebrities, to attempt to establish anchors of commonality and authenticity that leave them less vulnerable to brand-based reputational attacks. For example, Nike has established online social hubs for runners. Such companies are realizing that, just as it takes a community to raise a child or build a business, it also "takes a community, not a campaign, to create a brand."[90] They are also discovering that once such a brand is part owned by the community it is no longer under direct corporate control. Yet the brands of sharing companies—at least those with P2P, rather than B2P models—are even more genuinely co-produced and owned by the community that lies at the heart of the sharing process, and in that respect very different to conventional corporate brands, which remain the instrumental reason for the very existence of their brand communities.

For sharing economy companies, community cultures are incredibly valuable, and often valued by the entrepreneurs. For instance, Lyft's cofounder John Zimmer argues that his motivation is "making systems to connect people" because that "unlocks what people really want, which is a sense of belonging."[91] Michael Olson and Andrew Connor highlight the benefits sharing companies can obtain from their active stimulation of community-driven cultures. For example:

> Airbnb, CouchSurfing, HomeAway, Lyft and TaskRabbit all host regular community events, launch parties and milestone celebrations … [and] build compelling brands centered on a grassroots community spirit. Importantly, this community focus has helped [them] … overcome near-term regulatory setbacks. Through petitions, neighborhood rallies and word-of-mouth, the sharing economy community has influenced the regulatory policies of state and local governments and tipped the balance in favor of government acceptance and support.[92]

Such comments and activity might be dismissed as good corporate marketing, or even astroturfing. The blogger Mike Bulajewski even argues that Peers and leading companies in the commercial sharing economy are abusing positive perceptions of sharing to create "cult brands" (on the model of Apple, or Harley Davidson).[93] Cult brands develop distinctive ideologies and rituals that knit together their participant brand communities and "demonize the other"—in this case the entirety of the "old economy." Bulajewski particularly fears that these brand communities will be effectively mobilized to deliver political change that legitimizes the casualization of labor in the sharing economy.

Tonkinwise sees much the same problem from the other side. He suggests, that in engaging with commercial sharing platforms, participants may come to modify their personal identity project toward the creation of a personal "brand" that maximizes returns in the sharing economy. As a participant, he says:

I invest in cultivating a reputation that will attract "trust" from strangers-like-me. In so doing, I must make the least authentic role-play in service of others my most authentic self. And the only platforms for projecting that tradable reputation are in turn exploiting my unavoidable performance for their own data-based marketing manipulations.[94]

But things are not all so bleak. Juliet Schor highlights how Airbnb created its own organizing platform for guests, hosts, and employees, enabling the creation of local user groups, sharing advice and sometimes campaigning. She notes that:

The company wants these groups to push for favorable regulation. But they may develop agendas of their own, including making demands of the company itself, such as setting price floors for providers, pushing risk back onto the platforms, or reducing excessive returns to the entrepreneurs and the venture capitalists.[95]

And for the reasons offered above, we see something more authentic and less manipulable in the community cultures of the sharing economy. Similar communities have grown up around noncommercial sharing platforms like Freecycle, too. Such sharing platforms—regardless of whether they are commercial or not—rely on trustworthy interactions between real people. These inevitably build a sense of community and contribute to identities of belonging that are only superficially imitated in even the strongest cult-brand communities. As we discussed earlier in this chapter (in "Reciprocity Theory, Values, and Social Behavior"), monetary incentives can weaken such interactions and undermine trust. The idea that such identities could be entirely the product of commercial interests is therefore implausible. But we do need to beware that communities of trust could be abused in the interests of their commercial operators, or that some participants could become complicit in such abuse. All this hints yet again that organizational strategies that emphasize the financial aspects of sharing transactions—even as a benefit for the owners of the goods being shared—could be counterproductive not just for the sharing paradigm in general, but for the specific organization concerned, undermining efforts to build successful sharing communities.

And for building sharing communities, scale is also important. Too small a scale, and sharing cannot achieve critical mass, but at too large a scale—even online—any sense of community is inevitably diluted. We do not trust someone else simply because they are on Facebook any more than we trust someone because they wear the same brand of clothes. At larger scales more formal mechanisms of trust, or trustworthiness building, are required. Even Couchsurfing.com, for example, despite its noncommercial foundations, now requires ID verification, rich profile information, visitor and host ratings, and provides for a system of "vouching" in which users gain the right to vouch for others only when a certain number of so accredited users have already vouched for them. This has apparently reduced the organization's capacity to build social capital.[96] We would suggest this highlights that participants value surprise and adventure, too, as well as the psychological reality that trust needs some form of risk or vulnerability to be meaningful.[97] But if people we encounter shop at the same farmers' market or are part of the same local Freecycle list, a direct sense of commonality remains feasible and our empathy for them is likely to be strong. This is another reason why the overlap of virtual and urban space in sharing cities is so important, and why initiatives like Konnektid in Amsterdam are actively trying to integrate the local and the virtual in their user base.

Trust, Shame, and Stigma

The critical connection between empathy and trust is a powerful means of building inclusion. If we empathize with someone, we cannot help but understand them as our moral equals, meriting recognition and inclusion. Among other things this suggests that efforts to shame people into sharing would not work—at least in cultures where sharing is not already the social norm—and would even be counterproductive. Here we briefly explore the importance of avoiding shame, and the closely related sense of stigma, which can undermine trust and deter sharing.

Psychological research suggests that shame gets its power from denigrating the person, in contrast to guilt, which arises from criticizing the action, enabling a feeling of regret. Seeking to shame people for not sharing is likely to reinforce the negative effects of cognitive dissonance, encouraging them to further reject or suppress prosocial values. But in positive reinforcement the opposite may hold: to associate the desired action with good character has a deeper and longer lasting effect (on children) than to simply praise the action.[98] In this way the foregrounding of nonfinancial rewards for participation in sharing might help build people's sense of self-worth and strong moral identity, which would positively affect behavior in other arenas of life.

This sort of reinforcement of self-worth is remarkably powerful. For example, it affects how people deal with information that challenges political ideologies and identities. In psychological studies, self-affirmation consistently made people more open to accepting objectively correct information on contentious issues.[99] By implication, a secure positive self-identity makes us more resistant to misleading propaganda. This suggests that the undermining of identity by marketing and consumerism could also be, at least in part, to blame for the divisive and dangerous persistence of myths such as the ideas that vaccination might cause autism, or that climate change is a liberal conspiracy. A secure identity is also clearly essential to avoiding wider problems of bias, discrimination, and scapegoating that can arise from feeling one's identity to be under threat. And as we saw earlier, a secure identity means people are less likely to perceive encouragement to share as a threat.

This is perhaps reflected in the distinction between UK and Amsterdam survey evidence regarding the participation of ethnic minorities in mediated sharing.[100] In the UK, where interculturalism is weak, perhaps making racial minority identities more insecure, participation was low. In Amsterdam, where integration has been more consistently pursued, minority identities should be more secure, and enthusiasm for sharing was exceptionally high.

Before moving on, it is worth noting that the above conclusions are most clearly applicable in individualistic, Western societies like the US. In more collective and communitarian cultures, social pressure to comply, and the sense of shame in failing to do so,[101] may well be more effective in promoting sharing.

The strong and complex relationship between moral values and behavior has another important implication for the development of the sharing paradigm. Any association between sharing and moral inadequacy (like accusations that people relying on Social Security because of poverty are undeserving, feckless, or lazy) will make sharing unattractive. This can clearly apply to informal sharing—the shame associated with second-hand and hand-me-down clothes may mean that only families with other means to signal their non-poor status (or very brave pioneers of sharing) will dress their children that way. More generally reliance on informal and free sharing options might be seen and portrayed as inadequacy. Promotion of market-led development in the global South as essential to replace inefficient sociocultural sharing modes of provision of goods and services therefore stigmatizes entire cultures, regions, and continents!

Clearly, normalizing sharing and ensuring its uptake by peers, opinion formers and role models, too, can help lift such stigma. Even Airbnb got off to a slow start, due in part to the persistent social stigma around sharing. Airbnb's co-founder Brian Chesky says he was told time and again that renting to strangers was a "weird thing, a crazy idea."[102] Well-designed sharing approaches can also directly challenge stigma. For instance the forms of co-production of health and social services we discussed in chapter 2 build autonomy and self-respect among those using and co-producing those services[103] making them less vulnerable to stigma. At a city scale, real recognition of the contribution of all citizens to the urban commons could perhaps have a similar effect, as we saw in Medellín. A further example might lie in the way cities treat squatting of underused and vacant land and buildings. Miguel Martinez notes that typically—contrary to the popular image inculcated by much of the media and many politicians—squatters show "great effort to take care of the occupied places, to promote communal ways of living and to share their ideas with their surrounding neighborhoods."[104] In other words they help build the urban commons. For example, the Kukutza Gaztetxea in Bilbao, evicted in 2011, was a former factory "open to all who wanted to practice sports, learn foreign languages, create art, launch cooperative enterprises, organize meetings and engage in political campaigning."[105]

Martinez argues that—rather than criminalization and resistance from authorities—"squats have to be recognized and supported for what they are: vibrant social centers at the very heart of the 'commons', actively including the excluded."[106] For many years, but sadly not currently, as we saw in the case study, Amsterdam shared this view.

Cybertrust and Identity Online

As in the shared spaces of the city, trust is also critical in shared spaces online. Here the issue is complicated by the question of security of our online identities, which is essential for web-mediated sharing—both commercial and communal. Whether regarding our financial security or the reliability of our reputation, participants in sharing need to be able to trust the quality of the information available online. They also need to be able to trust that the online persona and the real person who turns up to borrow their car or look after their dog are one and the same.

We rightly distrust many online reviews—too many of them are paid for, or in some cases exemplify spoof humor run wild. In the travel industry for example, false reviews are prevalent.[107] A predominance of false reviews on sharing sites might not destroy the sharing economy, but could seriously undermine the trust it needs to function.

Worse, in the sharing economy a poor reputation for a participant could mean exclusion. Just as in the era when our sociocultural sharing nature evolved, humans remain keen to "punish" those who renege on their social commitments. The business writers Don Peppers and Martha Rogers suggest that the reason Wikipedia works is not so much the "prosocial" instinct to contribute, but the "punishment" instinct acting to ensure that misleading or inaccurate copy gets removed.[108] With sharing platforms determined to maintain trust, a few bad reviews—or even a personal vendetta—could mean a driver removed from Lyft, or a host from Airbnb, with serious consequences for their household economies.

The more we come to rely on sharing online, the more important our reputations will be—and if our activity is constantly linked to our Facebook or Google account, our reputations will follow us around, as will any predictions made about us by "big data" analysis (as discussed in "Surveillance, Anonymity, and Injustice" in chapter 4). "Reputation aggregators" are already emerging, intermediary companies like Trust Cloud[109] and e-rated,[110] which bring together personal reputation indicators from a set of different sharing platforms to provide transferable generic "reputation ratings." Benita Matofska has established a comparison site for the sharing economy—called *Compare and Share*—and is collaborating on a project to develop a "share-trade" mark for companies and platforms, which would guarantee certain standards, such as a code of conduct, background reputational checks, and insurance.[111]

The ability of key players such as Facebook and Google to securely manage the creation of multiple online identities becomes more important if the risk is that those identities will be used to abuse sharing platforms, rather than to re-spawn in online games, as hundreds of thousands of people have done, despite this being a breach of Facebook terms of service, for example. (We are not advocating here that people should be prevented from holding multiple accounts, and enjoying the rights to freedom of speech facilitated by such anonymity, merely that the capacity to connect the identity and the real person must be effectively and safely managed.)

At the same time, the right to be forgotten, or have certain personal online information deleted, is important. We might need "reputational bankruptcy" proceedings, as Zittrain has advocated,[112] to allow people a new chance if they get excluded on grounds of reputation. In 2014 the European Court of Justice judged that European data protection law already includes a form of "Right to be Forgotten," ruling that individuals should have an opportunity to insist that Google (and presumably other search engines) remove certain search results that come up in a search for their names, which violate "respect for private life" or the "right to protection of personal data."[113]

Some supporters of the ruling argue that it recognizes personal data as "property" and defends "property rights." Opponents, including the Global Network Foundation, claim it will only be used by rich and powerful to defend their interests. We might agree with neither extreme, seeing parallels with existing rights to bankruptcy and principles such as statutes of limitation, which seek to allow people to draw a line under their past errors and misdemeanors and reform themselves going forward. Such approaches must be open to all, however, and also respect the public interest in information and reputation.

Viktor Mayer-Schönberger highlights that even with such a ruling, our online legacies of data are still problematic, as they ignore change over time: "Without forgetting we ... risk misjudgment. As psychologists remind us, forgetting also is intimately linked to forgiving. If we can no longer forget, we may turn into an unforgiving society."[114]

Moreover, at the same time as being concerned about secure reputations, we also need the right to be anonymous online: otherwise the social and political freedoms that characterize urban space will not be available in cyber space. Manuel Castells notes, for example, how the "we are the 99%" meme spread particularly through Tumblr, which facilitates anonymity.[115] We need secure and protected online identities. Paul Bernal of the School of Law at the UK's University of East Anglia argues for a "privacy, identity, anonymity" model, key to which is the development of specific privacy rights: a right to roam the Internet with privacy, a right to monitor the monitors, a right to delete personal data, and a right to create, assert, and protect an online identity.[116] Assertion proves who you are, for democratic or economic purposes, but it must be linked to control over what aspects of your identity you need assert in different contexts. Appropriate protections extend to protection from exclusion, deletion (perhaps achieved by a right to transfer your data and profiles to a different platform[117]), impersonation, and defamation; and of the link between online and offline identity. Bernal points out that this last also enables the use of pseudonyms, protecting a right to anonymity. Sadly many in the cyber economy seem to believe that transparency is all. Facebook CEO Mark Zuckerberg for instance has said: "Having two identities for yourself is an example of a lack of integrity."[118]

Some sharing cities are engaging with issues of privacy and data protection, as well as reputation. Amsterdam's Smart City program says: "We strongly believe that the data is, in principle, owned by the person who generates it," promising to never share personal or group data without explicit approval from the subjects. They state that "users should also be able to get insight into the data that is being gathered wherever it is reasonably possible."[119] But the more complex issues of anonymity, identity, and broader online rights are only now emerging in debate.

In the absence of a framework of rights it is important to consider how "reputational" discrimination (exclusion from sharing) might interact with existing forms of discrimination. For instance, research has claimed that black hosts on Airbnb earn less from renting similar properties in New York.[120] The Harvard Business School professors Benjamin Edelman and Michael Luca suggest that this discrimination is in part enabled because of the required transparency that places a large picture of the host on the webpage.[121] By contrast, the relative anonymity of the Uber booking process means its users are less likely to be subject to racial profiling than those trying to hail a cab on the street.[122] But, Uber charges more than many taxi services, and as Latoya Peterson notes, "one shouldn't have to pay for premium service to get a racism-free ride experience."[123]

Edelman and Luca's work would suggest that the sharing economy might be more vulnerable to discrimination because it promotes interpersonal interaction that relies on a public open identity. But at a systemic level the evidence suggests that peer-to-peer contact and meaningful interaction with people of other races, for example, can contribute to reducing such discrimination. The demonstrated contextual effect of positive intercultural interactions in public spaces—reducing prejudice even among onlookers[124]—suggests that making such positive interactions in the public online realm visible as well would also help reduce discrimination.

These are not issues we can fully resolve here, but Bernal's work suggests some principles for how sharing platforms could promote inclusion and non-discrimination,[125] and we would also suggest that cities and the sharing sector should collaborate to gather and analyze data to allow such questions to be considered and solutions suggested if problems are found.[126]

A Habit-Forming Virus?

As we noted above, to meet its potential for social transformation, sharing needs to not only stimulate more prosocial values (whether through mainstream or countercultures), but also to be a sticky behavior that spreads and locks in those behaviors and values. In practice this suggests it must be both imitable and habitual. There is plenty of anecdotal evidence to support this. For example, Botsman and Rogers describe a Zipcar marketing campaign in 2009 that challenged 250 people who had not previously tried Zipcar to try it for a month.[127] At the end of the experiment participants had increased public transport usage, cycling and walking each by more than by 90 percent, and cut vehicle miles traveled by two-thirds. Sixty-one percent planned to continue to live without a private car, and a further 31 percent were considering it. In just one month, say Botsman and Rogers, shared travel had become a habit.

Habits appear to be central in stimulating pro-environmental, and by extension, prosocial behaviors, and are most easily established when intrinsic motivations and external incentives coincide.[128] Communications and campaign consultants, such as Chris Rose[129] and Solitaire Townsend, emphasize the importance of reaching out to people, beginning where they are. So rather than exhorting change in ways that imply someone needs to become a "better person," efforts to make green consuming "sexy" are more likely to increase participation.[130] Tom Crompton of the environmental charity WWF (popularly known in the US as the World Wildlife Fund) cautions, however, that attempts at behavior change which focus on such extrinsic motivations risk reinforcing values that run counter to the overall objective, even if the immediate goal is achieved.[131] For instance, selling solar PV panels as rooftop "bling" is of limited value if the owner consumes yet more electricity.

On the other hand, if the result of an extrinsically motivated action is a change in day-to-day behavior and habits, then there is potential for expressed values to subsequently adjust to match that behavior. Sharing is rooted in both a set of physical behaviors and cooperative values. Potential sharers might be attracted by either the extrinsic or intrinsic motivations, or by both. There are lots of reasons for people to adopt sharing behaviors for financial or convenience reasons, especially where the formal economy is struggling to deliver jobs and growth. But as we will see below, it seems that the adoption of new behaviors can also shift the values we express, and thus sharing can have impacts well beyond the specific arenas where it is practiced.

The underlying values in human societies are remarkably universal or "pan-cultural." Benevolence, stimulation, and security are valued in all societies but expressed to different degrees and in different ways in different cultures.[132] More individualistic cultures value stimulation and self-direction more than security and tradition; collective ones tend toward the opposite. Free-market consumerist cultures will emphasize hedonism, European social market cultures rather express universalism. Cutting across these tendencies is the general degree of "looseness" or "tightness" of cultural conformity (as discussed in chapter 2).

Figure 5.2 returns to the values circumplex of Shalom Schwartz.[133] In this model values adjacent to one another are likely to be expressed together, while those lying on opposite sides of the wheel are more inversely related,

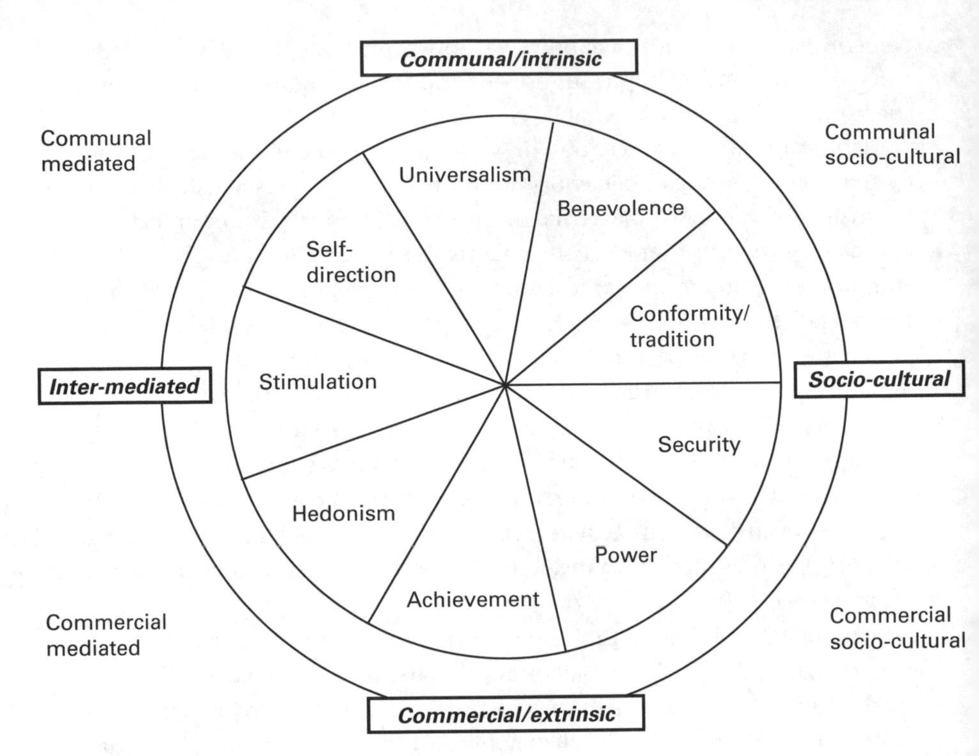

Figure 5.2
Sharing flavors as expressions of underlying values. (Drawing on the values circumplex of Schwartz [2006, see note 44 in chapter 2].)

although still expressed to some degree in everyone. In this presentation we have crudely overlaid our sharing matrix over the circumplex to indicate some of the ways in which different alignments of values might be expressed in different cultural flavors of sharing. For instance, societies marked by extrinsic hedonistic values might be expected to favor commercial mediated forms of sharing, while those where intrinsic self-direction and universalism are stronger might see more communal mediated peer-to-peer forms.

Different cultures can also be expected to respond to sharing technologies in different ways. The US response—dominated by commercialized sharing through secondary markets, underpinned by venture capital—is distinctive from the European, where communal sharing mediated by charitable or mutual platforms seems more prevalent. The Seoul case seems to suggest a third model—relatively non-monetized with emphasis on community, supported with city funds and legislation.[134] It is perhaps too early to distinguish how the uniquely Latin American blend of liberal and communitarian values is expressing itself in sharing models, although it too appears to have a strong role for state and city authorities as well as for community activism, as we saw in Medellín.

But what happens if behaviors stimulated in sharing are not coherent with dominant culturally expressed values? The commonplace assumption is that people's behaviors reflect their values, but things are not so simple. In childhood we accept more readily that action shapes character.[135] And many adults continue with addictive behaviors even when their values suggest they should stop. At a larger scale, technologies and structures can lock us all into behaviors that contradict our values and principles. For instance, we spend long hours at work, despite professing the importance of family life, and we have little choice but to commute by car in many cities regardless of our views on pollution and waste.

This gives rise to cognitive dissonance between our behavior and our values. Research has demonstrated that our reactions are complex, mediated by identity, and shaped by social norms disseminated in personal interactions.[136] We individually—or collectively, in businesses, political parties, and so forth—construct narratives that reframe our behaviors, their implications, and our motivations in ways that cast us in the best light possible and avoid the shame that can arise when apparent hypocrisy is exposed. Such ways of relieving cognitive dissonance have had serious consequences as societies have collectively wrestled with declining social capital, rapid technological change, and rampant consumer marketing. We have done damage to our own values and recast our identities in individualist forms.

But what if we could stimulate better behavior? Would values follow that as well and become more collective? The answer, unsurprisingly, is "it depends." It appears that values can be modified in this way, but only with difficulty. Instrumentally or extrinsically triggered marketized behaviors seem to have less influence on happiness and well-being than those that are intrinsically stimulated. Insofar as it is possible to stimulate sharing behavior by mobilizing inherent but under-expressed values of community and cooperation, then we could expect more widespread sharing practice to shift the expression of values away from hedonism and achievement, for example, and toward benevolence and universalism. Changed habits would be reinforced by the neurochemical happiness hit we appear to get from altruistic behavior.

Most importantly, because sharing is inherently collective, these values would begin to be expressed in social norms also—in a cultural as well as psychological process. Norms are internalized in part through cognitive dissonance, as we "trick ourselves into thinking the reality is what we ourselves might have chosen, or is what the right state of affairs should be."[137] In other words, even if sharing is initially forced upon us by convenience or cost, we can come to recognize it as preferable to ownership and consumerism, regardless of our previous stance on the matter.

But psychology also tells us that we can "learn to segment our lives so efficiently that normal well-adjusted adults can act differently in different settings"[138]—in terms of the moral norms we express. Benkler contrasts our moral behavior as parents with how we act in the office environment. So socially cooperative behavior from sharing will not necessarily spill over into other domains. But if we are exposed to sharing in multiple domains, the possibility of norm spillover would seem to grow. Then extrinsic and intrinsic motivations would reinforce one another, boosted by growing congruence between the widely shared values of community and environment and the new forms of collective consumption behavior.

Sharing Spreads through Networks

Rather than being a niche fashion, sharing is destined—and designed—to spread. In practice, sharing depends on bringing together owners and users, or mutual sharers. The larger the pool of sharers the more efficient and effective sharing is. Participants in sharing practices—and the managers of sharing platforms—therefore typically benefit from drawing in more and more participants. The same digital technologies that have enabled modern sharing—the Internet, mobile devices, online social networks—can also stimulate its spread.

Experience with networked and complementary products like faxes and mobile phones offers additional insights into sharing. Synergies arise from users sharing the same platforms. The snowball effect means, however, that suboptimal products can become locked in. And the need for critical mass means that the early stages of "market creation" are very difficult.

Sharing platforms share some of these characteristics. The more people involved in a time-bank, for example, the more likely new users will be able to access the skills they need, or find a user for the services they can offer. So there are good reasons for supporting early stage market development or consolidation of competing services in this space. As Jonathan Schifferes of London's Royal Society for the Encouragement of Arts, Manufactures and Commerce (RSA) notes: "Many sharing platforms struggle to reach critical mass" so we need "infrastructure to consolidate the sharing economy [such as] comparison websites."[139] But we should also beware of the possibility that a socially—or even economically—suboptimal sharing model might come to dominate a particular market, due to the power of networks. Airbnb, for example, is not necessarily superior to CouchSurfing.com in anything other than the sheer scale of its network of hosts and visitors. Intense battles between Lyft and Uber over ridesharing markets in many cities reflect the same insight, although neither riders nor even drivers are locked in to a particular platform.

The economist Paul Ormerod highlights the policy implications of the growing dominance of network effects in the modern world. He says governments and companies wrongly assume that individual incentives drive behavior.[140] Understanding network effects offers potential for much more effective policy interventions. Mathematical and economic network models suggest some critical factors for networks that spread behavior change effectively. These are the connectedness of early adopters to others and, most significantly, how easy it is to persuade others to change their behavior. The combination of Internet social networks and reputational markers, with highly convenient web and mobile applications for sharing, could scarcely meet these conditions better.

On this basis it is unsurprising how fast some forms of sharing are spreading. Yet there may be reasons beyond the economic and technical. Technological change can change tastes—or the expression of values—and lead to a greater increase in sharing than the technology alone implies.[141] In other words sharing behavior can spread like a cultural virus, infecting new participants and forming new habits as it goes.

From Counterculture to the Mainstream

Some sharing behaviors like squatting seem unlikely to ever become mainstream. But to focus exclusively on sharing as behavior change within mainstream culture would miss the importance of such insurgent countercultural sharing and its scope to synergize with art, protest, and politics, to spread through networks, and to lead wider values change (as we saw with filesharing online). Countercultural sharing behaviors happen in autonomous or "interstitial spaces"[142] beyond the reach of the powers that be. As a result they can be the birthplace of more far-reaching subversive or symbiotic approaches to social transformation.

Botsman and Rogers, however, are keen to portray collaborative consumption as part of mainstream culture and as a process, which offers even a "rational" self-interested consumer such excellent service that he will not "even realize he is doing something different or 'good.'"[143] We also see the potential for sharing services to meet needs this effectively, but believe that such a perspective rather misses the point. For us the sharing paradigm also embodies a countercultural ethos, but one with the potential to shift norms so far it transforms the mainstream. For this to happen, the *norms* of those who share must change, even if that process is only subconscious.

Botsman and Rogers imply that only the outcomes matter—the changed consumption behaviors with all their benefits for the environment and society. They are happy to see the trappings of consumerism used to sell a less damaging form of consumption. But the possibility for transformation in the sharing paradigm is so much greater, if only we can release its power to change norms and values in the cultural and political realms, not just the economic. By transmuting the lead of economic behavior change into the gold of political behavior change, sharing could be among the most powerful "symbiotic" or "subversive" strategies for social emancipation—supported not only by countercultural activists but also by existing elites, yet ultimately transforming values and norms, and redistributing power across a whole range of arenas.

The practical contribution to efficiency of resource use through sharing is (although substantial) of secondary importance in this respect. Without the transformation of cultural values, expressed politically in the public realm, such greater efficiency might be largely offset by rebound effects from growing overall consumption. Worse, it may even undermine justice—insofar as it exacerbates the integration of more people into the "always on" 24/7 commercial economy, without also developing their capabilities to participate there on their own terms. Fundamentally, insofar as the sharing economy treats underused labor as a surplus resource, it treats people instrumentally, as means to an end rather than as ends in themselves.

This section has explored how the adoption of a sharing paradigm could lead to social transformation: through the powerful effects of changed cues in language, culture, and practice on our behavior; reinforced by the synergistic trust-building effects of sharing on social capital and empathy; and locked in with new habits and norms first developed in interstitial countercultural niches and experiments, such as Occupy, Christiania, and integral cooperatives. As we have seen, sharing is potentially highly subversive, in the respect that it begins where people are today, yet introduces habits that can rebuild social norms and realign values with interculturalism and community solidarity. We believe that a virtuous circle is possible, between equitable sharing, social capital, the public realm, and collective political activity. As a result, where sharing builds social capital and transforms cultural values, it offers a real alternative to the dominance of neoliberal capitalism. In the next section we turn to the practical measures cities might take to launch themselves on that path.

Policy, Regulation, and Practice to Support the Sharing Paradigm

In this section we bring together lessons for cities that want to play a role in the sharing paradigm. We begin by stressing the importance of a broad approach whose scope covers sociocultural and communal sharing models, as well as commercial mediated ones, and recognizes the possibilities for sharing approaches to complement (or even replace) conventional market or public sector provision in some areas. We then outline some generic design principles for good sharing systems which apply to all four flavors of sharing. Recognizing the prevalence of online models in all four denominations, we then highlight approaches and measures that seem critical to the effective use of the digital domain, before considering in turn measures for scaling up and expanding sharing, extending it to new resources or domains, and strengthening infrastructures for sharing and collaboration. We conclude the section, and chapter, with some reflections on the challenges of governing a paradigm transition such as implied by sharing.

A Broad Approach

In practical governance it is easy to be distracted into focusing on the most salient issues, or the interests of those who can shout loudest (or otherwise hold the ear of decision makers), rather than providing policy measures driven by the public interest. In the case of sharing, this would mean focusing on the commercial mediated denomination, and in particular on the complaints of both incumbents and sharing businesses about regulation. We do not ignore such questions, but here we want to emphasize that cities can and should look to intervene across all denominations and domains of sharing, in ways that are alert to the context set by specific national and urban cultures.

In particular, their strategies must also include sectors in which sharing remains unrecognized, rare, or as yet nonexistent. For example, in current political debates around the sharing economy, co-production is almost invisible in comparison to collaborative consumption, yet some of the most exciting sharing opportunities recruit ordinary people into co-production of goods and services. Equally, a focus on products and services risks neglect of underpinning activities in community finance, energy supply, and materials recovery. Overarching all else, failure to recognize the city as a co-production of its citizens, rather than just as a venue for sharing, would critically handicap the new sharing paradigm.

In terms of policy, some suggest that a narrow focus is appropriate. Benkler for example argues:

> We do not need to focus consciously on improving the conditions under which friends lend each other a hand to move boxes, make dinner, or take kids to school, and we feel no need to reconsider the appropriateness of the production of automobiles by market-based firms.[144]

We think such a narrow approach would be wrong for all the same reasons that we think a purely economic analysis of sharing is misleading. Policy should also consciously seek to protect and strengthen sociocultural sharing in the gift economy and elsewhere and, indeed, to enhance the social dimensions of existing market-based transactions. This widens the case for intervention by understanding the cultural potential of sharing to promote justice and sustainability.

Principles for the Sharing Paradigm

In the first parts of this chapter we discussed a series of objections to the sharing paradigm, and explored how—understood as both a cultural and political process—sharing could be transformative. That highlighted five features or principles, all of which are fundamental to the design of sharing systems in a paradigm for just sustainabilities.

The first is that *trust and confidence* are required, both for mediated and sociocultural sharing. We vest trust in others whether we share a car with them, or a neighborhood park. The spaces and platforms involved should therefore be designed in ways that help to build trust, collectively or individually, and to enable users to obtain and signal their trustworthiness and reputation.

The second is that utilizing and stimulating *intrinsic motivations* is generally more effective for rebuilding communities through sharing than focusing on extrinsic rewards or sanctions. This has implications for the models of sharing we might choose, as well as for the incentives they offer or advertise. For instance, making the financial dimension of commercial systems simple and placing it in the background is advisable, while also favoring processes that stimulate personal interactions.

Third, systems must *empower users* both to control whether and how much they participate, and to influence the overall design and rules of the system. Obviously, cooperatively owned sharing systems can do this, but even public or business models can facilitate greater participation, following learning from commons-governance approaches.

Fourth, systems must achieve high standards of *protection of privacy* and security of personal data and as well enable anonymity where appropriate. Preventing the abuse of data or personal information for commercial or surveillance purposes is critical to trust. Protecting civil liberties is essential to enable wide participation.

Fifth, and most importantly, systems must be designed from the outset for *justice and inclusion* in intercultural societies. They must be equally accessible and attractive to those from different groups and cultures—especially those otherwise disadvantaged. They must allow participation on equal terms regardless of background, and do so openly and transparently. In other words they require both *designing in justice* and *justice in design*.

Understanding and Enabling the Move Online

Above we emphasized the importance of a broad approach to the various denominations of sharing, and previously we have stressed the ways in which sharing is part of a cycle of real world and virtual world interaction that enables participatory democracy. But we cannot ignore the growing prevalence of online platforms for sharing, and the challenges and opportunities the digital domain creates. Here we highlight the online disintermediation cycle, online tools for reputation, and the threats of constraints to civil liberties online, each of which requires some response by cities.

Students of the Internet early identified its potential to directly connect demand and supply. They also rapidly spotted the emergence of cycles of "disintermediation" as direct Internet marketing sidelined traditional intermediates like stockbrokers; followed by "re-intermediation" as new online enterprises—such as price-comparison websites—found profitable business models.[145] Disintermediation typically enhances transparency, which in turn can create new levels of P2P trust, even between strangers.[146] Trust may carry over to new intermediaries, depending on their motivations and management.

New intermediaries may be online subsidiaries of conventional businesses or entirely new enterprises. We see the same with sharing. As well as innovative peer-to-peer ridesharing services, we see car hire companies and even car makers buying into the car-sharing market. In similar terms, Airbnb is a novel online travel intermediary.

The sharing economy also exhibits cycles of dis- and re-intermediation. The sharing economy is essentially "enabled by platforms that connect supply and demand at the peer-to-peer level."[147] General-purpose listing websites such as Craigslist and Gumtree were disintermediators, enabling peer-to-peer transactions for all sorts of goods and services. Newer specialist sharing platforms are re-intermediators, aggregating demand and supply, offering benefits for both users and suppliers of resources and services. While the disintermediation step is a potentially profound expression of greater economic democracy, the nature of the re-intermediation process is much less clear-cut. It typically concentrates power, in the hands of the platform owner. But insofar as it further expands the sharing economy at the expense of conventional models it can maintain a democratizing momentum. More importantly, re-intermediating platforms can be designed, owned, and funded in different ways. If such platforms are owned and controlled in common by their users and suppliers, then they act to redouble the democratizing effects of sharing on the economy.

Culturally, norms may be changing more generally with respect to intermediation. For instance the growth of online dating sites suggests we are increasingly open to relying on an impersonal intermediate in a fundamentally personal part of our lives. This is not to suggest that such personal ads are completely new, just that they have become a more normal and accepted part of life as—in an Internet age—we have adopted horizons beyond our local and ethnic communities and accepted new norms, first online and then in the material world. And in this cycle, freedoms have typically grown: just think of the differences between intermediation in the form of dating websites and in the form of arranged marriages.

Cities need to tailor their interventions according to the point in the disintermediation cycle at which particular resources stand. They also need to recognize where city authorities are the conventional intermediaries under threat of disruption; and more positively, where they can play the role of new intermediary to the best interests of their citizens, by providing shared public resources and services, or investing on co-production models of service delivery. In particular, cities can be key intermediaries for political sharing, by creating and respecting physical and virtual public spaces for debate and protest.

Cities may also play an intermediary role with respect to reputation. We have noted the importance of reputational markers in sharing at several points in the book. Like trust, reputation is important in all different sharing domains. Online sharing businesses have invested heavily in mechanisms to help users build credible and verifiable reputations (through reviews, vouching systems, and interconnection with social media networks). City authorities might play important roles in overseeing such systems to help ensure their credibility and protect data against abuse. They might even play the role of a reputation intermediary or aggregator, rooting the credibility of the information in publicly held data on citizens (with adequate data protection of course). It is even possible to conceive of city authorities acting as some form of "reputation lender of last resort" or guarantor in extremis—allowing their citizens to participate in sharing services even before they have built a personal reputation online. This could be achieved by underpinning insurance-based approaches offered by sharing practitioners, or in a freestanding way. In either case such systems would require careful design to minimize moral hazard. Cities should use this role to promote justice by insisting on high standards of inclusion in any commercial or community-led sharing service that they are prepared to certify, brand or directly support. For example, in Helsinki's Kutsuplus mobility program, it was the city council as client that insisted on SMS (short message service) booking as well as smartphone access to help overcome the digital divide.

Online reputation is closely related to a secure online identity. And there is a tension here that must also be managed. The desirability of verifiable identity and reputation must not blind us to the importance of rights to anonymity, free speech, and participation in public domains (including the Internet). Cities and other authorities must collaborate to protect Internet rights and freedoms (as we discuss earlier in this chapter in the section "Cybertrust and Identity Online"), both to minimize civil liberties disincentives to sharing participation, and to ensure a strong, shared political demos.

Scaling Up: Expansion and Replication of Sharing Programs

Cities and city authorities can be key actors in the expansion and replication of sharing programs, with the potential to shape the development of sharing at a grand scale. San Francisco, Seoul, and Amsterdam are all attempting this, each in their own fashion. This is a key way in which cities might avoid a competitive economic race to the bottom by sharing—they are large enough to sustain a diverse economy, yet small enough to experiment with new approaches. In this section, we outline some of the opportunities and some of the benefits that might arise.

Yet even to mainstream sharing at the city scale implies very substantial expansion, and replication of sharing approaches. For many sharing commentators, the human scale of sharing is one of its great strengths, yet simultaneously, as Mike Jones—CEO of Science, a technology studio in Los Angeles—notes, the online setting gives "all sharing economy startups ... the ability to connect sellers with consumers on a scale that is astronomical in comparison with what was possible even a few years ago."[148] For Rachel Botsman, too, "'Big' does not have to be the foe." But, she suggests, "We have to be careful not to dilute the humanness and empowerment that lies at the core of collaborative consumption."[149]

This raises an obvious question: "How do you scale up while maintaining 'humanness?'"[150] Part of the answer resides in the mechanisms sharing operations have devised to build trust, reputation, and reciprocity across the huge expanses of cyberspace. For both evolutionary and technological reasons we no longer need to live in small communities to benefit from communal governance or communal sharing. Evolutionarily, empathy for others and our instincts to share and collaborate have been hardwired into us, while the modern technologies of the distributed web enable us to access the same levels of knowledge and reputation—and the power to hold others to account—at much larger scales.

Yet the materiality and connectedness of cities also helps such mechanisms work effectively to build communities that can spill out into urban space. The city scale is typically small enough to have a shared understanding of what is fair—and to resolve disagreements though public reasoning. It is large enough for critical mass, yet small enough for "critical trust." At this intermediate scale, density and real-life proximity enable rich and more complex webs of sharing and collaboration.

Despite notable exceptions like Airbnb, most sharing economy initiatives remain relatively small scale, and in many cases scaling-up might involve both niche expansion—sharing taking a greater share of the market for particular products or services—and geographical replication. Latitude suggest, for example, that sharing time/responsibilities, household items and appliances, money, cars, and living space all have significant potential for further growth in the US with relatively low current provision and high latent demand.[151]

But the sharing paradigm is not just about growing the sharing economy. It also offers a broader chance to reinvent economic models such that they treat us as whole people, not just consumers or shareholders, and thus to reshape our identities. Shared community production facilities, energy generation systems, and credit unions offer new models of co-production, further blurring the distinctions between work and consumption. If such models are nurtured and scaled up, then there are even better prospects for us to take our whole, sociable selves into our political interactions—rejecting one-dimensional dog-whistle politics in critical areas such as tax or immigration.

City Authorities as Leaders

City authorities are essential actors in such a transformation. Sharing advocates such as Janelle Orsi and her colleagues; April Rinne; and Rachel

Botsman and Roo Rogers have all suggested a range of actions cities might take.[152] Here we summarize and elaborate on their suggestions to paint a picture of the leadership role of the city authorities in a *sharing city*.

A sharing city will need a *public commitment to its vision*, backed by the necessary strategies, institutions, and resources.

It will need also to systematically *map the assets* owned by the city to identify those that could be shared, and opportunities for community engagement; and *review all operations and policies* to identify where it could utilize collaborative economy platforms to meet its needs or goals (also including shareability criteria in local procurement tenders and other municipal contracts).

It will actively invest in public services and *enable co-production in city-led services*, protecting and enhancing of public common resources, infrastructures, and services—from libraries to public spaces, and from health to education—paid for through taxation or insurance. It will support education and skill development that can build confidence and practice in sharing.

It will need to develop appropriate *indicators* for the city as a whole and its districts—such as a "neighborhood sharability index," and establish means of monitoring progress, perhaps piggybacking on smart city investments. It will *engage the public in governance*, for example, through participatory budgeting. It will also design and "police" the public realm in ways that enable participation and create physical, virtual and psychological spaces for insurgent countercultures and interculturalism.

It will *enable collaborative economy* operations in the city: reviewing and reforming regulation and policy as necessary across areas such as taxation, planning and zoning, insurance, and licensing; and investing its own capital in the sharing ecosystem, to support incubators and accelerators, research, capacity-building, and shared infrastructure; and particularly to provide finance for noncommercial approaches to sharing.

It will directly support *nonprofit, communal sharing* as a direct or enabling intermediary or facilitator in, for example, neighbor sharing initiatives, credit union–based peer-to-peer borrowing, and community involvement in providing housing, access to land, and community facilities.

And it will act as a "meta-intermediary" or *sharing hub*, publishing online directories, linking sharing activity operators together with one another and with citizens, perhaps also aggregating and guaranteeing reputation for its citizens, underpinned by enabling open, affordable high-speed (mobile) Internet access for all residents.

Using Existing Policy Levers

Some of these actions may require new policy or regulatory frameworks (which we will consider in the next section) but many—particularly those focused on enabling appropriate development of the commercial mediated sharing segment—simply imply more thoughtful application of existing tools like zoning, licensing, and local taxation. The legal researchers Daniel Rauch and David Schleicher highlight that local authorities such as cities do not face a binary choice between supporting the sharing economy or resisting it.[153] They argue that local authorities can and will develop mixed regulatory strategies including enhanced consumer protections, and mechanisms to incentivise sharing business to help deliver public goods and support redistribution.

Commercial models of collaborative consumption or co-production can be attractive to cities, as they don't need subsidies or grants to support their basic functioning and infrastructure. But before we consider how they can be enabled, we must note that this can also have a practical downside—as well as the cultural problems explained previously. The sharing sector would become vulnerable to the same sort of stock valuation bubbles that have particularly marked other generations of online enterprises. Current implied inflated valuations for Airbnb ($10bn) and Uber ($40bn)[154] suggest this process may well have begun. Cities and other authorities will need to be alert to this risk, especially to ensure that more socially oriented sharing platforms are not starved of capital, skills, and resources, and are not also lost when such a bubble bursts.

To facilitate sharing, Orsi and her colleagues suggest that reforms are particularly needed in rules on business registration, public health, planning and zoning, tax, insurance, and liability.[155] For example three states (California, Oregon, and Washington) have so far "passed laws relating to car-sharing, placing liability squarely on the shoulders of the car-sharing service and its own insurers, just as if it owned the car during the rental period. The laws also prohibit insurers from cancelling owners' policies."[156]

Orsi and her colleagues also argue that cities should "permit residents to use their homes for short-term renters or guests as a way to diversify local tourism opportunities and to help residents offset high housing costs."[157] Current planning, land use zoning, and permitted use rules were typically not written with sharing in mind. They were intended to separate out "incompatible uses" (e.g., a slaughter house next to homes). But they also tend to deter[158] "accessory uses" of homes as places of sharing business. In zoning parlance such uses are *customary* and *incidental*, and this needs explicit recognition in the rules, especially to ensure that commercial sharing is not driven into the gray economy.

Such measures need to be well calibrated to ensure that they do not instead enable excessive commercialization with its risks of gentrification or labor casualization. For instance, to ensure Airbnb delivers social benefits more cities might adopt

> nuanced permitting policies and fee structures to allow short-term guests. To prevent residential units from becoming too hotel-like, cities could adopt policies that limit the number of paid houseguests per year, limit the number of guest nights, or cap each household's gross income from short-term rentals at, for example, no more than 50 percent of the monthly costs associated with the unit.[159]

Such provisions would "recognize that the purpose of sharing is not necessarily to profit, but, rather, to offset the cost of housing."[160] Similar approaches might cap the charges ride sharers could levy (or donations they could solicit) at levels that help them cover the costs of journeys they intended to make anyway, but below amounts that would incentivize additional journeys.

Some city authorities appear to be responding positively. As we saw in the case study, Amsterdam limits rentals to 60 nights per year. Airbnb has revised its terms and conditions, highlighting hosts' obligations to declare what they earn, and to honor any conditions in their leases. It has also indicated that it is willing to collect hotel taxes in New York, Portland, and San Francisco. In New York the attorney general has targeted multiple-letters, demanded that Airbnb hand over information about hosts with more than one property listed on the site (allegedly such properties account for 30 percent of Airbnb's New York listings, and 102 letters have seven properties or more).[161] Airbnb has removed 2,000 hosts from its site, but is refusing to hand over information to the city.[162] San Francisco has adopted new rules (as we saw in the San Francisco case study).

Such approaches may not assuage the fears of incumbent businesses facing disintermediation or displacement by peer-to-peer letting, but they would help prevent the domination of previously affordable residential areas by "buy for (short) let" landlords, with negative consequences for social inclusion.

More broadly, Orsi and her colleagues suggest a series of regulatory reforms and other activities—notably using their procurement spending accordingly—that cities could implement to enable sharing of transport, food, housing, and economic opportunities. Among many practical suggestions they include: providing designated, discounted, or free parking for car-sharing; adopting a citywide public bikesharing program; updating the zoning code to make "food membership distribution points" a permitted activity throughout the city; subsidizing shared commercial kitchens; supporting the development of cooperative housing; allowing short-term rental in residential areas; expanding allowable home occupations to include sharing economy enterprise; and procuring goods and services from cooperatives.[163] In just one further example, Kirklees in the UK is developing a new city-run program that will "allow private citizens and businesses to offer up their unused resources—anything from vehicles to skills—to trade or barter, in a sort of time-banking system for the entire city."[164]

Smart Regulation for Sharing

The essence of smart regulation is simply to maximize effectiveness in delivering intended effects while minimizing burdens on those needing to comply. Arun Sundararajan argues for more self-regulation. "There's a real danger," he says, "that today's misalignment between newer peer-to-peer business models and older regulations will impede economic growth". He suggests delegating "more regulatory responsibility to the marketplaces and platforms while preserving some government oversight."[165] Mediated peer-to-peer marketplaces have built-in digital reputation systems, identity verification systems, supplier-screening protocols, and instant and transparent user feedback mechanisms that might all reduce the need for direct regulation. From this perspective moves such as Houston's new cumbersome 40-step regulatory process for "anyone who drives for pay"[166] are simply overkill.

As Sundararajan acknowledges, "Self-policing isn't a universal panacea." Government mandates will still be needed "for, say, providing accessible vehicles and ensuring disaster preparedness—things that markets don't easily self-provide."[167] Stokes and her colleagues similarly suggest a hybrid:

> Peer review and self-regulation tools ... do not and should not supersede the role of government. However, a combination of self-regulation—for activities where peer review is well suited—and government regulation could prove to be more powerful and efficient than today's regulatory norms.[168]

But Sundararajan is also sanguine about self-regulation in areas where we might be skeptical: notably finance and real estate, as well as in law and medicine. This does not necessarily argue for command-and-control regulation, but a smarter approach which involves regulators in setting requirements that genuinely empower other stakeholders—notably the peer users and providers in sharing systems.

So we would argue that in smart regulation, governments should not simply "proactively partner with or delegate responsibility to the platforms" for provision of a social safety net, as Sundararajan suggests.[169] They also need to empower users and providers, and support collective efforts through bodies like the Freelancers Union. Rather than trusting that competition between the platforms for good quality providers will mean they provide social protections, governments should guide and harness that competition. This might involve active support for unionized or cooperative alternatives at the city scale.

The sharing entrepreneur Debbie Wosskow's review of the sharing economy for the UK government—whose recommendations led to the choice in 2015 of Leeds and Manchester as pilot sharing cities in the UK—calls for clear minimum standards for health and safety, systems to ensure workers are paid at least the living wage, and for tax requirements to be clear and firmly regulated.[170] But it also suggests regulation proportionate to scale, arguing for instance that "someone renting out a spare room for a few nights [should not be] subject to the same level of regulation as a business renting out 100 rooms all year-round," while a skillsharing platform which plays only "a passive role in matching users" should not be classed as an employment agency.[171]

Albert Cañigueral of Spain's ConsumoColaborativo.com also sees the degree of transparency and involvement of users as naturally suiting sharing activities to a degree of self-regulation.[172] Sharing activities, he says, are rarely comparable enough with their conventional "competitors" for existing regulatory approaches to be directly transferable as a legal framework to ensure safety and security to users and platforms. Moreover, because of the diversity of different sharing activities, specialized approaches are generally needed. But he recommends two exceptions: a general framework for insurance to protect owners or lenders; and a generic approach to the treatment of income or benefits for the individuals involved in sharing. Cañigueral also argues that where possible, regulation should be developed at the local or city level, contrasting those administrations (at least in Europe) with national governments more subject to regulatory capture by incumbent business interests.

The treatment of justice concerns probably also requires specialized approaches, tailored to the political context, specific resources being shared and the models applied. But this does not mean that design for justice becomes a secondary, bolt-on concern. It also has to be considered at a system level, in generic ways. We noted above the scope for capping charges for or frequency of commercial sharing to avoid negative impacts that can arise from excess commercialization. This is the sort of principle that could helpfully guide regulators in this area. In addition, and perhaps more importantly, regulators need to learn from commons governance and enable participation by users and owners of shared resources—on an equal basis.

Similarly we would suggest that sharing sector regulators need to pay particular attention to labor protections and liabilities whenever the user of the service is an institution or business, rather than an individual. This might leave TaskRabbits able to undertake tasks for other householders on negotiated terms, but provide minimum wages, and other protections and security for those doing things for companies, and also prevent companies from abusing such sharing platforms to deliberately casualize their labor force. And, as we suggested earlier, supporting cooperative ownership models for sharing economy companies—Rabbits owning TaskRabbit, Zipsters owning Zipcar, hosts owning Airbnb and so forth—would offer a robust and sustainable solution.

There are good reasons (as we saw above) to expect larger benefits for communities from larger sharing networks. But that does not mean that sharing businesses should be permitted to operate monopolies. We would suggest that a commercial sharing business should not be permitted to build a monopoly anywhere antitrust or antimonopoly rules would prevent conventional business from doing so. So, for example, Uber or Lyft should not be permitted to dominate the taxi-cum-ride-share business in a given city. More generally, this implies that regardless of the enthusiasm of Cañigueral and Sundararajan, consumer protection regulation will not be completely replaceable with self-regulation and user-monitoring mechanisms.[173] It might even suggest that in the places where network benefits suggest natural monopolies the equivalent of nationalization might need to be considered to avoid abuses of monopoly power. In the US, the policy analyst Richard Eskow suggests that the level of past public research and investment that underpins platform monopolies like Google and Amazon, their social significance, and the high risks they pose to civil liberties might all argue for them to be turned into public utilities.[174] Meanwhile in France the influential think tank Centre Nationale de Numerique (National Digital Council) argues for intervention to prevent platform monopoly, so as to protect both consumer and citizen interests.[175] Such proposals, and related suggestions—for public interest duties for social media platforms[176] or Zittrain's "information fiduciaries" idea[177]—seem set to proliferate in the face of ongoing controversies around the exercise of power by Facebook and Google.

Strengthening Infrastructures and Enablers of Sharing

Our survey of sharing activity suggests there are few resources as yet untouched by the sharing economy, but that there are still many opportunities for wider application of the sharing paradigm in service co-production and in the political domain. But, we would suggest, truly effective and powerful models for sharing have perhaps yet to be identified in two areas. First in finance, where the potential synergies between local currencies, crowd financing, and credit unions may prove substantial in creating what we might call a citywide financial commons. Such a financial commons would simultaneously enable enhanced social investment and undermine the unelected power of bankers and financiers.

The second area to consider is land, where land reform and land taxation measures might unlock vast potential for enhanced sharing of scarce urban land. Getting land sharing right underlies the idea of the genuinely sharing city. It is not only a means to efficient use of perhaps the scarcest resource in cities, but inflated land and property values otherwise exclude poorer and disadvantaged groups from the other shared facilities and opportunities of the city. Delivering a "right to the city" requires cities to overturn the domination of land use and development by speculative interests in favor of diverse land sharing through, for example, land rights, community ownership, community titling of squatted land, more multiple use, recognition of informal uses, rent control, and land value taxation.

Enhanced sharing of land and money strengthen the foundations for sharing in many other dimensions—supporting everything from shared workplaces to community facilities. They would also ensure that cities continued to house dense diverse populations, providing the underlying market for effective sharing.

In parallel with reforms to maintain urban space, measures are also needed to protect and extend the shareability of cyberspace. Cities need to continue investing in ICT infrastructure, such as high-capacity broadband connections, and in ensuring access for all through, for example, free Wifi in public spaces and public buildings, and access to training and, where necessary, finance and resources to overcome the digital divide. Many of these measures overlap with the prescriptions of the "smart cities" discourse, but sharing cities would place much greater emphasis on ensuring equal participation, and civil liberties in cyberspace. Mechanisms to regulate and sensibly enable sharing platforms that are commercial enterprises, yet free to users through revenues from advertising or data provision are likely to be needed in this context.

Sharing cities would also recognize the importance of continuing to invest in soft infrastructures that enable collaboration and sharing, such as education, skills, capabilities, and encounter spaces in streets and parks.

All these underlying infrastructures can also be understood as commons, for which we can draw useful lessons from commons management.[178] The new virtual commons of the Internet may raise novel challenges because the spaces or infrastructure of these new commons are privately owned, while the content and activity are managed in common. But the urban commons of streets and public spaces; the educational commons of knowledge, skills and learning; the cultural commons of art and music: all these too, and more, are shared in cities.

Garret Hardin, in his 1968 article "The Tragedy of the Commons," suggests that in the absence of effective rules or norms for communication, users of common property tend to overconsume it rather than share it effectively, degrading its overall quality and productiveness. But this only applies to forms of property that can be degraded in use, rather than forms whose value is enhanced by the numbers using them—such as communication networks or social capital. More generally effective common ownership provides a mechanism where individual interests in immediate consumption can be moderated in the interests of the collective. With deliberative negotiation and communication, long-term and future interests are *more* likely to be considered.[179] Fair involvement of users is critical in effective commons management. Effective shared institutions will need to be part of the essential infrastructure of urban and virtual commons: institutions both to defend both the new commons and to protect civil liberties, especially where the underlying infrastructure is privately owned.

Governing Transition or Stimulating Transformation

Most of the policy recommendations of sharing advocates are entirely reasonable. But as a whole they take for granted a governance system that can translate objective policy into broadly predictable outcomes. Unfortunately, in the real world we know this is too great a simplification.

We need to consider urban governance in context as part of multidimensional city systems. This reminds us that while modern cities do enjoy a significant degree of political autonomy and executive power, they are particularly constrained by the economics of urban development, inter-urban competition, and economic globalization. To realize the sort of autonomy needed to deliver policies for the sharing paradigm, cities will need to build networked, distributed autonomy, involving citizens and communities in both the "how" of policy formulation and its implementation.[180] This will help enable sharing solutions to emerge from interstitial niches and countercultures, and symbiotically or subversively transform mainstream economic and political structures.[181]

Recognizing cities as systems can be helpful in two further respects. First it reminds us that cities are also nodes in physical flows of materials and energy with environmental consequences. Sharing policy can, and should, seek to intervene in these flows, closing loops and reducing environmental impacts. Second, and critically, it raises the question of what sort of systems cities are. Are they primarily sociotechnical systems, political-economic systems; or cultural-ecological systems; or indeed, in different ways, all three? The researchers Phillip Späth and Harald Rohracher note that the existing configurations in urban systems are very obdurate.[182] They highlight the importance of creating new visions or discourses at a regional level, shared among stakeholders, to deliver transformations in such contested sociopolitical domains. This is one reason we have emphasized sharing as a paradigm shift, rather than as an economic practice.

Such perspectives are essential if the urban system is also an economic-political one, and not merely sociotechnical in nature. The latter assumes that (to some degree) planned interventions can be assembled and implemented, while the former emphasizes political conflicts between interests which lead to economic expressions in urban form. The sharing paradigm recognizes that exploiting technological change—whether to build smart cities or sharing cities—is not a simple managerial process, even though technological changes can help overcome many practical (and even some psychological) obstacles to sharing. Technology is always embedded in sociotechnical systems in which practice is structured by social norms and other cultural factors as well as technological ones. In effect, much of this book has been about the importance of understanding the sociocultural side of sociotechnical systems.

But our simultaneous focus on culture as a political phenomenon with implications for justice means that we also recognize cities as economic-political systems. In this model, the physical and economic development and redevelopment of urban systems is seen in part as a product of the interests of the finance and property sectors.[183] The commercial emanations of the sharing economy may similarly reflect business and financial interests. In the face of disruptive business models, as with disruptive technology, incumbents can be expected to simultaneously *resist* new entrants—using political power—and *negotiate* with them. Some will change strategies to better complement the new intermediaries: for example manufacturing more durable products better suited to intensive shared use, or to hold resale value; adding features that make the product more easily sharable (such as in-car systems for keyless sharing); or more easily (re)customizable to multiple sharers' needs. But most will—as is typical in the face of disruption—resist.

So *policies* for sharing cities—which can help deliver the changes in technology and practices—must be complemented with a *politics* of the sharing city, which engages with norms and values, both driving and driven by the *people* in terms of the cultural transformation that the sharing paradigm involves.

Summary

In our case study of Amsterdam, we saw the city actively using existing policy levers as well as developing new policy and planning around quality of life and the public realm (like Copenhagen), exhibiting some of the same sharing-city ambition of Seoul, but in the perhaps more challenging setting of a Western intercultural city. Amsterdam is welcoming of both innovation and immigration, and offers high levels of participation in service evaluation and design that can be understood as the early stages of co-production. Enhancing sharing appears to be an important purpose of Amsterdam's "Smart City" projects, not just an incidental outcome.

In this chapter we saw sharing as being ultimately about developing human capabilities and delivering well-being. We argued that our thinking about sharing should start with the question of how it contributes to those capabilities, consequently, sharing approaches and shared resources that more directly deliver capabilities for all are the most important to encourage.

We considered a range of objections to sharing—and sources of opposition from "it's not human nature" to "sharing is a threat to consumer based identities"—not because we want to give them publicity, but because we want the reader to fully understand some of arguments that skeptics and sharing pessimists might use to obfuscate efforts to embed the sharing paradigm. Despite these objections and oppositional arguments, we posited the power of sharing as a vehicle for cultural transformation, based in explorations of reciprocity theory, trust, and empathy, and in the power of positive feedback in networks that help sharing spread as a cultural meme. We also saw the possibilities for insurgent countercultural forms of sharing to stimulate new norms.

We concluded the chapter by considering practical planning and policy options for the development of the sharing city, including a set of five principles: designing in justice, civil liberties, trust, user empowerment, and releasing intrinsic motivations. We also highlighted the outstanding potential for cities to enable a sharing infrastructure while strengthening the urban commons by developing synergies between local currencies, crowd financing, and credit unions as well as through land reform and land taxation that could unlock the huge potential for enhanced sharing of scarce urban land.

In effect, we have argued for a grounded, simultaneously *countercultural* and *intercultural* redefinition of consumption and development that is a collaborative, shared, co-production of services and products supplying fundamental human needs. This will in turn present a challenge to identities rooted in consumerism, not merely by offering different products or services, but different activities and—fundamentally—different relationships, responsibilities, and capabilities. Just another version of Herb Caen's eternal conflict producing eternal truths, perhaps?

6 Case Study: Bengaluru

Our final case study explores a global South city poised between the temptations of the global smart city and the needs of a rapidly growing population seeking new freedoms, yet facing severe environmental constraints—something of a microcosm of the challenges facing our urban futures. Bengaluru, still widely known as Bangalore, is the capital city of the southwest Indian state of Karnataka, and as we shall see, exhibits tensions common to the development of sharing and smart cities the world over.

As the third most populous city in India, its estimated 2014 population is just over 10 million people. Between the census years of 2001 and 2011, Bengaluru's population grew by 46.7 percent (to 9.6 million), the highest rate in the country.[1] Partly this rapid growth can be attributed to Bengaluru's rise as India's IT (information technology) capital, with both international and domestic technology companies settling in the city and attracting a surge of young Indian engineers and tech professionals. In the past two decades, Bengaluru has seen an overwhelming influx of this professional technology class, triggering something of an identity crisis for the city.[2]

The novelist Bharati Mukherjee describes Bengaluru as once the pleasantest city in India, a former British Army cantonment boasting wide boulevards, parks, and a perfectly inviting climate. Its recent population surge has brought traffic congestion and rapidly expanding suburbs. As Mukherjee notes, "Bangalore" is not only a city, but also a concept. For young Indian women in particular, who otherwise face limited socioeconomic prospects and educational opportunities, Bengaluru is a promise of a new life. Among the vast Indian middle class this promise might imply self-expression, money, and freedom; but for many American and European multinational corporations it has meant cheap outsourced operations.[3]

But Bengaluru is surpassing its reputation as an outsourcing hub, with more indigenous entrepreneurial startup companies, several of which are becoming global businesses in their own right.[4] It is also becoming known for pioneering research at its academic institutions and technology labs.[5] Co-working has taken off, as many startups are sharing office space to bring down operational costs. In fact, India's own Silicon Valley, as Bengaluru has been called, has become one of the largest technology innovation clusters in the world.[6]

At 3,000 feet above sea level, Bengaluru boasts a pleasant high-altitude climate throughout the year, adding to its attractiveness to businesses.[7] The business journalist Anirudha Dutta argues that in cities' race for attracting talent, "you just need 'place,' and 'place' is what Bengaluru brings to the residents who live in it." The city's ability to draw and retain young professionals reflects its ability to outcompete Mumbai with more affordable real estate, cleaner air, and stronger feeling of safety.[8] Yet poor infrastructure, pollution, and water scarcity are still serious problems that the city is struggling to keep up with as its population grows.

Bangalore Water Supply and Sewage Board (BWSSB) chief Gaurav Gupta warns that it will be at least 10 years before water and underground drainage facilities can be supplied in the newly developing suburbs of Bengaluru.[9] The Hesaraghatta and Tippagondanahalli reservoirs have been drained or are drying up, and the city is already drawing more water than allocated from the Cauvery River basin, the only remaining reliable water supply. V. Balasubramanian, the former additional chief secretary of Karnataka, suggests that as much as half the city may become uninhabitable due to water scarcity and contamination in 10 years if urgent measures are not taken.[10] Alternative sources of water, such as recycling or harvesting of rainwater, are being discussed as potential solutions.

As Bengaluru's population explodes, the city's waste management is also reaching a critical point. The political journalist Aravind Gowda highlights the growing consternation in the nearby village of Mandur that serves as Bengaluru's landfill.[11] Residents are up in arms against the 1,800 tons of garbage being dumped in their backyard on a daily basis. A one-year deadline on the Karnataka government's promise to the villagers of Mandur to find a different dumping location has now passed.[12] Villagers complain of various health ailments believed to be caused by the garbage, and in June 2014 they began staging demonstrations that blocked waste-hauling trucks from entering the village. The Greater Bangalore City Corporation (GBCC) has failed to find a solution, and has even hired police protection for the dumping grounds. It has been suggested that corruption may be contributing to delays in construction of new more efficient waste processing facilities.[13] These challenges contrast starkly with the co-produced approaches to waste we saw in Seoul, as well as in Buenos Aires (see chapter 2), and they emphasize the desperate need to build trust and solidarity among the city's citizens and stakeholders.

In this context the entry of some of Bengaluru's IT professionals into local politics is ambiguous.[14] These tech professionals are typically viewed by longtime residents as incomers, insulated from the city's struggles by their affluence, contributing to traffic congestion and the strain on resources while driving up prices. On the other side, the newcomers are frustrated by the city's collapsing infrastructure and apparent lack of political will to address it. They campaign under a promise of clean, responsive governance, presented as a break with the status quo, which has allegedly failed to support grassroots initiatives for sustainable city transport and waste management.[15]

The potential advantages of Bengaluru's most prominent citizens engaging with city government, investing both time and wealth, are clear. But so are the risks. One engineer who has lived in the city for 20 years fears this new political force is strictly self-interested, arguing that "they want to show their clients that Bangalore is a world-class city, whatever that means. The rest of Bangalore might as well not exist."[16]

The more these challenges come to a head, the more Bengaluru presents itself as a laboratory to explore alternative and shared solutions. In 2012 a local nonprofit research group called the Institute for the Future (IFTF) organized a workshop at Jaaga, a temporary gathering location, for an expanding hub of artists, designers and cultural entrepreneurs around Bengaluru, to help develop a work program for mapping information, and data on access to public services and infrastructure to assist future city planning.[17] The participants prioritized efforts to enable poor groups to participate in and benefit from the process, and highlighted the need to focus on key resources such as water and land and on crowdsourcing of public services.

One appealing route for Bengaluru is to pursue its potential to become India's first smart city. It is one of a hundred Indian cities slated for modernization and investment as "smart cities."[18] The reporter N. V. Krishnakumar suggests Bengaluru already meets the first three of six broad characteristics: smart economy, smart people, and smart mobility; and the city has good potential to also deliver a smart environment, smart living, and smart governance. "Amongst all cities of India, [Bengaluru] has the pedigree to evolve into a smart city the fastest," he concludes.[19]

The city is investing heavily in improved, smarter transport, including extension of the Namma Metro by 72 kilometers (44 miles) by 2019,[20] in addition to the 42 kilometers (27 miles) currently due for completion in 2016.[21] A proposed Bangalore Metropolitan Transport Corporation (BMTC) smart card will be accepted by Namma Metro and could extend to transport associations, such as taxis and auto rickshaws, which could be integrated using exclusively open source software.[22] The metro is also involved in Namma Wifi, which provides limited free Wifi access in 5 public locations in the city with 10 more proposed. The first, launched in January 2014, was based at the MG Road Metro Station.[23]

Free Wifi access is a small step toward more participatory decision making in order to allow for smart governance and living—as urged by Krishnakumar, who also advocates measures such as establishing an online portal for citizen feedback on public services. Krishnakumar also recognizes the importance of social infrastructure and public services in an inclusive smart city. He says:

> Social infrastructure needs to be upgraded. ... Government primary schools must become e-learning centers while health clinics are required in every ward that can provide quality healthcare including remote treatment and tele-assistance. Making affordable housing available for the poor and marginalized as well as slum redevelopment plans requires substantial investment.[24]

But the smart city buzz in Bengaluru and elsewhere is often less inclusive and less participative than this might suggest. Dholera—"India's twenty-first-century utopian urban experiment"—is one of 24 entirely new smart cities planned across the country in order to accommodate India's rapidly expanding population. But the geographer Ayona Datta highlights how villagers and small-scale subsistence farmers who currently inhabit the proposed site have been staging peaceful protests with support from the grassroots land rights movement Jameen Adhikar Andolan Gujarat (JAAG). In addition to the displacement effects of the planned city, it is also being criticized for the unaccounted engineering expenses that will arise with the high risk of flooding in the chosen location on an expanse of salt flats.[25]

The proposals for Dholera echo the development of Masdar in the United Arab Emirates. Sennett describes Masdar as "a half-built city rising out of the desert, whose planning ... comprehensively lays out the activities of the city, the technology monitoring and regulating the function from a central command center." Such smart cities, he says, are "over-zoned, defying the fact that real development in cities is often haphazard, or in between the cracks of what's allowed."[26]

The urbanist Adam Greenfield of the London School of Economics also targets Masdar in his critique of the smart city paradigm. The idea of a "turnkey installation" of "a collection of technologies that, once deployed, will function consistently and uniformly" is mistaken. Ignoring the specifics of place and social milieu, and above all the inhabitants of cities, is a recipe for disaster.[27] Greenfield calls instead for recognition of the value of democracy, "citizen cunning and unglamorous technology."[28] Mathieu Lefevre, the director of the New Cities Foundation, similarly highlights how promotional images of smart cities are "entirely devoid of human life," but he argues that in reality "cities succeed not because of how 'smart' they are, but because of how human they are."[29]

The urban researchers Hug March and Ramon Ribera-Fumaz point out that the "smart cities" ideology also acts to depoliticize urban planning and development.[30] The architect Michele Provoost also highlights the anti-democratic nature of "smart" cities. She argues that they have a deliberate social dark side; that smart infrastructure is being marketed to cities intentionally to construct a privatized, commercial platform for services, health, and education, intentionally replacing sociocultural sharing and enclosing the existing commons.[31] From this perspective the smart city is platform capitalism on steroids, with every citizen as part of its captive market. Even worse, in the stand-alone model of new smart cities built from scratch, following the model of Songdo in South Korea for example, they provide exclusive private services only for the rich, enabling the elite to flee the megacities.

Provoost is also scathing about the design of stand-alone smart cities: often highway oriented, car-based—with exclusive spaces and design drawing on US cul-de-sacs rather than inclusive ones learning from European neighborhoods. She highlights a deep divide in architecture between "starchitect" advocates of commercially motivated, privately financed, so-called sustainable smart cities; and those—like Jamie McGuirk, who we met in chapter 4—working with "self-organized cities," in which development possibilities are emerging from the slums and favelas through collaboration.

Sennett similarly sees more potential in self-organization:

> A more intelligent attempt to create a smart city comes from work currently under way in Rio de Janeiro. Rio has a long history of devastating flash floods, made worse socially by widespread poverty and violent crime. In the past people survived thanks to the complex tissues of local life; the new information technologies are now helping them, in a very different way to Masdar and Songdo. ... The technologies have been applied to forecasting physical disasters, to co-ordinating responses to traffic crises, and to organizing police work on crime. The principle here is co-ordination rather than ... prescription.[32]

Sennett also asks: "But isn't this comparison unfair? Wouldn't people in the favelas prefer, if they had a choice, the pre-organized, already planned place in which to live?" No, he answers, research reveals that "once basic services are in place people don't value efficiency above all; they want quality of life. ... If they have a choice, people want a more open, indeterminate city in which to make their way; this is how they can come to take ownership over their lives."[33]

Bengaluru has the opportunity to become an inclusive, participatory, emergent smart city of the sort Sennett praises. But that means rejecting the lure of the "global city" and instead adopting the new sharing paradigm, learning as much from Medellín and Seoul as from San Francisco, and as much from Amsterdam as from New Amsterdam (now known as New York). If an alliance can be built between the IT crowd and the urban citizenry, Bengaluru could build a shared urban commons extending to land reform and financial innovation, not just a shared virtual commons—or worse still, an exclusive virtual commons for the elite, part of a global city. Such an elite virtual commons would add separation and exclusion in cyberspace to the separation and exclusion in physical space that arise where development capital is allowed to direct urban development. If the IT skills of Bengaluru can be harnessed to social objectives it will have a head start for developing sharing infrastructures such as: peer-to-peer marketplaces; neighborhood networks for co-production; platforms for repair, upgrading, and customization services; a mainstream virtual complementary currency; and indeed a whole ecosystem of apps for phones and tablets enabling sharing of any kind of product, skill, time, or service, virtually anywhere[34].

The challenges Bengaluru currently faces present an incredible opportunity to develop a smart *and* sharing vision for the future, one that can serve as a model for other cities in the global South.

Synthesis

In our book we believe we've successfully made the case that rewiring our minds and our cities toward the sharing paradigm is the single most important task for urban governance and urban futures in the twenty-first century. To briefly synthesize some of our thoughts, we revisit and reflect upon the case laid out in our introduction for both *understanding cities as shared spaces, and acting to share them fairly*.

Our hope is that through our arguments, literature reviews, case studies, and other examples, we have contributed to both an *understanding* of cities as historically shared spaces, and set out policy and planning strategies on how to *act* upon this understanding with solid ideas for implementing policies and plans at the municipal level. We have shown what some of the more progressive cities are doing to deepen our understanding of the potential for sharing. But at present, for every Seoul with an explicit, proactive, and multidimensional strategy for sharing, there are hundreds of cities simply *reacting* to sharing trends, with no strategy, no policy, no coordination, and presumably little or no understanding.

But there are also many cities that do not use the explicit (and very fashionable) language of sharing yet are developing policy and planning that *contributes* toward the sharing paradigm. Much of this activity, however, is disconnected or even incidental to other "economic" policy goals. These cities need help in understanding their roles, responsibilities, and the benefits they will accrue: the emerging Sharing Cities Network, supported by Shareable (www.shareable.net/sharing-cities), aims to get 100 cities to emulate Seoul as a formal "sharing city" by 2015; and existing national and international associations of cities such as ICLEI-Local Governments for Sustainability—who took an early and impressive lead in galvanizing Cities for Climate Protection in 1993—and the National League of Cities should help to rectify this as a matter of urgency.

The specific points in our case were as follows:

- Humans are natural sociocultural sharers.

We believe the evidence for sharing and cooperation as evolved, sociocultural traits, summarized in chapter 2, but permeating throughout the book, is compelling. We have argued that, as a result of commercialization and consumerism in modern economies, these natural traits, and associated norms and values have, literally taken a back seat as our competitive, individualized selves took over, particularly in the era of neoliberalism. We have highlighted the contrasts between commercial discourses of the sharing economy that are easily co-opted by neoliberalism, or are at best reformist, and the potentially transformative communal sharing discourses that are fundamental to our sharing paradigm.

In this sense we are arguing for a cultural shift toward sharing as a part of a political reawakening of our natural sharing tendencies. In doing so, however, we have been careful not to fall into the "post-political trap." Sharing is also about changing values in political space, not just about changing individual behaviors in consumer space. And in stressing the need for a cultural shift and changed behaviors, values, and norms, we also see the need for the building of new public institutions as both a product of the new norms, and as a mechanism to fully reflect them in society.

We are also not seeking to privilege traditional sociocultural sharing over the new mediated models that are spreading in both commercial and communal flavors across the digital commons of the Internet. We have seen that different incentives to share will work in different cultures, and that sharing models at the commercial end of the spectrum can change norms (and even values) in ways that are probably an essential precursor to delivering political support for the necessary shifts. But we argue for awareness that socioculturally evolved sharing traits underpin the new mediated forms, and despite the growth of commercial platforms like Airbnb, still constitute the vast majority of all sharing behavior around the world. They are the hidden mass of the iceberg of sharing, essential, but easy to miss. Moreover, commercial sharing alone cannot restore the shared public realm nor the urban commons that we have described, nor the hard and soft infrastructures and resources that are the foundation of all successful urban economies.

- The future of humanity is urban, which necessitates sharing: of resources, infrastructures, goods, services, experiences, and capabilities.

The globe is becoming ever more urban, yet even the most economically successful cities face a host of social and environmental challenges. We argue that the sharing paradigm offers a framework and a robust set of understandings and actions that can yield myriad opportunities for rediscovering and reinvesting in the public realm and urban commons, leading ultimately to urban transformation. The future of sharing and the future of cities are increasingly and inextricably interlinked, and sharing cities could be the trigger for new economic models that deliver "just sustainabilities."

The mediated, commercial spaces of collaborative consumption have a role to play in the emanation of the sharing city as a platform for sharing goods, services, and experiences, with potential to massively increase the efficiency of resource use (which is already higher in cities than in suburban sprawl). But it is in the revival of sharing and co-production of the multiple dimensions of the urban commons—through co-produced services of health and education, through co-creation of the cultural and financial commons of the city, through collaborative efforts to build gray and green infrastructures of transport, power, water, sanitation, clean air, green space, and waste reduction—that the sharing paradigm can deliver the very fabric of future cities. In losing sight of the shared nature of these urban commons, cities have devalued them, and contributed to both social and spatial inequalities and injustices as a result of selective degradation (especially in low income neighborhoods), and privatization of the public realm. And with that has come declining trust, and eroded social capital.

In devaluing the urban commons, cities have also undermined their potential to support creative and productive economies, and they have been forced to compete for footloose development capital, as well as to give up political autonomy to corporate and financial interests. Yet within the sharing paradigm, co-production can extend to the productive economy as well as to the reproductive economy of shared public services. Our focus on "the urban" helps reveal the commonalities in these various different expressions of collaboration and co-production. It also reveals the scale of the challenge if cities are to adopt truly transformative sharing approaches.

- New opportunities for sharing will give new opportunities to enhance trust and rebuild social capital.

But, from acorns grow mighty oaks. Every sharing interaction, even mediated and commercial ones, is interpersonal, and an opportunity for trust building. As sharing begets trust, so trust begets more sharing. And every sharing organization, every platform, every human interaction is another institutional deposit in a new city bank of social capital. Where social capital is weak and consumerism dominant, commercial mediated models with formal mechanisms for building trust and reputation will be needed to extend sharing beyond family and close friends. But once that barrier is overcome, sharing offers the chance to experience empathy with, and build trust in the different others with whom we share the physical realm of the city and its underlying infrastructures and commons.

And alongside investing in mechanisms to enhance trust and reputation among individual sharers, even commercial mediated models of sharing actively seek to build cosmopolitan sharing communities. And as we saw, these are inevitably more authentic forms of social capital than the brand communities of global corporations. But commercial models also seem more vulnerable to the pressures of financial capital—especially venture funds—which push them to redefine their missions away from the social purposes sharing entrepreneurs often exhibit; this results in the potential commodification of yet more areas of life, and threatens the casualization of labor in the sharing economy. In such ways the focus of commercial models can further exacerbate the very precariousness of urban life that they might have been established to confront.

• Sharing with equity and justice can naturally shift cultural values and norms toward trust and collaboration.

We have illustrated throughout our book, various shortcomings of sharing without equity or justice, or where both have been retrofits once the economic or environmental benefits of sharing were confirmed. We saw dangers such as co-option of the sharing economy, exclusive sharing for the privileged, farmers markets for white liberals, gated communities, Airbnb for whites, "hobby" making, and self-provisioning and self-build for a wealthy elite.

Instead we advocated supporting cooperative models of sharing business rather than market- or venture-funded models. We advocate as well active city support to enable gift-based communal sharing to replace market transactions, and to supplement public services, in ways that build social inclusion and release people from the constraints of inability to pay.

Such models—free of excessive commercial pressures—are inherently better suited to building cultural norms of trust and collaboration. If participants feel they are being treated instrumentally, or even exploited as casual labor or advertising fodder for example, then trust will grow slowly if at all. These risks of commercialized, exclusive, unfair sharing mean we cannot realize the benefits of sharing without active intervention to ensure sharing systems are designed around justice and equality.

At this point we must restate that while reciprocity is potentially a very powerful tool to change behavior and influence values, it does not apply directly to corporations and other institutions. The case for an explicit framework of human rights, and for interventions to guarantee capabilities—especially where the threats to those arise largely from the actions of institutions and corporations—therefore remains. The hope of the sharing paradigm is that our political capacity to agree on such rights and design new institutions will be enhanced in sharing cities.

• An enhanced public realm or urban commons establishes a precondition and motivation for collective political debate that recognizes the city as a shared system.

Fundamental to our arguments throughout the book has been that the current sharing trend must be understood and developed politically and culturally, not just technologically and behaviorally, if we are to get more just, inclusive, and environmentally sustainable sharing. Rebuilding social capital in sharing could also help rebuild the public square of collective politics. Without this, extending sharing city action to the infrastructures, urban commons, and public realm of the city would be difficult. But the ways in which social capital can be rebuilt (and cultural norms shifted) by sharing with justice reveal another virtuous cycle: as the public square of collective politics is strengthened, so investments in sharing infrastructures are more easily agreed, and those investments in turn lead to more sharing and even stronger social capital. The domains of sharing and support for sharing can expand hand in hand until they encompass the urban commons and indeed the whole city.

However, throughout the book we have seen potential challenges to sharing from incumbent interests (in both commerce and politics) and obstacles resulting from the co-option of the sharing economy by commercial and financial interests. So the cultural effects of sharing in the political realm will be resisted and contested.

Yet sharing can build the political constituency for change. We have seen how shared knowledge—particularly through web access—is critical to inclusion and potentially transformative. Libraries (such as Medellín's Parque Biblioteca), self-organized and popular education, and online creative and knowledge commons all offer access to knowledge and learning opportunities that can underpin political participation. Political inclusion is also fundamental to democracy, while in turn genuine democracy offers the essential means to manage injustice through public reasoning and deliberation. Moreover, insofar as sharing helps get commerce, consumerism, and wealth out of politics it becomes a way to make democracy more meaningful.

• "Sharing the whole city" should become the guiding purpose of the future city.

A vision of the sharing city then, must extend way beyond bikesharing, free Wifi and a supportive view of Airbnb. It must extend to the whole city physically, the whole city socially and—crucially in our increasingly different and diverse cities—the whole city interculturally. It implies integrating policies to support sharing with public deliberation to shift power, politically, over the key resources of land and finance. In these respects the sharing city stands in counterpoint to Saskia Sassen's idea of the global city: driven by financial interests; and structured by the development industry into two exclusive tiers, a central zone tied into elite global markets and a periphery of surplus and casual labor. The sharing city equally challenges the discourse of the smart city, competing for inward investment by building up a high-tech core. Yet our vision offers a way to disconnect the smart city concept from an exclusive, competitive agenda.

Adopting what we are calling the "sharing paradigm" offers cities the opportunity to lead a transition to just sustainabilities. This offers a radically different vision compared with a global race to the bottom to attract footloose investment capital. It redefines what "smart cities" of the future might really mean—harnessing smart technology to an agenda of sharing and solidarity, rather than one of competition, enclosure, and division.

In conclusion therefore, as we understand it, sharing offers both a sustainable foundation for participatory urban democracy and a transformative approach to urban futures. The emerging sharing paradigm as we've described it, echoes and helps fulfill the basic tenets of the Right to the City as an idea, a movement, and manifesto:

> Right to the City was born out of desire and need ... [for] a stronger movement for urban justice. But it was also born out of the power of an idea of a new kind of urban politics that asserts that everyone, particularly the disenfranchised, not only has a right to the city, but as inhabitants, have a right to shape it, design it, and operationalize an urban human rights agenda. (www.righttothecity.org)

But as citizens we need to go further in claiming that right to make and remake the city, not simply to access it. That process of making and remaking is fundamentally collaborative and collective, as a shared endeavor among all citizens and a project for the sharing city.

As we have seen, sharing does not just offer this potential in cities in the rich world—it also offers a new strategy and direction for cities in the developing world. The sharing paradigm has broad application to urban challenges including slum improvement and infrastructure development—which might be met partly through communally shared and co-produced sanitation, water and power supply, and distribution investments, for example. And perhaps even more importantly, a sharing culture offers the potential to build greater empathy and solidarity between rich and poor neighborhoods, rich and poor cities, and the rich and poor worlds. We cannot overstate the importance of building a sense of global community in face of breakdowns in global governance, especially regarding climate change, where cities have been the only real leaders for the past 25 years. In this way, sharing our "urban living space" can be seen as a metaphor for, and a step toward, sharing our "global living space" in justice and sustainability.

Beyond the Sharing City: Co-creating the Future

We close with a handful of themes that have arisen in the course of writing our book, which strike us as being of wider import. We share them here in a spirit of collaboration, as incomplete and emergent thoughts and questions, seeking to incite transdisciplinary consideration and investigation.

First, we see potential in integrating thinking that is being done about empathy largely as a tool for social inclusion with the more theoretical concept of recognition and the political concept of dignity, and exploring the potential for them to influence policy and planning. Recognition of an individual or group's identity and moral standing would appear to be an expression of empathy. And the recognition of one's own privilege in light of growing inequality is perhaps an essential precursor to developing empathy.

Such explorations of empathy and justice would require not only *inter*disciplinary attention, linking psychology, philosophy, and justice, but also *trans*disciplinary approaches as we recognize the interlinked nature of our urban (and societal) problems. These problems are not resolvable by academics, professionals, or experts in isolation.

Secondly, further development and exploration of the sharing paradigm might offer ways to transcend existing tensions in notions of "development" and "progress." Sociocultural sharing approaches such as we have described them can seem the antithesis of "development," and even a little "backward." Yet by examining the different flavors of sharing, and how they bleed into one another, and critically how they may impact in different cultures, we saw ways of offering mediated sharing approaches that combine the communal values of sociocultural sharing with the freedoms of individualized markets. Such thinking, we suggest, might have wider application to discourses of development and progress.

Third, we see much potential in the further development of reciprocity theory as a way of thinking about influencing behavior, values, and norms. Particularly, we hope that the way we have applied it here with respect to sharing behaviors and flavors offers a way to reconsider those debates over behavior change for sustainability that have focused either on the intrinsic motivations of values, or on the extrinsic motivations for behavior. Moreover, in the way we have applied reciprocity theory, it raises the prospect that sharing (and other similar interventions) might be seen not only as an instrumental, utilitarian way of getting better outcomes (efficiency or equality for example), nor indeed only as something shifting individual values, but also as a much more transformative way of simulating the emergence of cultures that are tolerant, respectful, and fair, for example. We are acutely aware that this could be perceived as social engineering or patronizing, but would note that we support participatory and deliberative decision-making processes for the adoption of any such interventions.

Fourth, and connected to the case for participatory decision making, we believe firmly that real solutions to the problems of cities and society can only emerge if they are co-produced by informed community members and enlightened city officials—as they are currently in many experiments in empowered participation around the world. In this context we were struck by the potential power of self-organized learning across the various educational systems we considered in chapter 2. It seems to us that of all the specific forms of sharing and collaboration we document in this book, the ramifications of self-organized learning are potentially the most far reaching for the development of capabilities and for the development of genuinely participatory politics, in which people engage not only in debate and the setting of agendas, but also in the construction of the processes and institutions involved.

Fifth and finally, we believe we have helpfully extended the concept of the urban commons beyond its usage as a description of the co-created physical and cultural spaces of the city, to also encompass the potential for co-created and co-governed commons across multiple dimensions of urban systems: cultural, financial, infrastructural, and more. In the light of commons governance literature—in which it is almost universally agreed that in the right conditions, effective communal governance of commons can emerge—we suggest that reconceptualizing urban and even wider global systems as commons could be morally beneficial, could decrease inequality, and decrease resource use.

Afterword

As we complete our text, we see a battle raging on blogs, websites and newspaper pages, a battle over the definition and meaning of sharing, and the sharing economy. Should sharing be understood narrowly, or broadly? Should the booming mediated forms be called sharing at all? Should its supporters concede that the term sharing has been co-opted by commercial interests, or continue to fight for its inherent communal meaning?

Our position is clear. As veterans of decades-long battles over the term "sustainability" we have never seen circumstances so ripe with potential:

The global economic system is still fragile and hasn't stopped stuttering since the financial crisis of 2008.

A whole generation has emerged of young people who are questioning the values of consumerism, especially but not only in the US—the heartland of consumer capitalism.

Contemporary technologies and skills provide an unparalleled capacity to build a just and sustainable sharing society.

And the fast-growing mediated "sharing economy" itself is absolutely loaded with what we have called subversive or symbiotic potential to transform values and norms even as it wins support from existing elites.

In these circumstances, why wouldn't you fight?

Notes

Introduction

1. David Harvey, *Social Justice and the City* (London: Edward Arnold, 1974).

2. Manuel Castells, *The Urban Question: A Marxist Approach* (Cambridge, MA: MIT Press, 1977).

3. Manuel Castells, *The Rise of the Network Society* (London: Wiley-Blackwell, 1996).

4. Saskia Sassen, *The Global City: New York / London / Tokyo* (Princeton, NJ: Princeton University Press, 2001).

5. Leonie Sandercock, *Cosmopolis II: Mongrel Cities of the 21st Century* (London: Continuum, 2003).

6. Richard Florida, *The Rise of the Creative Class: And How It's Transforming Work, Leisure, Community and Everyday Life* (New York: Perseus Book Group, 2002).

7. Charles Landry, *The Creative City* (London: Earthscan/Comedia, 2008).

8. Jeb Brugmann, *Welcome to the Urban Revolution: How Cities Are Changing the World* (New York: Bloomsbury Press, 2009).

9. Susan Fainstein, *The Just City*. (Ithaca and London: Cornell UniversityPress, 2011).

10. Edward Glaeser, *The Triumphant City* (New York: The Penguin Press, 2011).

11. David Harvey, *Rebel Cities: From the Right to the City to the Urban Revolution* (London and New York: Verso, 2012).

12. World Bank, "Urban Development" (2012). Available at http://data.worldbank.org/topic/urban-development.

13. *The Economist,* "Open-Air Computers," 25 October 2012. Available at http://www.economist.com/news/special-report/21564998-cities-are-turning-vast-data-factories-open-air-computers.

14. By 2025, there will be at least 24 megacities of 10 million people or more but more than 50 online "communities" of 10 million or more (and at least 10 times more objects connected to the grid). McKinsey, cited by James McRitchie, "Pay It Forward for the World of 2025," *Corporate Governance*, 21 October 2014. Available at http://linkis.com/corpgov.net/2014/10/cQacS.

15. Will Steffen, Katherine Richardson, Johan Rockström, et al., "Planetary Boundaries: Guiding human development on a changing planet." *Science*, 347(6223) (2015).

16. Kate Raworth, *A Safe and Just Operating Space for Humanity: Can We Live Within the Doughnut?* (Oxford: Oxfam Discussion Papers, 2012).

17. Hans (JB) Opschoor and Rob Weterings, *Environmental Utilization Space* (Amsterdam: Boom, 1994); Joachim Spangenberg et al. *Toward Sustainable Europe.* (Brussels: Friends of the Earth Europe, 1995); Sara Larrain, "La línea de dignidad como indicador de sustentabilidad socioambiental. Avances desde el concepto de vida mínima hacia el concepto de vida digna." *Polis* 3 (2002). Available at http://polis.revues.org/7695#text.

18. Raworth, *A Safe and Just Operating Space*, 4.

19. Duncan McLaren, "Environmental Space, Equity and the Ecological Debt," in *Just Sustainabilities: Development in an Unequal World,* ed. Julian Agyeman, Robert D. Bullard, and Bob Evans (London: Earthscan, 2003), 25.

20. Harvey, *Rebel Cities.*

21. Julian Agyeman, Robert D. Bullard, and Bob Evans, "Exploring the Nexus: Bringing Together Sustainability, Environmental Justice and Equity." *Space and Polity* 6(1) (2002): 70–90, 78.

22. Susan Shaheen, Stacey Guzman, and Hua Zhang, "Bikesharing in Europe, the Americas, and Asia: Past, Present, and Future." Institute of Transportation Studies Report (2010). Available at http://escholarship.org/uc/item/79v822k5.

23. One honorable exception is the Bicycle Coalition of Greater Philadelphia and the Philadelphia Mayor's Office of Transportation and Utilities, which is holding focus groups with minority and low income populations to influence the design, structure, marketing, and outreach of their planned bikeshare system. The tender for research is available at http://bicyclecoalition.org/wp-content/uploads/2014/08/RFQ-Bike-Share-Perceptions-2014.pdf.

24. Geoffrey West notes that as cities grow, so does the scope for such social and economic interactions: doubling of city size means 15 percent more interaction per capita and also 15 percent less infrastructure per capita. See Geoffrey West, "Scaling: The Surprising Mathematics of Life and Civilization." *Foundations & Frontiers: Santa Fe Institute Bulletin* 28(2) (2014). Available at https://medium.com/sfi-30-foundations-frontiers/scaling-the-surprising-mathematics-of-life-and-civilization-49ee18640a8.

25. Erik Swyngedouw, "Impossible/Undesirable Sustainability and the Post-Political Condition," in *The Sustainable Development Paradox,* ed. J. Rob Krueger and David Gibbs (New York: Guilford Press, 2010), 13–40.

26. Richard Belk, "Why Not Share Rather Than Own." *Annals of the American Academy of Political and Social Science* 611 (2007): 126–140.

27. Don Tapscott and Anthony D Williams, *Wikinomics: How Mass Collaboration Changes Everything* (London: Atlantic Books, 2006); *Macrowikinomics: Rebooting Business and the World* (London: Atlantic Books, 2010).

28. Tapscott and Williams, *Macrowikinomics.*

29. Viktor Mayer-Schönberger and Kenneth Cukier, *Big Data: A Revolution That Will Transform How We Live, Work and Think* (London: John Murray, 2013).

30. Cheyanna Weber, "Sharing Power: Building a Solidarity Economy," *Shareable,* 5 October 2011. Available at http://www.shareable.net/blog/sharing-power-building-a-solidarity-economy.

31. Ethan Miller, "Solidarity Economy: Key Concepts and Issues," in *Solidarity Economy I: Building Alternatives for People and Planet,* ed Emily Kawano, Tom Masterson, and Jonathan Teller-Ellsberg (Amherst, MA: Center for Popular Economics, 2010), 25.

32. Ethan Miller, "Solidarity Economy: Key Concepts and Issues."

33. Jeremy Rifkin, *The Zero Marginal Cost Society* (New York: Palgrave Macmillan, 2014).

34. P2PValue. "What is P2Pvalue?" Available at http://www.p2pvalue.eu.

35. Rifkin, *Zero Marginal Cost Society.*

36. David Harvey, *The Enigma of Capital and the Crises of Capitalism* (London: Profile Books, 2011).

37. Neal Gorenflo, quoted in Sharon Ede, "Transactional Sharing, Transformational Sharing," Share Adelaide, 21 October 2011. Available at http://www.shareadelaide.com/transactional-sharing-transformational-sharing.

38. Rachel Botsman, "The Sharing Economy Lacks a Shared Definition: Giving Meaning to the Terms," *Fast Company Exist,* 21 November 2013. Available at http://www.fastcoexist.com/3022028/the-sharing-economy-lacks-a-shared-definition.

39. Rachel Botsman and Roo Rogers, *What's Mine Is Yours: The Rise of Collaborative Consumption* (London: Harper Business, 2010).

40. Susan Wise, "New Survey Reveals Disownership Is the New Normal." Sunrun Power Forward blog, 3 April 2013. Available at http://blog.sunrun.com/new-survey-reveals-disownership-is-the-new-normal.

41. Jeremy Rifkin, *The Age of Access: How the Shift from Ownership to Access Is Transforming Modern Life*. (London: Penguin, 2000).

42. Fleura Bardhi and Giana M. Eckhart, "Access-Based Consumption: The Case of Car Sharing." *Journal of Consumer Research* 39 (2012): 881–898.

43. Fleura Bardhi and Giana Eckhardt, "The Sharing Economy Isn't about Sharing at All." *Harvard Business Review,* 28 January 2015. Available at https://hbr.org/2015/01/the-sharing-economy-isnt-about-sharing-at-all.

44. Botsman and Rogers, *What's Mine Is Yours*.

45. Rob Walker, "Like Minds: A Conversation with Rob Walker," Archive Global (undated), http://archiveglobal.org/unconsumption; and Unconsumption website, http://unconsumption.tumblr.com.

46. Yochai Benkler, "Sharing Nicely: On Shareable Goods and the Emergence of Sharing as a Modality of Economic Production." *The Yale Law Journal*, 114 (2) (2004): 273–358.

47. Kathleen Stokes, Emma Clarence, Lauren Anderson, and April Rinne, *Making Sense of the UK Collaborative Economy* (London: Collaborative Lab and NESTA, 2014).

48. Botsman and Rogers, *What's Mine Is Yours*.

49. Matthew Yglesias, "There Is No 'Sharing Economy,'" *Slate*, 26 December 2013. Available at http://www.slate.com/blogs/moneybox/2013/12/26/myth_of_the_sharing_economy_there_s_no_such_thing.html.

50. Sven Eberlein, "Sharing for Profit—I'm Not Buying it Anymore," *Shareable*, 20 February 2013. Available at http://www.shareable.net/blog/sharing-for-profit-im-not-buying-it-anymore.

51. Benkler, "Sharing Nicely."

52. Cathy Goodwin, Kelly L. Smith, and Susan Spiggle, "Gift Giving: Consumer Motivation and the Gift Purchase Process," in *NA—Advances in Consumer Research Volume 17*, ed. Marvin E. Goldberg, Gerald Gorn, and Richard W. Pollay (Provo, UT: Association for Consumer Research, 1990), 690–698.

Chapter 1, Case Study: San Francisco

1. Alison Vekshin, "San Francisco Beating Silicon Valley with Growth in Jobs," Bloomberg.com, 25 March 2014. Available at http://www.bloomberg.com/news/2014-03-25/san-francisco-beating-silicon-valley-with-growth-in-jobs.html.

2. Inbal Orpaz, "San Francisco: The New Golden Gateway for Israeli High-Tech Startups," *Haaretz News*, 17 March 2014. Available at http://www.haaretz.com/business/.premium-1.580154.

3. Botsman and Rogers, *What's Mine Is Yours*, 169.

4. Sara Horowitz, "What is New Mutualism?" *Freelancers Broadcasting Network*, 5 November 2013. Available at https://www.freelancersunion.org/blog/2013/11/05/what-new-mutualism.

5. Jung-Hoon Lee and Marguerite Gong Hancock, *Toward a Framework for Smart Cities: A Comparison of Seoul, San Francisco & Amsterdam* (Yonsei University, Seoul, and Stanford Program on Regions of Innovation and Entrepreneurship, 2012): 13. Available at: http://iis-db.stanford.edu/evnts/7239/Jung_Hoon_Lee_final.pdf.

6. Noreen Malone, "How San Francisco's Latest Gold Rush Has Transformed the City. *New Republic*, 25 November 2013. Available at http://www.newrepublic.com/article/115731/what-tech-wealth-has-done-real-estate-san-francisco.

7. Michael Stoll, "San Francisco Pitched as Beacon of 'the Sharing Economy,'" *Shareable*, 4 April 2012. Available at http://www.shareable.net/blog/san-francisco-pitched-as-beacon-of-'the-sharing-economy'; See also Janelle Orsi, Yaisi Eskandari-Qajar, Eve Weissman, Molly Hall, Ali Mann, and Mira Luna, *Policies for Shareable Cities: A Sharing Economy Policy Primer for Urban Leaders* (Oakland, CA: Shareable and the Sustainable Economies Law Center, 2013). Available at http://www.shareable.net/blog/policies-for-a-shareable-city.

8. Patrick Hoge, "Critics Slam Mayor Lee's Phantom 'Sharing Economy' Working Group," *San Francisco Business Times* TechFlash blog, 1 May 2014. Available at http://www.bizjournals.com/sanfrancisco/blog/techflash/2014/05/mayor-lees-sharing-economy-task-force-never-met.html?page=all.

9. San Francisco Office of the Mayor, "Mayor Lee Also Announces First in the Nation Appointment of Chief Innovation Officer in Mayor's Office to Make City Government More Engaged & Responsive with New Approach & Technology." Press release, 6 January 2012. Available at http://www.sfmayor.org/index.aspx?page=643.

10. Jay Nath quoted in Stoll, "San Francisco Pitched as Beacon."

11. San Francisco Office of the Mayor, "Mayor Lee & Board President Chiu Announce New Sharing Economy Emergency Preparedness Partnership." Press release, 11 June 2013. Available at http://www.sfmayor.org/index.aspx?recordid=333&page=941.

12. Alice Gillet, "Resiliency: San Francisco Leverages Sharing Economy to Prepare for Disasters," *L'Atelier*, 17 June 2013. Available at http://www.atelier.net/en/trends/articles/resiliency-san-francisco-leverages-sharing-economy-prepare-disasters_421314.

13. San Francisco Office of the Mayor, "Mayor Lee Announces Selected Startups for Entrepreneurship-in-Residence Program." Press release, 13 March 2014. Available at http://www.sfmayor.org/index.aspx?recordid=537&page=846.

14. Jake Levitas, quoted in John Metcalfe, "San Francisco is Rolling out High Tech Playgrounds for Adults," *The Atlantic CityLab*, 5 November 2013. Available at http://www.citylab.com/tech/2013/11/san-francisco-rolling-out-high-tech-playgrounds-adults/7474.

15. Ariel Schwartz, "A New Way to Explore Some of San Francisco's Hidden Public Spaces." *Fast Company Exist*, 2 May 2013. Available at http://www.fastcoexist.com/1681937/a-new-way-to-explore-some-san-franciscos-hidden-public-spaces.

16. Elise Hu, "Bay Area's Steep Housing Costs Spark Return to Communal Living," *National Public Radio*, 19 December 2013. Available at http://www.npr.org/blogs/alltechconsidered/2013/12/19/250548681/bay-areas-steep-housing-costs-spark-return-to-communal-living.

17. Jordan Grader, quoted in Nellie Bowles, "Tech Entrepreneurs Revive Communal Living," *San Francisco Chronicle*, 18 November 2103. Available at http://www.sfgate.com/bayarea/article/Tech-entrepreneurs-revive-communal-living-4988388.php#page-1.

18. Nellie Bowles, "Tech Entrepreneurs Revive Communal Living."

19. David Streitfeld, "Some Setbacks for the Sharing Economy," Bits blog, *New York Times*, 23 April 2014. Available at http://bits.blogs.nytimes.com/2014/04/23/some-setbacks-for-the-sharing-economy/.

20. Carolyn Said, "Window into Airbnb's hidden impact on S.F.," *San Francisco Chronicle*, 16 June 2014. Available at http://www.sfchronicle.com/business/item/airbnb-san-francisco-30110.php; San Francisco Board of Supervisors, Budget and Legislative Analyst, "Policy Analysis Report: Analysis of the impact of short-term rentals on housing," 13 May 2015. Available at http://www.sfbos.org/Modules/ShowDocument.aspx?documentid=52601.

21. Carolyn Said, "Supes back 'Airbnb law,'" *SFGate*, 8 October 2014. Available at http://www.sfgate.com/news/article/Supervisors-approve-Airbnb-law-5807858.php.

22. Lizzy Acker, "Why San Francisco's Tech Community Is Creating Problems, Not Solving Them." *Policy Mic*, 22 October 2013. Available at http://www.policymic.com/articles/68969/why-san-francisco-s-tech-community-is-creating-problems-not-solving-them.

23. Nancy Scola, "The Battle over San Francisco's Bus Stops," *The Next City*, 13 December 2013. Available at http://nextcity.org/daily/entry/the-battle-over-san-franciscos-bus-stops.

24. Jeff J. Roberts, "Cabbies Sue to Drive Car Service Uber out of San Francisco," Gigaom blog, 14 November 2012. Available at http://gigaom.com/2012/11/14/cabbies-sue-to-drive-car-service-uber-out-of-san-francisco.

25. *The Economist*, "All Eyes on the Sharing Economy," 9 March 2013. Available at http://www.economist.com/news/technology-quarterly/21572914-collaborative-consumption-technology-makes-it-easier-people-rent-items.

26. Verne Kopytoff, "Airbnb's Woes Show How Far the Sharing Economy Has Come," *Time*, 7 October 2013. Available at http://business.time.com/2013/10/07/airbnbs-woes-show-how-far-the-sharing-economy-has-come.

27. Ibid.

28. Ibid.

29. Arun Sundararajan, "Trusting the 'Sharing Economy' to Regulate Itself," Economix blog, *New York Times*, 3 March 2014. Available at http://economix.blogs.nytimes.com/2014/03/03/trusting-the-sharing-economy-to-regulate-itself/.

30. Cat Johnson, "Profiles in Sharing: Janelle Orsi—The Sharing Economy Lawyer," *Shareable* [blog], 23 July 2013. Available at http://www.shareable.net/blog/profiles-in-sharing-janelle-orsi-the-sharing-economy-lawyer.

31. Carolyn Said, "S.F. Conference Brings Sharing Economy into Spotlight," *SFGate*, 13 May 13, 2014. Available at http://www.sfgate.com/technology/article/S-F-conference-brings-sharing-economy-into-5475596.php.

32. Janelle Orsi, quoted in Carolyn Said, "Cooperatives Give New Meaning to Sharing Economy," *SFGate*, 24 May 2014. Available at http://www.sfgate.com/business/article/Cooperatives-give-new-meaning-to-sharing-economy-5502879.php.

33. Josh Danielson, quoted in Said, "Cooperatives Give New Meaning to Sharing Economy."

Chapter 1, Sharing Consumption: The City as Platform

1. Robert D. Putnam, *Bowling Alone: The Collapse and Revival of American Community* (New York: Simon & Schuster, 2000).

2. Jean M. Twenge, W. Keith Campbell, and Nathan T. Carter, "Declines in Trust in Others and Confidence in Institutions Among American Adults and Late Adolescents, 1972–2012." *Psychological Science* 25 (10) (2014): 1914–1923.

3. Richard Sennett, *Together: The Rituals, Pleasures and Politics of Cooperation* (London: Penguin, 2013).

4. Jim Edwards, "Airbnb Takes New Funding at a $10 Billion Valuation." *Business Insider*, 18 April 2014. Available at http://www.businessinsider.com/airbnb-takes-new-funding-at-a-10-billion-valuation-2014-4.

5. PWC, "Five Key Sharing Economy Sectors Could Generate £9 Billion of UK Revenues by 2025." *PWC Press Room,* 15 August, 2014. Available at: http://pwc.blogs.com/press_room/2014/08/five-key-sharing-economy-sectors-could-generate-9-billion-of-uk-revenues-by-2025.html. This provides a conservative estimate of value, as it focuses on just five sectors and on monetized revenues, rather than the full value of goods and services shared.

6. Tomio Geron, "Airbnb and the Unstoppable Rise of the Share Economy." *Forbes,* 11 February 2013.

7. Latitude, *The New Sharing Economy: A Study by Latitude in Collaboration with Shareable Magazine* (2013). Available at http://latdsurvey.net/pdf/Sharing.pdf.

8. Harald Heinrichs and Heiko Grunenberg, *Sharing Economy: Toward a New Culture of Consumption?* (Lüneburg: Centre for Sustainability Management, 2013), 12.

9. Latitude, *The New Sharing Economy.*

10. Despite the relative anonymity of the web, many sharing platforms provide facilities for user reviews, which (like eBay seller ratings) have become a key tool for trust building, while direct links to Facebook and other social media sites can allow users of sharing services to check out potential lenders/borrowers in advance.

11. Nielsen, "68% of Global Survey Respondents Willing to Share or Rent Personal Items; Two-Thirds Likely to Use Products and Services from Others." Press release, 28 May 2014. Available at http://www.nielsen.com/apac/en/press-room/2014/global-consumers-embrace-the-share-economy.html.

12. Jeremiah Owyang, *Sharing Is the New Buying* (3 March 2014). Available at http://www.web-strategist.com/blog/2014/03/03/report-sharing-is-the-new-buying-winning-in-the-collaborative-economy.

13. Latitude, *The New Sharing Economy*; Sunrun, "Sunrun Survey Finds Nearly 92 Million Americans Plan to Participate in "Disownership" this Summer." Press release, 15 June 2013. Available at http://www.sunrun.com/why-sunrun/about/news/press-releases/sunrun-survey-finds-nearly-92-million-americans-plan-to-participate-in-disownership-this-summer.

14. Heinrichs and Grunenberg, *Sharing Economy.*

15. Rachel Griffiths, *The Great Sharing Economy: A Report into Sharing Across the UK* (Manchester: Cooperatives UK, 2011). Available at http://www.uk.coop/sites/storage/public/downloads/sharing_0.pdf; The People Who Share, *The State of the Sharing Economy: Food Sharing in the UK* (2013). Available at: http://www.thepeoplewhoshare.com/tpws/assets/File/TheStateoftheSharingEconomy_May2013_FoodSharingintheUK.pdf; Stokes et al., *Making Sense of the UK Collaborative Economy.*

16. Chris Diplock, *The Sharing Project: A Report on Sharing in Vancouver.* The Sharing Project, 2012. Available at http://thesharingproject.ca/TheSharingProject_Report.pdf.

17. Griffiths, *The Great Sharing Economy*, 3.

18. Ibid.

19. Wise, "New Survey Reveals Disownership."

20. Latitude, *The New Sharing Economy*, 2.

21. Neilsen, "Global Survey Respondents Willing to Share."

22. Diplock, *The Sharing Project*, 15.

23. Ibid., 16.

24. Stokes et al., *Making Sense of the UK Collaborative Economy*.

25. Sunrun, "Sunrun Survey—Disownership."

26. Latitude, *The New Sharing Economy*, 3.

27. Sunrun, "Sunrun Survey—Disownership."

28. Latitude, *The New Sharing Economy*, 2.

29. Griffiths, *The Great Sharing Economy*, 3.

30. Owyang, *Sharing is the New Buying*.

31. Chris Riedy, "The Sharing Economy Spooking Big Business." *The Conversation*, 14 November 2013. Available at http://theconversation.com/the-sharing-economy-spooking-big-business-19541.

32. US Chamber of Commerce Foundation, *Millennials Report* (2012). Available at http://www.uschamberfoundation.org/MillennialsReport.

33. Derek Thompson and Jordan Weissmann, "The Cheapest Generation: Why the Millennials Aren't Buying Cars or Houses and What That Means for the Economy." *The Atlantic Magazine*, September 2012. Available at http://www.theatlantic.com/magazine/archive/2012/09/the-cheapest-generation/309060.

34. Seth Schneider, "Gen Y and the Sharing Economy." *Shareable*, 30 May 2012. Available at http://www.shareable.net/blog/gen-y-and-the-sharing-economy.

35. Anthony Flint, "What Millennials Want—And Why Cities Are Right to Pay Them So Much Attention." *Atlantic Cities*, 5 May 2014. Available at http://m.theatlanticcities.com/housing/2014/05/what-millennials-wantand-why-cities-are-right-pay-them-so-much-attention/9032.

36. Nick Hiebert, "Why Driving for Lyft Is Good for the Soul." *Shareable*, 10 July 2013. Available at http://www.shareable.net/blog/why-driving-for-lyft-is-good-for-the-soul.

37. Ibid.

38. Victoria Hurth, David Jackman, Jules Peck, and Enrico Wensing, "Reforming Marketing for Sustainability," *Big Ideas Thinkpiece* (London: Friends of the Earth, 2015). Available at http://www.foe.co.uk/sites/default/files/downloads/reforming-marketing-sustainability-full-report-76676.pdf.

39. Ibid.

40. Dara O'Rourke, "New Millennial Consumers and Sustainability," 23 January 2014, *Guardian Sustainable Business Blog: Davos 2014: Business and Sustainability—Day Two as It Happened,* ed. Jo Confino and Caroline Holtum. Available at http://www.theguardian.com/sustainable-business/2014/jan/23/davos-2014-climate-change-resource-security-sustainability-live.

41. Ibid.

42. Jamelle Bouie, "Why Do Millennials Not Understand Racism?" *Slate Magazine,* 16 May 2014. Available at http://www.slate.com/articles/news_and_politics/politics/2014/05/millennials_racism_and_mtv_poll_young_people_are_confused_about_bias_prejudice.html.

43. Mike Brown, personal communication.

44. Jonathan Schifferes, "Sharing Our Way to Prosperity; Profiting from Sharing," RSA Blog entries, 6 August 2013. Available at http://www.rsablogs.org.uk/2013/social-economy/sharing-prosperity.

45. Estimate from Rachel Botsman cited in *The Economist,* "The Rise of the Sharing Economy," 9 March 2013. Available at http://www.economist.com/news/leaders/21573104-internet-everything-hire-rise-sharing-economy.

46. Estimate reported in "The People Who Share," *State of the Sharing Economy.*

47. Bill Jacobson, personal communication.

48. Rifkin, *Zero Marginal Cost Society.*

49. McEntee, Jesse, "Realizing Rural Food Justice: Divergent Locals in the Northeastern United States," in *Cultivating Food Justice. Race, Class and Sustainability,* ed. Alison H. Alkon and Julian Agyeman (Cambridge, MA: MIT Press, 2011), 254.

50. Alison H. Alkon, and Julian Agyeman, *Cultivating Food Justice: Race, Class and Sustainability.* (Cambridge, MA: MIT Press, 2011), citing Winson 1993.

51. Ibid., 10.

52. Casserole Club website. https://www.casseroleclub.com.

53. Foodcycle website, http://foodcycle.org.uk; Fareshare website, http://www.fareshare.org.uk.

54. Patrycja M. Długosz, "The Rise of the Sharing City: Examining Origins and Futures of Urban Sharing." (MSc thesis, International Institute for Industrial Environmental Economics (IIIEE), Lund University, 2014).

55. Kristina Dervojeda, Diederik Verzijl, Fabian Nagtegaal, Mark Lengton, Elco Rouwmaat, Erica Monfardini, and Laurent Frideres, *The Sharing Economy: Accessibility Based Business Models for Peer-to-Peer Markets* (Price Waterhouse Coopers for Business Innovation Observatory, 2013). Available at http://ec.europa.eu/enterprise/policies/innovation/policy/business-innovation-observatory/files/case-studies/12-she-accessibility-based-business-models-for-peer-to-peer-markets_en.pdf.

56. Eatwith website, http://www.eatwith.com/brand/about.

57. Rachel Botsman, "Collaborative Economy Services Changing the Way We Travel." Collaborative Consumption, 25 June 2014. Available at http://www.collaborativeconsumption.com/2014/06/25/collaborative-economy-services-changing-the-way-we-travel.

58. Noah Karesh, quoted on Feastly website, https://www.eatfeastly.com/info/about/story.

59. Ibid.

60. Feastly website, https://www.eatfeastly.com/t_and_c.

61. Grubclub website, http://grubclub.com.

62. *The Independent,* "Everyone Back to Mine: Pop-Up Restaurants in Private Homes Are the Latest Foodie Fad." 4 June 2009. Available at http://www.independent.co.uk/life-style/food-and-drink/features/everyone-back-to-mine-popup-restaurants-in-private-homes-are-the-latest-foodie-fad-1696262.html.

63. League of Kitchens website, https://www.leagueofkitchens.com.

64. Gordon W. Allport, *The Nature of Prejudice* (Cambridge, MA: Addison-Wesley, 1954); Oliver Christ, Katharina Schmid, Simon Lolliot, Hermann Swart, Dietlind Stolle, Nicole Tausch, Ananthi Al Ramiah, Ulrich Wagner, Steven Vertovec, and Miles Hewstone, "Contextual effect of positive intergroup contact on outgroup prejudice." *PNAS*, March 18 2014, 111 (11) 3996–4000.

65. *The Independent,* "Everyone Back to Mine."

66. Mariaana Nelimarkka, "Restaurant Day Is a Carnival of Food." Visit Helsinki, (undated). Available at http://www.visithelsinki.fi/en/stay-and-enjoy/eat/restaurant-day-carnival-food.

67. Todd Barnard, "New Trend: Pop-Up Restaurants Selling Tickets." Thundertix, 20 September 2013. Available at http://www.thundertix.com/restaurant-ticketing-software/new-trend-pop-up-restaurants-selling-tickets.

68. For example, Richard Belk, "Sharing." *Journal of Consumer Research* 5 (2010): 715–734.

69. US Chamber of Commerce Foundation, *Millennials Report.*

70. Joseph Pine and James Gilmore, *Authenticity: What Consumers Really Want.* (Cambridge MA: Harvard Business Press, 2007), 136.

71. Rachel Botsman, "The Case for Collaborative Consumption." TEDx Sydney, May 2010. Available at http://www.ted.com/talks/rachel_botsman_the_case_for_collaborative_consumption.html.

72. Rifkin, *Zero Marginal Cost Society*, 11.

73. Ibid., 17.

74. Michael J. Olson and Andrew D. Connor, "The Disruption of Sharing: An Overview of the New Peer-to-Peer 'Sharing Economy' and the Impact on Established Internet Companies" (Piper Jaffray Investment Research, 2013). Available at https://piper2.bluematrix.com/docs/pdf/35ef1fcc-a07b-48cf-ab80-04d80e5665c4.pdf.

75. Matt Hamblen, "Verizon Auto Share Service to Launch Later This Year." *Computerworld*, 8 September 2014. Available at http://www.computerworld.com/article/2602872/verizon-auto-share-service-to-launch-later-this-year.html.

76. Evgeny Morozov, "Don't Believe the Hype, the 'Sharing Economy' Masks a Failing Economy." *The Guardian,* 28 September 2014. Available at http://www.theguardian.com/commentisfree/2014/sep/28/sharing-economy-internet-hype-benefits-overstated-evgeny-morozov.

77. Tapscott and Williams, *Macrowikinomics*, citing Hendershott.

78. Olson and Connor, *The Disruption of Sharing*, 8.

79. Ibid.

80. Morozov, "Don't believe the hype."

81. Latitude, *The New Sharing Economy.*

82. David Weinberger, "Millennials Are Committed to a Multidimensional Approach to Saving the Environment." Next New Deal—Blog of the Roosevelt Institute, 9 April 2012. http://www.nextnewdeal.net/millennials-are-committed-multidimensional-approach-saving-environment.

83. Botsman and Rogers, *What's Mine Is Yours.*

84. US Today 2006 poll, cited by Botsman and Rogers, *What's Mine Is Yours,* 54.

85. Olson and Connor, *The Disruption of Sharing*, 6.

86. Tim Jackson, *Prosperity Without Growth: Economics for a Finite Planet* (Oxford and New York: Routledge, 2009).

87. Nigel Piercy, David Cravens and Nikala Lane, "Marketing out of the Recession: Recovery Is Coming, but Things Will Never Be the Same Again." *Marketing Review* 10 (1) (2010) 3–23, 3.

88. Thompson and Weissman, *"The Cheapest Generation."*

89. Botsman, "The Case for Collaborative Consumption."

90. Belk, "Sharing"; Botsman and Rogers, *What's Mine Is Yours.*

91. Christopher L. Weber, Jonathan G. Koomey, and H. Scott Matthews, "The Energy and Climate Change Impacts of Different Music Delivery Methods." *Final report to Microsoft Corporation and Intel Corporation* (2009). Available at http://download.intel.com/pressroom/pdf/cdsvsdownloadsrelease.pdf.

92. Botsman and Rogers, *What's Mine Is Yours.*

93. Ibid.

94. Swiss Federal Office of Energy. *Evaluation Car-Sharing* (Bern: SFOE, September 2006).

95. Elliot W. Martin and Susan A. Shaheen, *Greenhouse Gas Emission Impacts of Carsharing in North America.* Mineta Transportation Institute Report 09–11 (2010). Available at http://tsrc.berkeley.edu/sites/tsrc.berkeley.edu/files/Greenhouse%20Gas%20Emission%20Impacts%20of%20Carsharing%20in%20North%20America%20(final%20report).pdf; Elliot W. Martin and Susan A. Shaheen, "The Impact of Carsharing on Household Vehicle Ownership." *Access* 38 (2011): 22–27. Also, note that a US ton equals .91 tonnes, and that carbon dioxide equivalents also take account of changed emission of other greenhouse gases.

96. Shaheen et al. "Bikesharing."

97. Bobby Magill and Climate Central, "Is Bike Sharing Really Climate Friendly?" *Scientific American*, 19 August 2014. Available at http://www.scientificamerican.com/article/is-bike-sharing-really-climate-friendly.

98. Shaheen et al., "Bikesharing."

99. Corinne Kisner, *Integrating Bike Share Programs into a Sustainable Transportation System.* National League of Cities—City Practice Brief (2011). Available at http://www.nlc.org/documents/Find%20City%20Solutions/Research%20Innovation/Sustainability/integrating-bike-share-programs-into-sustainable-transportation-system-cpb-feb11.pdf.

100. Joe Gebbia, "AirBNB: How Sofa-Surfing Helps the Environment." *Greenbiz*, 7 August 2014. Available at http://www.greenbiz.com/blog/2014/08/07/airbnb-how-sofa-surfing-helps-environment.

101. Sarah Burch, Yuill Herbert, and John Robinson, "Meeting the Climate Change Challenge: A Scan of Greenhouse Gas Emissions in BC Communities." *Local Environment: The International Journal of Justice and Sustainability* 19(5) (2014): 1–19, 17.

102. David Owen, "Greenest Place in the U.S.? It's Not Where You Think." *Yale e360,* 26 October 2009. Available at http://e360.yale.edu/feature/greenest_place_in_the_us_its_not_where_you_think/2203.

103. Ibid.

104. Ibid.

105. Ibid.

106. Repair Café website, http://repaircafe.org/about-repair-cafe.

107. Cited in *The Economist,* "3D printing: The printed world." 10 February 2011. Available at http://www.economist.com/node/18114221.

108. Harald Heinrichs, "Sharing Economy: A Potential New Pathway to Sustainability." *GAIA* 22/4 (2013): 228–231; Marc Gunther, "Is Sharing Really Green?" *Ensia,* 29 July 2014. Available at http://ensia.com/voices/is-sharing-really-green.

109. Juliet Schor, "Debating the Sharing Economy." Great Transition Initiative, October 2014. Available at http://www.greattransition.org/publication/debating-the-sharing-economy.

110. Botsman and Rogers, *What's Mine Is Yours*; Rifkin, *Zero Marginal Cost Society.*

111. Benkler, "Sharing Nicely," 278.

112. Latitude, *The New Sharing Economy,* 2.

113. Botsman and Rogers, *What's Mine Is Yours,* 59–60.

114. Tapscott and Williams, *Macrowikinomics.*

115. Liat Clark, "Pirate Bay Traffic Has Doubled Post-ISP Blocks." *Wired,* 18 July 2014. Available at http://www.wired.co.uk/news/archive/2014-07/18/pirate-bay-traffic-doubles.

116. Tapscott and Williams, *Macrowikinomics,* 236.

117. Ibid.; Rifkin, *Zero Marginal Cost Society.*

118. Tapscott and Williams, *Macrowikinomics.*

119. Nicholas A. John, "The Social Logics of Sharing." *Critical Studies in Media Communication.* 16 (3) (2013) 113–131.

120. Jeremy Rifkin, *The Empathic Civilization: The Race to Global Consciousness in a World in Crisis* (Cambridge: Polity Press, 2010).

121. From a letter from Zuckerberg that accompanied Facebook's initial public offering (IPO) in May 2012, cited in John, "The Social Logics of Sharing," 117.

122. John, "The Social Logics of Sharing."

123. Nicholas A. John, "Sharing and Web 2.0: The emergence of a keyword." *New Media & Society* 15(2) (2013): 167–182; 171.

124. John, "Sharing and Web 2.0," 174–175.

125. Ibid.

126. Kester Brewin, *2012. Mutiny! 8.*

127. David Graeber, *The Democracy Project: A History. A Crisis. A Movement* (London: Penguin, 2013), 179–180.

128. Hom 2008 cited in John, "Sharing and Web 2.0," 177.

129. Renaud Garcia-Bardidia, Jean-Philippe Nau, and Eric Rémy, "Consumer Resistance and Anti-consumption: Insights from the Deviant Careers of French Illegal Downloaders." *European Journal of Marketing* 45 (2011, 11/12): 1789–1798.

130. John, "The Social Logics of Sharing," 125.

131. Aaron Balick, "The Digital Economy of Recognition: The Psychology of the Socially Networked Self." Mindswork blog, 22 June 2013. Available at http://www.mindswork.co.uk/wpblog/the-digital-economy-of-recognition-the-psychology-of-the-social-networked-self.

132. Tapscott and Williams, *Macrowikinomics*, 368.

133. Ibid., 369.

134. Mayer-Schönberger and Cukier, *Big Data*.

135. Rifkin, *Zero Marginal Cost Society*, 76.

136. Tapscott and Williams, *Macrowikinomics*.

137. Cameron Tonkinwise, "Sharing you can Believe in: The Awkward Potential within Sharing Economy Encounters." *Medium*, July 1st 2014. Available at https://medium.com/@camerontw/sharing-you-can-believe-in-9b68718c4b33.

138. Latitude, *The New Sharing Economy;* The People Who Share, *State of the Sharing Economy*.

139. Stokes et al., *Making Sense of the UK Collaborative Economy*.

140. Tapscott and Williams, *Wikinomics*.

141. Michael Porter, "Clusters and the New Economics of Competition." *Harvard Business Review*. November–December 1998: 77–90.

142. Tapscott and Williams, *Wikinomics.*

143. Richard Florida and Gary Gates, *Technology and Tolerance: The Importance of Diversity to High-Technology Growth.* The Brookings Institution Center on Metropolitan and Urban Policy, 2001.

144. Mariana Mazzucato, *The Entrepreneurial State: Debunking Public vs. Private Sector Myths* (London: Anthem, 2013); Ha-Joon Chang, *23 Things They Don't Tell You about Capitalism* (London: Penguin, 2010).

145. Ahmedabad University website, http://emba.ahduni.edu.in/cambridgeecosystem.shtml.

146. Clive Thompson, *Smarter Than You Think: How Technology Is Changing Our Minds for the Better* (London: William Collins Books, 2014).

147. Roberta Capello and Alessandra Faggian, "Collective Learning and Relational Capital in Local Innovation Processes." *Regional Studies* 39(1) (2005): 75–87.

148. Manuel Castells, *Networks of Outrage and Hope* (Cambridge: Polity Press, 2012).

149. Tapscott and Williams, *Macrowikinomics.*

150. Dan M. Kahan, "The Logic of Reciprocity: Trust, Collective Action, and Law" John M. Olin Center for Studies in Law, Economics, and Public Policy Working Paper 281: 31 (2002). Available at http://digitalcommons.law.yale.edu/cgi/viewcontent.cgi?article=1007&context=lepp_papers.

151. Erik Olin Wright, *Envisioning Real Utopias* (London: Verso, 2010).

152. Botsman and Rogers, *What's Mine Is Yours*; Heinrichs, "Sharing Economy"; Rifkin, *Zero Marginal Cost Society.*

153. Botsman and Rogers, *What's Mine Is Yours.*

154. Benkler, "Sharing Nicely," 305.

155. Beate Volker and Henk Flap, "Sixteen Million Neighbors: A Multilevel Study of the Role of Neighbors in the Personal Networks of the Dutch." *Urban Affairs Review* 43 (2007): 256.

156. David A. Crocker and Toby Linden, eds. *Ethics of Consumption: The Good Life, Justice and Global Stewardship* (Oxford: Rowman & Littlefield Publishers, 1998).

157. Benkler,"Sharing Nicely."

158. Both cited in Botsman and Rogers, *What's Mine Is Yours,* 115.

159. Bardhi and Eckhardt, "Access-Based Consumption."

160. Zygmunt Bauman, *Liquid Life* (Cambridge: Polity Press, 2005).

161. Cait P. Lamberton and Randall L. Rose, "When Is Ours Better than Mine? A Framework for Understanding and Altering Participation in Commercial Sharing Systems." *Journal of Marketing,* 76 (4) (2012): 109–125.

162. Benkler, "Sharing Nicely."

163. Thompson, *Smarter Than You Think*, 152.

164. Mike Jones, "How Capitalism and Regulation Will Reshape the Sharing Economy." *Forbes*, 9 October 2013. Available at http://www.forbes.com/sites/ciocentral/2013/10/09/how-capitalism-and-regulation-will-reshape-the-sharing-economy

165. Botsman and Rogers, *What's Mine Is Yours,* citing Cialdini.

166. Mark Pagel, *Wired for Culture: The Natural History of Human Cooperation* (London: Allen Lane, 2012).

167. Schor, "Debating the Sharing Economy," 5.

168. Ibid.

169. Lamberton and Rose, "When Is Ours Better."

170. Ibid.

171. Botsman and Rogers, *What's Mine Is Yours.*

172. Ibid., 101.

173. Josh Steimle, "Slicify Lets You Sell the Spare Computing Power of Your Desktop." *Forbes*, 25 November 2013. Available at http://www.forbes.com/sites/joshsteimle/2013/11/25/slicify-lets-you-sell-the-spare-computing-power-of-your-desktop.

174. Benkler, "Sharing Nicely," 343.

175. Kahan, "The Logic of Reciprocity," 1.

176. David Graeber, "On the Invention of Money—Notes on Sex, Adventure, Monomaniacal Sociopathy and the True Function of Economics." *Naked Capitalism,* 13 September, 2011. http://www.nakedcapitalism.com/2011/09/david-graeber-on-the-invention-of-money-%E2%80%93-notes-on-sex-adventure-monomaniacal-sociopathy-and-the-true-function-of-economics.html.

177. Ibid.

178. Stephen Sinofsky, "The Four Stages of Disruption." LinkedIn blog, 7 January 2014. Available at https://www.linkedin.com/today/post/article/20140107190944-2293107-four-stages-of-disruption.

179. Julie Wernau, "Enterprise Buying Chicago's I-Go Car Service." *Chicago Tribune*, 28 May 2013. Available at http://articles.chicagotribune.com/2013-05-28/business/chi-igo-enterprise-enterprise-buys-chicagos-igo-car-service-20130528_1_feigon-phillycarshare-enterprise-holdings.

180. Peugeot Group website, http://www.psa-peugeot-citroen.com/en/inside-our-industrial-environment/a-socially-responsible-business/setting-the-standard-in-sustainable-mobility-article.

181. Olson and Connor, *The Disruption of Sharing.*

182. Stokes et al., *Making Sense of the UK Collaborative Economy.*

183. Olson and Connor, *The Disruption of Sharing.*

184. Jeremiah Owyang, "The Six Scenarios for Hotels to Address the Collaborative Economy." Web Strategist blog, 22 July 2014. Available at http://www.web-strategist.com/blog/2013/07/22/the-six-scenarios-for-hotels-to-address-the-collaborative-economy.

185. David Pérez, "Guerra total de la industria contra el consumo colaborativo en España." *Noticias de Tecnología* 18 March 2014. Available at http://www.elconfidencial.com/tecnologia/2014-03-18/guerra-total-de-la-industria-contra-el-consumo-colaborativo-en-espana_103273/.

186. José A. Navas, "El Gobierno limita el crowdfunding: un millón máximo con aportaciones de 3.000€." *Noticias de Tecnología,* 28 February 2014. *Available at* http://www.elconfidencial.com/tecnologia/2014-02-28/el-gobierno-limita-el-crowdfunding-un-millon-maximo-con-aportaciones-de-3-000_94989.

187. Pérez, "Guerra Total."

188. Donna Tam, "NY Official: Airbnb Stay Illegal; Host Fined $2,400." *CNET,* 21 May 2013. Available at http://www.cnet.com/uk/news/ny-official-airbnb-stay-illegal-host-fined-2400.

189. David Golumbia quoted in Rebecca Burns, "The 'Sharing' Hype: Do Companies Like Lyft and Airbnb Help Democratize the Economy?" *In These Times*, 27 January 2014. Available at http://inthesetimes.com/article/16111/the_sharing_economy_hypeBurns.

190. Slavoj Žižek, "Fat-Free Chocolate and Absolutely No Smoking: Why Our Guilt about Consumption Is All-Consuming." *The Guardian*, 21 May 2014. http://www.theguardian.com/artanddesign/2014/may/21/prix-pictet-photography-prize-consumption-slavoj-zizek.

191. Danielle Sacks, "The Sharing Economy." *Fast Company*, 18 April 2011. Available at http://www.fastcompany.com/1747551/sharing-economy.

192. Christopher Maag, "The Sharing Economy Grows Up: The Emerging Models & Investment Trends." *CleantechIQ*, 23 April 2014. Available at http://cleantechiq.com/2014/04/the-sharing-economy-goes-from-gawky-adolescent-to-prom-queen.

193. Jeremiah Owyang, "How Investors Are Sharing Their Money into the Collaborative Economy." *Web Strategist*, 17 March 2015. Available at: http://www.web-strategist.com/blog/2015/03/17/how-investors-are-sharing-their-money-into-the-collaborative-economy/.

194. Janelle Orsi, "The Sharing Economy Just Got Real." *Shareable*, 16 September 2013. Available at http://www.shareable.net/blog/the-sharing-economy-just-got-real.

195. Quoted in Burns, "The 'Sharing' Hype."

196. Rifkin, *Zero Marginal Cost Society*.

197. Tapscott and Williams, *Wikinomics*; and *Macrowikinomics*.

198. Cited by Danielle Sacks, "How Silicon Valley's Obsession with Narrative Changed TaskRabbit." *Fast Company*, 17 June 2013. Available at http://www.fastcompany.com/3012593/taskrabbit-leah-busque.

199. Ibid.

200. Colleen Taylor, "TaskRabbit Confirms Layoffs As It Realigns to Focus on Mobile and Enterprise." *TechCrunch*, 8 July 2013. Available at http://techcrunch.com/2013/07/08/taskrabbit-confirms-layoffs-as-it-realigns-to-focus-on-mobile-and-enterprise.

201. Andrew Leonard, "You're Not Fooling Us, Uber! 8 Reasons Why the "Sharing Economy" Is All about Corporate Greed." *Salon*, 17 February 2014. Available at http://www.salon.com/2014/02/17/youre_not_fooling_us_uber_8_reasons_why_the_sharing_economy_is_all_about_corporate_greed; Mike Bulajewski, "The Cult of Sharing." Blog post, 5 August 2014. Available at http://www.mrteacup.org/post/the-cult-of-sharing.html; Yglesias, "There Is No 'Sharing Economy'"; Morozov, "Don't Believe the Hype."

202. Yglesias, "There Is No 'Sharing Economy.'"

203. Sascha Lobo, "Die Mensch-Maschine: Auf dem Weg in die Dumpinghölle." *Spiegel Online*, 3 September 2014. Available at http://www.spiegel.de/netzwelt/netzpolitik/sascha-lobo-sharing-economy-wie-bei-uber-ist-plattform-kapitalismus-a-989584.html.

204. Sebastian Olma, "Never Mind the Sharing Economy: Here's Platform Capitalism." *Institute of Network Cultures Blog*, 16 October 2014. Available at http://networkcultures.org/mycreativity/2014/10/16/never-mind-the-sharing-economy-heres-platform-capitalism.

205. Tonkinwise, "Sharing You Can Believe In."

206. Anthony Kalamar, "Sharewashing is the New Greenwashing." *Op Ed News*, 13 May 2014. Available at http://www.opednews.com/articles/Sharewashing-is-the-New-Gr-by-Anthony-Kalamar-130513-834.html.

207. Anya Kamenetz, "Is Peers the Sharing Economy's Future or Just a Great Silicon Valley PR Stunt?" *Fast Company*, 9 December 2013. Available at http://www.fastcompany.com/3022974/tech-forecast/is-peers-the-sharing-economys-future-or-just-a-great-silicon-valley-pr-stunt.

208. Burns, "The 'Sharing' Hype."

209. Quoted in Burns, "The 'Sharing' Hype."

210. Amartya Sen, *Development as Freedom* (Oxford: Oxford University Press, 2001).

211. Quoted in Burns, "The 'Sharing' Hype."

212. Ibid.

213. Ibid., our emphasis.

214. Ibid.

215. Freelancers Union website. https://www.freelancersunion.org/about.

216. Adam Parsons, "Putting the 'Sharing' Back in to the Sharing Economy." *Share the World's Resources*, 24 April 2014. Available at http://www.sharing.org/information-centre/articles/putting-%E2%80%98sharing%E2%80%99-back-sharing-economy/.

217. Michael Sandel, *Justice: What's the Right Thing To Do?* (London: Penguin, 2010); Michael Sandel, *What Money Can't Buy: The Moral Limits of Markets* (London: Penguin, 2012).

218. Jesse C. McEntee and Elena N. Naumova, "Building Capacity Between the Private Emergency Food System and the Local Food Movement: Working Towards Food Justice and Sovereignty in the Global North." *Journal of Agriculture, Food Systems, and Community Development* 3(1) (2012): 235–253; 248.

219. Jamie Peck and Adam Tickell, "Neoliberalizing Space." *Antipode* 34 (3) (2002): 380–404.

220. Peck and Tickell, "Neoliberalizing Space"; Jamie Peck and Nik Theodore, "Variegated Capitalism." *Progress in Human Geography* 31(6) (2007): 731–772; Neil Brenner, Jamie Peck, and Nik Theodore, "Variegated Neoliberalization: Geographies, Modalities, Pathways." *Global Networks* 10(2) (2010): 182–222.

221. Lita Kurth, "Destroying Labor Law in the 'Sharing Economy.'" Classism, 3 September 2013. Available at http://www.classism.org/destroying-labor-law-sharing-economy.

222. Shontell, Alyson, "My Nightmare Experience as a TaskRabbit Drone." *Business Insider*, 7 December 2011. Available at http://www.businessinsider.com/confessions-of-a-task-rabbit-2011-12#ixzz35bKIcYd5.

223. SolidarityNYC, quoted in Burns, "The 'Sharing' Hype."

224. Ibid.

225. Janelle Orsi, Yassi Eskandari-Qajar, Eve Weissman, Molly Hall, Ali Mann, and Mira Luna, *Policies for Shareable Cities: A Sharing Economy Policy Primer for Urban Leaders* (Oakland CA: Shareable and the Sustainable Economies Law Center, 2013). Available at http://www.shareable.net/blog/policies-for-a-shareable-city; 6.

Chapter 2, Case Study: Seoul

1. Cat Johnson, "Is Seoul the Next Great Sharing City?" *Shareable*, 16 July, 2013. Available at http://www.shareable.net/blog/is-seoul-the-next-great-sharing-city.

2. Frederico Guerrini, "How Seoul Became One of the World's Sharing Capitals." *Forbes Tech*, 25 May 2014. Available at http://www.forbes.com/sites/federicoguerrini/2014/05/25/how-seoul-became-one-of-the-worlds-sharing-capitals.

3. Matt Phillips, "It Takes $290,000 in Cash to Rent an Apartment in Seoul." *Quartz*, 10 March 2014. Available at http://qz.com/183412/koreas-crazy-system-for-renting-apartments-is-driving-the-country-deeper-into-debt.

4. Guerrini, "How Seoul Became"; Zak Stone, "Why the Sharing Economy Is Taking Off in Seoul." *Fast Company Exist*, 17 July 2013. Available at http://www.fastcoexist.com/1682623/why-the-sharing-economy-is-taking-off-in-seoul.

5. Ken Lee, "20 Cultural Mistakes to Avoid in Korea." Seoulistic blog (undated). Available at http://seoulistic.com/korean-culture/20-cultural-mistakes-to-avoid-in-korea.

6. Korean, The, "Super Special Korean Emotions?" Ask a Korean, Blogspot, 25 April 2008. Available at http://askakorean.blogspot.co.uk/2008/04/super-special-korean-emotions.html.

7. Ibid.

8. Ejang, "'The Sharing City, Seoul' Project." Co-up/Share blog, 24 January 2013. Available at http://co-up.com/share/archives/32498.

9. Johnson, "Is Seoul the Next Great Sharing City?"

10. Quoted in Cat Johnson, "Sharing City Seoul: A Model for the World." *Shareable*, 3 June 2014. Available at http://www.shareable.net/blog/sharing-city-seoul-a-model-for-the-world.

11. Johnson, "Is Seoul the Next Great Sharing City?"

12. Johnson, "Sharing City Seoul: A Model."

13. Johnson, "Is Seoul the Next Great Sharing City?"

14. Ibid.

15. Yewon Kang, "Sharing Markets Growing among Young Consumers." *Yonhap News,* 3 September 2012. Available at http://english.yonhapnews.co.kr/n_feature/2012/08/30/90/4901000000AEN20120830003000315F.HTML.

16. Tim Draimin, "Taking the Seoul Train to the Sharing Economy Part III." Social Innovation Generation (SiG) blog, 3 October 2013. Available at http://www.sigeneration.ca/taking-seoul-train-sharing-economy-part-iii.

17. Johnson, "Sharing City Seoul: a Model."

18. Quoted in Kang, "Sharing Markets Growing."

19. Ibid.

20. Johnson, "Sharing City Seoul: a Model."

21. Ibid.

22. Neal Gorenflo, "Why Banning Uber Makes Seoul Even More of a Sharing City." *Shareable,* 25 July 2014. Available at http://www.shareable.net/blog/why-banning-uber-makes-seoul-even-more-of-a-sharing-city.

23. Tim Draimin, "Taking the Seoul Train to the Sharing Economy Part I." Social Innovation Generation (SiG) blog, 17 September 2013. Available at http://www.sigeneration.ca/taking-seoul-train-sharing-economy-part; Kang, "Sharing markets growing."

24. Johnson, "Sharing City Seoul: a Model."

25. David Satterthwaite, "The Ten and a Half Myths That May Distort the Urban Policies of Governments and International Agencies" (UCL, 2003). Available at: http://www.ucl.ac.uk/dpu-projects/drivers_urb_change/urb_infrastructure/pdf_city_planning/IIED_Satterthwaite_Myths_complete.pdf.

26. Tim Smedley, "The New Smart City—from High-Tech Sensors to Social Innovation." *The Guardian,* 26 November 2013. Available at http://www.theguardian.com/sustainable-business/smart-cities-sensors-social-innovation.

27. Quoted in Nicole Fray, "South Korean City of Seoul Works toward Becoming a 'Smart City' in 2015." ExecutiveGov, 15 January 2014. Available at http://www.executivegov.com/2014/01/south-korean-city-of-seoul-works-toward-becoming-a-smart-city-in-2015/.

28. Ibid.

29. Sunhyuk Kim and Jeongwon Kim, "Emerging Patterns of Co-production in South Korea: Strengthening Democracy and Constructing (Not Complementing) a Welfare State." Discussion paper for European Group of Public Administration (EGPA) Conference, Madrid, Spain, 19–22 September 2007. Available at https://www.researchgate.net/publication/228433971_Emerging_Patterns_of_Co-Production_in_South_Korea_Strengthening_Democracy_and_Constructing_%28Not_Complementing%29_a_Welfare_State; 1.

30. Ibid., 3.

31. Ibid., 1.

32. Jung Hoon Kim, "Sustainable Urban Waste Management System in Metropolitan Seoul, South Korea," in *The Sustainable City III*, ed. N. Marchettini, C. A. Brebbia, E. Tiezzi and L. C. Wadhwa (Southampton: WIT Press, 2004), 718–726.

33. Ibid., 719.

34. Ibid., 725.

35. Park Won-soon, "Forging Ahead with Cross-Sector Innovations." *Stanford Social Innovation Review*, Supplement, Summer 2013. Available from http://www.ssireview.org/articles/entry/forging_ahead_with_cross_sector_innovations.

36. UNDP (United Nations Development Program), "Bright Young Minds Share Social Innovation Ideas in Seoul." UNDP Seoul Policy Center for Global Development Partnerships press brief, 6 October 2013. Available at http://www.undp.org/content/seoul_policy_center/en/home/presscenter/articles/2013/10/06/bright-young-minds-share-social-innovation-ideas-in-seoul.

37. Melissa J. Rowley, "Smart City Seoul." Cisco, The Network, 14 January 2014. Available at http://newsroom.cisco.com/feature/1309662/Smart-City-Seoul.

38. Johnson, "Sharing City Seoul: a Model."

39. Rowley, "Smart City Seoul."

40. Renjie Butalid, "Lessons from Seoul Mayor Park Won-soon." Center for International Governance Innovation (CIGI) blog post, 14 November 2013. Available at http://www.cigionline.org/blogs/sharing-local-knowledge-globally/lessons-seoul-mayor-park-won-soon.

41. Park Won-soon, "Mayor's Story: Why Is There a Giant Ear in Front of the City Hall?" *Mayor Park Won-soon's Hope Journal*, 104, 2013. Available at http://english.seoul.go.kr/mayor-park-won-soon%E2%80%99s-hope-journal-104-why-is-there-a-giant-ear-in-front-of-the-city-hall.

42. Creative Commons Korea, "Seoul City Government Is Sharing Documents with the Public." *CC News in English*, 14 November 2013. Available at http://www.cckorea.org/xe/?document_srl=647861.

43. Park Won-soon, "Forging Ahead."

44. Ibid.

45. Johnson, "Is Seoul the Next Great Sharing City?"

46. Johnson, "Sharing City Seoul: a Model."

47. Quoted in ibid.

48. Johnson, "Is Seoul the Next Great Sharing City?"

49. Ibid.

50. UNDP (United Nations Development Program) "Seoul forum shows how social business can cut poverty." UNDP Seoul Policy Center for Global Development Partnerships press brief, 6 November 2013. Available at http://www.ge.undp.org/content/seoul_policy_center/en/home/presscenter/articles/2013/11/06/seoul-forum-highlights-how-social-business-can-cut-poverty.

Chapter 2, Sharing Production: The City as Collective Commons

1. Benkler, "Sharing Nicely," 337.

2. Diplock, *The Sharing Project,* 26.

3. Stokes et al., *Making Sense of the UK Collaborative Economy.*

4. Marcela Basch, "20 + Latin American Cities Celebrate Collaborative Economy This Week." *Shareable,* 5 May 2014. Available at http://www.shareable.net/blog/20-latin-american-cities-celebrate-collaborative-economy-this-week.

5. Ahmad Sufian Bayram, "The Sharing Economy Is Gaining Momentum in the Arab World." *Ouishare,* 17 October 2013. Available at http://magazine.ouishare.net/2013/10/the-sharing-economy-is-gaining-momentum-in-the-arab-world/.

6. Marco Schmidt and Jessica Sommerville, "Fairness Expectations and Altruistic Sharing in 15-Month-Old Human Infants." *PLoS ONE* 6(10) (2011): e23223. doi:10.1371/journal.pone.0023223.

7. Matt Ridley, *The Origins of Virtue* (London: Viking, 1996).

8. Michael Tomasello and Felix Warneken, "Share and Share Alike: The Happy Tendency to Share Resources Equitably." *Nature* 454.7208 (2008): 1057.

9. Pagel, *Wired for Culture.*

10. Ibid.

11. Ibid.

12. Ibid., 78.

13. Ibid., 368–369.

14. Kahan, "The Logic of Reciprocity."

15. Robert Boyd, Herbert Gintis, and Samuel Bowles "Coordinated Punishment of Defectors Sustains Cooperation and Can Proliferate When Rare." *Science* 328 (5978) (2010): 617–620.

16. Peter Richerson and Robert Boyd, *Not by Genes Alone: How Culture Transformed Human Evolution* (Chicago: University of Chicago Press, 2005); Martin Nowak with Roger Highfield, *Supercooperators: Altruism, Evolution, and Why We Need Each Other to Succeed* (New York: Free Press, 2012); Samuel Bowles and Herbert Gintis, *A Cooperative Species: Human Reciprocity and Its Evolution* (Princeton NJ: Princeton University Press, 2013).

17. Bowles and Gintis, *A Cooperative Species.*

18. Richerson and Boyd, *Not by Genes Alone.*

19. Sennett, *Together*, 9.

20. Ibid., 11.

21. Ibid., 13.

22. Schmidt and Sommerville, "Fairness Expectations and Altruistic Sharing."

23. Jonathan Haidt, "The New Synthesis in Moral Psychology." *Science* 316 (5827) (2007): 998–1002. doi: 10.1126/science.1137651.

24. Roman Krznaric, *Empathy: A Handbook for Revolution* (London: Random House, 2013).

25. Rifkin, *The Empathic Civilization.*

26. Sennett, *Together*, 38.

27. Allport, *The Nature of Prejudice*; Christ et al. "Contextual Effect of Positive Intergroup Contact."

28. Franjo Weissing quoted in Science Daily, "Evolution of competitiveness: Scientists explain diversity in competitiveness," *Science Daily*, 29 October 2014. Available at http://www.sciencedaily.com/releases/2014/10/141029084021.htm.

29. Wright, *Envisioning Real Utopias*, 49.

30. Ibid.

31. Margaret Heffernan, *A Bigger Prize: Why Competition Isn't Everything and How We Do Better* (New York: Simon & Schuster, 2014).

32. Robert Frank and Philip Cook, *The Winner-Take-All Society: Why the Few at the Top Get So Much More Than the Rest of Us* (London: Penguin Books, 1996).

33. Rami Gabriel, *Why I Buy: Self, Taste, and Consumer Society in America* (Chicago: University of Chicago Press, 2013).

34. Geert Hofstede, *Culture's Consequences: International Differences in Work-Related Values* (Beverly Hills, CA: Sage Publications, 1980); Harry Triandis, "Self and Social Behavior in Differing Cultural Contexts." *Psychological Review* 96 (1989): 506–520; Harry Triandis, *Individualism & Collectivism* (Boulder, CO: Westview Press, 1995); Gabriel, *Why I Buy*.

35. Thomas Talhelm, X. Zhang, S. Oishi, C. Shimin, D. Duan, X. Lan, and S. Kitayama, "Large-Scale Psychological Differences within China Explained by Rice Versus Wheat Agriculture." *Science* 344 (6184) 9 May 2014: 603–608.

36. Christopher Chung and Samson Cho, *Significance of "Jeong" in Korean Culture and Psychotherapy*. Harbor-UCLA Medical Center, undated. Available at http://www.prcp.org/publications/sig.pdf, 2.

37. Putnam, *Bowling Alone*.

38. Described in Sennett, *Together*.

39. Chung and Cho, *Significance of "Jeong,"* 2.

40. Ibid., 3.

41. Ibid., 3.

42. Wendi Gardner, Shira Gabriel, and Angela Y. Lee, "'I' Value Freedom, but 'We' Value Relationships: Self-Construal Priming Mirrors Cultural Differences in Judgment." *Psychological Science* 10(4) (1999): 321–326.

43. Michele J. Gelfand, Jana L Raver, Lisa Nishii, et. al, "Differences Between Tight and Loose Cultures: A 33-Nation Study." *Science* 332 (6033) (2011): 1100–1104.

44. Shalom H. Schwartz, "Les valeurs de base de la personne: Théorie, mesures et applications." *Revue française de sociologie* 42 (2006): 249–288.

45. Amitai Etzioni, *The Spirit of Community: The Reinvention of American Society* (New York: Simon and Schuster, 1994); "Communitarianism," in *Encyclopedia of Community: From the Village to the Virtual World, Vol. 1*, eds. Karen Christensen and David Levinson (Thousand Oaks, CA: Sage Publications, 2003), 224–228.

46. Etzioni, "Communitarianism."

47. Sandel, *Justice*.

48. Maria N. Ivanova, "Consumerism and the Crisis: Wither 'the American Dream'?" *Critical Sociology*, 37 (3) (2011): 329–350.

49. Gabriel, *Why I Buy*.

50. Belk, "Why Not Share."

51. Volker and Flap, "Sixteen Million Neighbors."

52. Gabriel, *Why I Buy*. 123.

53. Ridley, *The Origins of Virtue*.

54. Adam Grant, *Give and Take: A Revolutionary Approach to Success* (New York: Viking, 2013).

55. Schor "Debating the Sharing Economy," 7.

56. Mark Granovetter, "The Strength of Weak Ties." *American Journal of Sociology* 78 (6) (1973): 1360–80.

57. Putnam, *Bowling Alone*.

58. Ibid.

59. Harvey, *Rebel Cities*.

60. Lewis Mumford, *The City in History: Its Origins, Its Transformations, and Its Prospects* (New York: Harcourt, Brace & World, 1961).

61. Christopher Friedrichs, *The Early Modern City, 1450–1750* (London: Longman, 1995).

62. Tapscott and Williams, *Macrowikinomics*, 31–32.

63. Rifkin, *Zero Marginal Cost Society*, 17.

64. Adam M. Brandenburger and Barry J. Nalebuff, *Co-opetition* (New York: Currency Doubleday, 1997).

65. *The Economist*, "The Trillion Dollar Gap." 20 March 2014. Available at http://www.economist.com/news/leaders/21599358-how-get-more-worlds-savings-pay-new-roads-airports-and-electricity.

66. Quoted in Christina Romano, "How Bus Rapid Transit Will Transform City Transportation." Metrofocus, 19 February 2014. Available at http://www.thirteen.org/metrofocus/2014/02/how-bus-rapid-transit-will-transform-city-transportation.

67. Ibid.

68. Mary Skelton Roberts, "Why Not Go for Gold in Boston?" Barr Foundation News, 14 October 2014. Available at: http://www.barrfoundation.org/news/why-not-go-for-gold-in-boston.

69. Milton Mueller, John Mathiason, and Hans Klein, "The Internet and Global Governance: Principles and Norms for a New Regime." *Global Governance* 13 (2007): 237–254.

70. Cited in Tapscott and Williams, *Macrowikinomics*.

71. Mathis Wackernagel and William Rees, *Our Ecological Footprint: Reducing Human Impact on the Earth* (Gabriola Island, BC: New Society Publishers, 1998); Herbie Girardet, *Regenerative Cities*. World Future Council and HafenCity University Hamburg (HCU) Commission on Cities and Climate Change, 2010. Available at http://www.worldfuturecouncil.org/fileadmin/user_upload/papers/WFC_Regenerative_Cities_web_final.pdf.

72. Mark Huxham, Sue Hartley, Jules Pretty, and Paul Tett, "No dominion over nature: Why treating ecosystems like machines will lead to boom and bust in food supply." *Big Ideas Thinkpiece* for Friends of the Earth, London, 2014. Available at http://www.foe.co.uk/sites/default/files/downloads/no-dominion-over-nature-why-treating-ecosystems-machines-will-lead-boom-bust.pdf.

73. Kristie Robinson, "How Formalizing a Network of 'Waste-Pickers' Is Making Buenos Aires Greener." *Atlantic Citylab,* 17 October 2014. Available at http://www.citylab.com/work/2014/10/how-formalizing-a-network-of-waste-pickers-is-making-buenos-aires-greener/381626/; Adrian Smith, "A day with Argentina's 'Street Engineers.'" STEPS Center blog, 22nd August 2014. Available at http://steps-centre.org/2014/blog/urban-infrastructure-day-argentinas-street-engineers.

74. Recycle Together website, undated, http://www.recycletogether.com/cities/new-york/new-york-new-york.

75. McLaren, "Environmental Space, Equity."

76. Girardet, *Regenerative Cities*.

77. John Ehrenfield and Nicholas Gertler, "Industrial Ecology in Practice: The Evolution of Interdependence at Kalundborg." *Journal of Industrial Ecology* 1(1) (1997): 67–79; The Ellen MacArthur Foundation, *Toward the circular economy: Economic and Business Rationale for an Accelerated Transition*. 2012. Available at: http://www.ellenmacarthurfoundation.org/business/reports.

78. Jo Confino and Caroline Holtum, "Davos 2014: Business and Sustainability—Day Two as It Happened," *Guardian* sustainable business blog. Available at http://www.theguardian.com/sustainable-business/2014/jan/23/davos-2014-climate-change-resource-security-sustainability-live.

79. Schifferes, "Sharing Our Way to Prosperity."

80. Rajesh Makwana, "Financing the Global Sharing Economy," Share the World's Resources, 2012. Available at http://www.stwr.org/downloads/pdfs/financing-global-sharing-report-final.pdf.

81. Wright, *Envisioning Real Utopias*.

82. Harriet Bulkeley, Andres Luque, Colin McFarlane, and Gordon MacLeod, *Enhancing Urban Autonomy: Toward a New Political Project for Cities* (Friends of the Earth and Durham University, 2013). Available at http://www.foe.co.uk/sites/default/files/downloads/autonomy_briefing.pdf.

83. Richelle H. Plesse, "Parisians Have Their Say on City's First €20m 'Participatory Budget.'" *The Guardian*, 8 October 2014. Available at http://www.theguardian.com/cities/2014/oct/08/parisians-have-say-city-first-20m-participatory-budget.

84. Peter Levine, "You Can Add Us to Equations but They Never Make Us Equal: Participatory Budgeting in Boston." Open Democracy: Transformation, 4 August 2014. Available at https://www.opendemocracy.net/transformation/peter-levine/you-can-add-us-to-equations-but-they-never-make-us-equal-participatory-b.

85. Tony Bovaird, "Beyond Engagement and Participation: User and Community Coproduction of Public Services." *Public Administration Review*. September/October 2007: 846–860.

86. Lucie Stephens, Josh Ryan-Collins, and David Boyle, *Co-production: A Manifesto for Growing the Core Economy* (London: New Economics Foundation, 2008).

87. Stephens et al. in *Co-production: A Manifesto*, credit the term "core economy" to the environmental economist Neva Goodwin.

88. Tapscott and Williams, *Macrowikinomics*, 268.

89. Ibid.

90. Stephens et al. *Co-production: A Manifesto*, 12–13.

91. Ibid., 18.

92. Ibid.

93. Philip A. Stephenson, "Text Messages Are Saving Swedes from Cardiac Arrest." *Quartz*, 23 October 2013. Available at http://qz.com/138334/text-messages-are-saving-swedes-from-cardiac-arrest.

94. Marina Sitrin, "The Commons and Defending the Public." Telesur, 17 November 2014. Available at http://www.telesurtv.net/english/opinion/The-Commons-and-Defending-the-Public-20141117-0020.html.

95. Tapscott and Williams, *Macrowikinomics*.

96. Tapscott and Williams, *Macrowikinomics*, 182, citing Pincock.

97. Sen, *Development as Freedom*.

98. Richard Wilkinson and Michael Marmot eds., *Social Determinants of Health: The Solid Facts*, 2nd ed. (WHO Europe, 2003), 9.

99. Orsi et al. *Policies for Shareable Cities*, 6.

100. Ken Robinson "Changing Education Paradigms." Lecture to the RSA, 2010. Available at http://www.thersa.org/events/video/archive/sir-ken-robinson.

101. Danny Dorling, *Injustice: Why Social Inequality Persists* (Bristol: Policy Press, 2011).

102. Stokes et al., *Making Sense of the UK Collaborative Economy*.

103. Ibid., 14.

104. Matthew Davis, "How Collaborative Learning Leads to Student Success." Edutopia, 5 December 2012. Available at http://www.edutopia.org/stw-collaborative-learning-college-prep?page=15.

105. Ibid.

106. Sennett, *Together*.

107. Paul Dolan, Paul, David Leat, Laura Mazzoli Smith, Sugata Mitra, Liz Todd, and Kate Wall, "Self-Organised Learning Environments (SOLEs) in an English School: An Example of Transformative Pedagogy?" *The Online Educational Research Journal*, (2013) paper 109, www.oerj.org, 13.

108. Ibid., 14.

109. Ibid., 16.

110. Budd Hall and Rajesh Tandon, "No More Enclosures: Knowledge Democracy and Social Transformation." Open Democracy, 20 August 2014. Available at https://opendemocracy.net/transformation/budd-hall-rajesh-tandon/no-more-enclosures-knowledge-democracy-and-social-transformat.

111. Ibid.

112. Jane E. Barker, Andrei D. Semenov, Laura Michaelson, Lindsay S. Provan, Hannah R. Snyder, and Yuko Munakata, "Less-Structured Time in Children's Daily Lives Predicts Self-Directed Executive Functioning." *Frontiers of Psychology* 5 (593), 17 June 2014. doi: 10.3389/fpsyg.2014.00593.

113. Ivan Illich, *Deschooling Society* (New York: Harper and Row, 1971).

114. Charles Eisenstein, "The Deschooling Convivium," 2013. Available at http://charleseisenstein.net/the-deschooling-convivium.

115. Cited in Lee Bryant, "Pierre Bourdieu." History Learning website, http://www.historylearningsite.co.uk/pierre_bourdieu.htm.

116. Paulo Freire, *Pedagogy of the Oppressed* (New York: Continuum, 2007).

117. Ettore Gelpi, *A Future for Lifelong Education* (Manchester: University of Manchester Press, 1979).

118. Dave Watson, "Self-Directed Care—Reality Doesn't Always Match the Rhetoric." 1 April, 2014. Available at http://unisondave.blogspot.co.uk/2014/04/self-directed-care-reality-doesnt.html.

119. Jenny Shank, "Utah Is on Track to End Homelessness by 2015 with This One Simple Idea." NationSwell, 19 December 2013. Available at http://nationswell.com/one-state-track-become-first-end-homelessness-2015.

120. Sennett, *Together*, 252.

121. Stephens et al. *Co-production: A Manifesto,* 15.

122. Ibid., 15.

123. Ibid., 16.

124. Tapscott and Williams, *Macrowikinomics*, 274.

125. Tapscott and Williams, *Macrowikinomics,* 195–196, citing Robinson and Ginsburg.

126. Wright, *Envisioning Real Utopias*, 207.

127. See for example Giraff website, http://www.giraff.org/?lang=en.

128. Alissa Walker, "These Five Ideas for Smarter Cities Just Won Millions in Funding." *Gizmodo,* 19 September 2014. Available at http://gizmodo.com/these-five-ideas-for-smarter-cities-just-won-millions-i-1636858366.

129. Tapscott and Williams, *Macrowikinomics*, 69.

130. *Boston Consulting Group,* "Chris Andersen on Why Community-Driven Companies Will Always Win (in conversation with Ralf Dreischmeier)," 9 July 2014. Available at https://www.bcgperspectives.com/content/videos/technology_strategy_innovation_chris_anderson_why_community_driven_companies_will_always_win.

131. Gijs Hillenius, "'Open Source Crucial for Cutting Edge Industry'—South Korea ICT Ministry." Open Source Observatory, 16 June 2014. Available at https://joinup.ec.europa.eu/community/osor/news/open-source-crucial-cutting-edge-industry-south-korea-ict-ministry.

132. Yochai Benkler, *The Penguin and the Leviathan. The Triumph of Cooperation Over Self-Interest* (New York: Crown Business, 2011).

133. Benkler, *Penguin and Leviathan*, 182, citing Lakhani and Wolf.

134. Thompson, *Smarter Than You Think,* 169.

135. James B. Quilligan, "People Sharing Resources: Towards a New Multilateralism of the Global Commons." *Kosmos*, Fall/Winter 2009: 36–43.

136. Kiley Kroh, "Arizona May Impose Unusual New Tax on Customers Who Lease Solar Panels." Think Progress, 30 April 2014. Available at http://thinkprogress.org/climate/2014/04/30/3432172/arizona-solar-property-tax.

137. Laurie Guevara-Stone, "The Rise of Solar Co-ops." RMI blog, 22 April 2014. Available at http://blog.rmi.org/blog_2014_04_22_the_rise_of_solar_coops.

138. Brentin Mock, "Black Lawmakers Push Back against Coal Utilities' New Trick." *Grist*, 27 October 2014. Available at http://grist.org/climate-energy/black-lawmakers-can-push-back-against-coal-utilities.

139. Kristin Shrader-Frechette, *Environmental Justice: Creating Equality, Reclaiming Democracy* (Oxford: Oxford University Press, 2002).

140. Tapscott and Williams, *Macrowikinomics*.

141. Tobias Timm, "How Germany Delivers." *Presentation at 'Great British Refurb' Conference*, London School of Economics, 8 December 2009. Available at: http://sticerd.lse.ac.uk/textonly/LSEHousing/Events/Great_British_Refurb/Session4/Tobias_Timm_Session_4.pdf

142. Ibid.

143. Cited in Kiley Kroh, "Germany Sets New Record, Generating 74 Percent of Power Needs From Renewable Energy." Think Progress, 13 May 2014. Available at http://thinkprogress.org/climate/2014/05/13/3436923/germany-energy-records.

144. Caroline Julian and Rosie Olliver, *Community Energy: Unlocking Finance and Investment—The Way Ahead* (London: ResPublica, 2014), 5.

145. Tom de Castella, "Self-Build: Should People Build Their Own Homes?" *BBC News magazine*, 19 July 2011. Available at http://www.bbc.co.uk/news/magazine-14125196.

146. Self Build Homes, "Sluggish Self-Build Housing Statistics Reveal Government Needs to Do More to Promote the Industry." Undated. Available at http://www.selfbuildhomesmag.com/sluggish-self-build-housing-statistics-reveal-government-needs-to-do-more-to-promote-the-industry.

147. Reforesting Scotland campaign website, http://www.thousandhuts.org/?page_id=2.

148. Yvonne Rydin, *The Future of Planning: Beyond Growth Dependence* (Bristol: Policy Press, 2013).

149. Oliver Wainwright, "The Future's Communal: Meet The UK's Self-Build Pioneers." *The Guardian*, 7 May 2013. Available at http://www.theguardian.com/artanddesign/2013/may/07/uk-self-build-pioneers.

150. Habitat for Humanity, "What We Do." *Habitat World*, March 2009. Available at http://www.habitat.org.nz/HW_March09.pdf.

151. Bulkeley et al., *Enhancing Urban Autonomy*.

152. Colin McFarlane, *Learning the City: Knowledge and Translocal Assemblage* (Oxford: Wiley-Blackwell, 2011).

153. SDI (Shack / Slum Dwellers International) Annual Report 2013–14. 2014. Available at http://www.sdinet.org/media/upload/documents/SDI_Annual_Report_2013-14.pdf.

154. Ivette Arroyo and Johnny Åstrand, "Organized Self-Help Housing: Lessons from Practice with an International Perspective," in *Proceedings of the 19th International CIB World Building Congress*, ed. S. Kajewski, K. Manley, & K Hampson (Brisbane: Queensland University of Technology, 2013).

155. Castells, *Networks of Outrage and Hope*; Rifkin, *Zero Marginal Cost Society*; Rachel Botsman, "Collaborative Finance: By the People, For the People." Collaborative Consumption, 31 July 2014. Available at http://www.collaborativeconsumption.com/2014/07/31/collaborative-finance-by-the-people-for-the-people/.

156. Mark DeCambre, "Why Crowd-Funding Is Set to Explode in Size over the Next Few Years." *Quartz*, 23 April 2014. Available at http://qz.com/202090/why-crowd-funding-is-set-to-explode-in-size-over-the-next-few-years/.

157. Rifkin, *Zero Marginal Cost Society*.

158. Arash Massoudi and Richard Waters, "Lending Club Valuation Rises to $3.76bn." FT.com, 17 April 2014. Available at http://www.ft.com/intl/cms/s/0/fe75cbd6-c651-11e3-9839-00144feabdc0.html.

159. Naiwen Zhang and Yangjie She, "The Evolution of Peer to Peer Lending in China." Crowdfund Insider, 12 September 2014. Available at http://www.crowdfundinsider.com/2014/09/48954-evolution-p2p-lending-china.

160. Ibid.

161. Felix Oldenburg, "Say Hello to Hybrid Finance." *Forbes*, 8 January 2014. Available at http://www.forbes.com/sites/ashoka/2014/08/01/say-hello-to-hybrid-finance.

162. Cited in Deborah M. Todd, "Pittsburgh pilot program to help finance entrepreneurs morphs into Kiva City." *PostGazette.com*, 26 March 2014. Available at http://www.post-gazette.com/business/2014/03/27/Pittsburgh-pilot-program-to-help-finance-entrepreneurs-morphs-into-Kiva-Cities/stories/201403260150.

163. Kiva website, http://www.kiva.org/about.

164. Tapscott and Williams, *Macrowikinomics*, 301.

165. Ibid., 301–302.

166. Maren Duvendack, Richard Palmer-Jones, James G. Copestake, Lee Hooper, Yoon Loke, and Nitya Rao, *What Is the Evidence of the Impact of Microfinance on the Well-Being of Poor People?* (London: EPPI-Centre, Social Science Research Unit, Institute of Education, University of London, 2011). Available at https://eppi.ioe.ac.uk/cms/LinkClick.aspx?fileticket=8rhQ5Dp_ceQ%3d&tabid=3178&mid=5928.

167. David M. Roodman, *Due Diligence: What Social Investors Should Know about Microfinance* (Washington, DC: Center for Global Development, 2011).

168. Chang, *23 Things.*

169. Cited in Claire Provost, A Voice of Reason Amid the Sound and Fury of the Microfinance Debate. *The Guardian,* Poverty Matters blog, 6 January 2012. Available at http://www.theguardian.com/global-development/poverty-matters/2012/jan/06/david-roodman-reasoned-microfinance-debate.

170. Kamal Munir, "Microfinance Has Been a Huge Disappointment around the World." *Business Insider,* 17 February 2014. Available at http://www.businessinsider.com/microfinance-has-been-a-huge-disappointment-around-the-world-2014-2.

171. Ibid.

172. Amanda Palmer, quoted in Kickstarter, "Seven Things to Know about Kickstarter," undated. Available at https://www.kickstarter.com/hello?ref=footer.

173. Kickstarter, "Stats." Available at https://www.kickstarter.com/help/stats.

174. Kickstarter, "Seven things to know about Kickstarter," undated. Available at https://www.kickstarter.com/hello?ref=footer.

175. Alex Hern, "Kickstarter: A Platform for Investment, Philanthropy or Shopping?" *The Guardian,* 23 June 2014. Available at http://www.theguardian.com/technology/2014/jun/23/kickstarter-crowdfunding-start-ups-pebble.

176. Cited in ibid.

177. Sarah Wakefield, Fiona Yeudall, Carolin Taron, Jennifer Reynolds, and Ana Skinner, "Growing Urban Health: Community Gardening in South-East Toronto." *Health Promotion International* 22 (2) (2007): 92–101.

178. Landshare website, http://www.landshare.net/about.

179. Tuintjedelen website, http://tuintjedelen.nl/over-tuintjedelen.

180. Beacon Food Forest website, http://www.beaconfoodforest.org.

181. Co-operatives UK, *Setting Up a Cider Co-Operative at Work: Get Fruity With Your Colleagues!* (Manchester: Co-operatives UK, 2011). Available at http://www.uk.coop/sites/storage/public/downloads/cider_co-op.pdf.

182. Not Far From the Tree website, http://notfarfromthetree.org.

183. Orsi et al. *Policies for Shareable Cities.*

184. Juliet Schor, *True Wealth* (London: Penguin, 2011).

185. Alkon and Agyeman, *Cultivating Food Justice.*

186. Guthman, Julie, "Bringing Good Food to Others: Investigating the Subjects of Alternative Food Practice." *Cultural Geographies* 15(4) (2008): 431–47.

187. Schor, "Debating the Sharing Economy," 8.

188. Jimiliz M. Valiente-Neighbors, "Mobility, Embodiment, and Scales: Filipino Immigrant Perspectives on Local Food." *Agriculture and Human Values* 29 (2012): 531–41, 539.

189. Branden Born and Mark Purcell, "Avoiding the Local Trap: Scale and Food Systems in Planning Research." *Journal of Planning Education and Research* 26(2) (2006): 195–207.

190. Ibid., 195–196.

191. L. V. Anderson, "Review—*Labor and the Locavore: The Making of a Comprehensive Food Ethic*, by Margaret Gray." *Dissent Magazine*, Spring 2014. Available at http://www.dissentmagazine.org/article/limits-of-the-locavore.

192. Penn Loh, "Land, Co-ops, Compost: A Local Food Economy Emerges in Boston's Poorest Neighborhoods." *YES! Magazine,* November 2014. Available at http://www.yesmagazine.org/commonomics/boston-s-emerging-food-economy.

193. Tapscott and Williams, *Wikinomics.*

194. Rifkin, *Zero Marginal Cost Society.*

195. Tapscott and Williams, *Macrowikinomics.*

196. Ibid.

197. Gorenflo, cited in Ede, "Transactional Sharing, Transformational Sharing."

198. Tapscott and Williams, *Macrowikinomics,* 63.

199. Fab Lab Seoul website, https://www.fablabs.io/fablabseoul.

200. Rifkin, *Zero Marginal Cost Society.*

201. Tapscott and Williams, *Wikinomics.*

202. Sennett, *Together.*

203. Pat Kane, "Radical Animal: Innovation, Sustainability and Human Nature." Radical Animal, 1 May 2011. Available at http://radicalanimal.ning.com/profiles/blogs/radical-animal-the-general.

204. Orsi et al. *Policies for Shareable Cities,* 31.

205. Orsi, "The Sharing Economy Just Got Real."

206. Rifkin, *The Empathic Civilization.*

207. Graeber, "On the Invention of Money."

208. Margaret Attwood, *Payback: Debt and the Shadow Side of Wealth* (Toronto: House of Anansi Press, 2008); Graeber, David. *Debt: The First 5000 Years* (New York: Melville House Publishing, 2013).

209. Tonkinwise, "Sharing You Can Believe In."

210. Botsman and Rogers, *What's Mine Is Yours,* 133.

211. Benkler, *Penguin and Leviathan,* 43.

212. Sennett, *Together.*

213. Freecycle website, https://www.freecycle.org/about/background.

214. Dervojeda et al., *The Sharing Economy: Accessibility Based Business Models.*

215. Streetclub website, https://www.streetclub.co.uk/our-values.

216. Wright, *Envisioning Real Utopias,* 80.

217. Tonkinwise, "Sharing You Can Believe In."

218. *Žižek!* 2005. Documentary film on Slavoj Žižek, directed by Astra Taylor.

219. Giles Tremlett, "Mondragon: Spain's Giant Co-Operative Where Times Are Hard but Few Go Bust." *The Guardian,* 7 March 2013. Available at http://www.theguardian.com/world/2013/mar/07/mondragon-spains-giant-cooperative.

220. Wright, *Envisioning Real Utopias.*

221. Jessica Gordon Nembhard, "The Deep Roots of African American Cooperative Economics." Interviewed by Mira Lund. *Shareable,* 28 April 2014. http://www.shareable.net/blog/interview-the-deep-roots-of-african-american-cooperative-economics.

222. Ibid.

223. Orsi et al. *Policies for Shareable Cities,* 34.

224. Wright, *Envisioning Real Utopias.*

225. Ibid.

226. CIC (Cooperativa Integral Catalana), "Links in English to Know More about Us." CIC, 8 June 2013. Available at http://cooperativa.cat/en/welcome-to-cooperativa-integral-catalana-2.

227. Patricia Manrique, "From Critique to Construction: The Integrated Cooperative in Catalonia." Por Diagonal English, 6 August 2012. Available at https://www.diagonalperiodico.net/blogs/diagonal-english/from-critique-to-construction-the-integrated-cooperative-in-catalonia.html.

228. Mike Gilli, "Integral CoOps, 15M and Occupy. An embryo for Post Capitalism?" Infoshop News, 25 June 2012. http://news.infoshop.org/article.php?story=20121130120028363.

229. World Council of Credit Unions, "What is a Credit Union?" WCCU, undated. Available at http://www.woccu.org/about/creditunion.

230. Castells, *Networks of Outrage and Hope.*

231. In the United States, community banks are defined as independent banks and savings institutions holding companies with aggregate assets less than $1 billion.

232. Jim Blasingame, "A community Bank Is Not a Little Big Bank." *Forbes,* 5 April, 2013. Available at http://www.forbes.com/sites/jimblasingame/2013/04/08/a-community-bank-is-not-a-little-big-bank.

233. Dean Baker, *Work Sharing: The Quick Route Back to Full Employment.* Center for Economic and Policy Research (CEPR), 2011. Available at http://www.cepr.net/documents/publications/work-sharing-2011-06.pdf; ILO (International Labour Organisation). Global Jobs Pact Policy Brief No 18: Work Sharing: Working Time Adjustments as a Job Preservation Strategy. ILO (undated). Available at http://www.ilo.org/wcmsp5/groups/public/---dgreports/---integration/documents/publication/wcms_146815.pdf.

234. Anna Coote, "Reduce the Working Week to 30 Hours." The NEF blog, 12 March 2014. Available at http://www.neweconomics.org/blog/entry/reduce-the-working-week-to-30-hours.

235. Anna Coote, "Introduction: A New Economics of Work and Time," in *Time on Our Side: Why We All Need a Shorter Working Week,* eds. Anna Coote and Jane Franklin (London: New Economics Foundation, 2013) xi–xxii.

236. Joseph Stiglitz, *Toward a General Theory of Consumerism: Reflections on Keynes' Economic Possibilities for our Grandchildren.* Working paper, 2007, 18. Available at: http://www0.gsb.columbia.edu/faculty/jstiglitz/download/papers/2007_General_Theory_Consumerism.pdf.

237. Wright, *Envisioning Real Utopias.*

238. Russell W. Belk, "Possessions and the extended self." *Journal of Consumer Research* 15(2) (1988): 139–168.

239. Harvey, *Rebel Cities.*

240. Elinor Ostrom, "A General Framework for Analyzing Sustainability of Social-Ecological Systems," *Science* 325 (5939) (2009): 419–422, doi:10.1126/science.1172133.

241. Harvey, *Rebel Cities,* 70.

242. Ibid., 71.

243. Brewin, *Mutiny!*

244. Sennett, *Together.*

245. Harvey, *Rebel Cities.*

246. Sennett, *Together.*

247. Teresa Mares and Devon Peña, "Environmental and Food Justice: Toward Local, Slow and Deep Food Systems," in *Cultivating Food Justice. Race, Class and Sustainability,* ed. Alison Alkon and Julian Agyeman (Cambridge, MA: MIT Press, 2011), 197–219.

248. Ibid., 209.

249. Harvey, *Rebel Cities,* 74.

250. Ibid., 78.

251. Cited in Moira Jeffrey, "Don't Call Glasgow's Contemporary Art Scene a Miracle." *The Guardian,* 10 May 2014. Available at http://www.theguardian.com/artanddesign/2014/may/10/dont-call-glasgow-contemporary-art-scene-a-miracle.

252. Ibid.

253. Carl Grodach, Elizabeth Currid-Halkett, Nicole Foster, and James Murdoch III, "The Location Patterns of Artistic Clusters: A Metro- and Neighborhood-Level Analysis." *Urban Studies* (2014): 1–22, 2. doi:10.1177/0042098013516523.

254. Wilson Sherwin, Maria-Valerie Schegk, and Olivia Sandri, "Cultural Creative Clusters For Dummies." MCR Digipaper, 7 March 2011. Available at: http://mcrdigipaper.wordpress.com/2011/03/07/cultural-creative-clusters-for-dummies.

255. Miguel Martinez, "Squatting for Justice, Bringing Life to the City." *ROARMAG* 13 May 2014. Available at http://roarmag.org/2014/05/squatting-urban-justice-commons.

256. Ibid.

257. *Column, The,* "The Tragic Birth of an Artist's Village." 8 February 2014. Available at http://thecolumn.net/2014/02/08/the-tragic-birth-of-an-artists-village.

258. Ibid.

259. Holt, "Why the Sustainable Economy Movement Hasn't Scaled."

260. David Byrne, "If the 1% Stifles New York's Creative Talent, I'm Out of Here." *The Guardian,* 7 October 2013. Available at http://www.theguardian.com/commentisfree/2013/oct/07/new-york-1percent-stifles-creative-talent.

261. Ibid.

262. Ibid.

263. Harvey, *Rebel Cities.*

264. Sassen, *The Global City.*

265. Harvey, *Rebel Cities,* 110–111.

266. Ibid., 110.

267. Julia Elyachar, "Next Practices: Knowledge, Infrastructure, and Public Goods at the Bottom of the Pyramid." *Public Culture* 24(1 66) (2012): 109–129.

268. Lewis Hyde, cited in Brewin, *Mutiny!*: 42.

Chapter 3, Case Study: Copenhagen

1. The Happy Planet Index does not concur. It looks at long and happy lives in relation to the ecological footprint of the country and puts Costa Rica top and Denmark in 93rd place.

2. Cited in Jay Walljasper, "Public Spaces Make the World Go 'Round." *Shareable,* 18 July 2012. Available at http://www.shareable.net/blog/public-spaces-make-the-world-go-round.

3. McKenzie Funk, *Windfall: The Booming Business of Global Warming* (New York: Penguin, 2014): 67.

4. Cited by Rich Heap, "Copenhagen Puts Edges at Heart of Planning." *UBM's Future Cities,* 3 December 2013. Available at http://www.ubmfuturecities.com/author.asp?section_id=242&doc_id=526256.

5. Maryam Omidi, "Anti-Homeless Spikes Are Just the Latest in 'Defensive Urban Architecture.'" *The Guardian,* 12 June 2014. Available at http://www.theguardian.com/cities/2014/jun/12/anti-homeless-spikes-latest-defensive-urban-architecture.

6. City of Copenhagen, *A Metropolis for People: Visions and Goals for Urban Life in Copenhagen 2015.* The Technical and Environmental Administration with consultant Gehl Architects, 2008. Available at http://www.engageliverpool.com/uploads/default/files//Copenhagen_2008_A_Metropolis_for_People.pdf.

7. Sustainia, *Guide to Copenhagen 2025*. (Copenhagen: Sustainia, 2013). Available at http://issuu.com/sustainia/docs/cph2025.

8. City of Copenhagen, *A Metropolis for People,* 14.

9. Jordan Lewis, "Reflecting on Urban Play in Denmark." *Gehl Architects—Cities for People* [blog], 28 December 2013. Available at http://gehlarchitects.com/blog/reflecting-on-urban-play-in-denmark.

10. John Metcalfe, "Please, Take a Load Off on Copenhagen's Free Street Hammocks." *The Atlantic CityLab,* 6 August 2013. Available at http://www.citylab.com/design/2013/08/copenhagen-encourages-public-laziness-street-hammocks/6430/.

11. Official Website of Denmark, "Art and Culture in Public Spaces—In Denmark and in the Middle East." Available at http://denmark.dk/en/lifestyle/architecture/superkilen-celebrates-diversity-in-copenhagen/art-and-culture-in-public-spaces.

12. Borghi, Antonio, "Superkilen, an Innovative Public Space in Copenhagen." *Well Designed and Built* [blog], 10 June 2013. Available at http://welldesignedandbuilt.com/2013/06/10/superkilen-an-innovative-public-space-in-copenhagen.

13. Official Website of Denmark, "Art and Culture in Public Spaces."

14. Tom Freston, "You Are Now Leaving the European Union." *Vanity Fair,* 12 September 2013. Available at http://www.vanityfair.com/politics/2013/09/christiana-forty-years-copenhagen.

15. Ibid.

16. Ibid.

17. Helen Jarvis, "Saving Space, Sharing Time: Integrated Infrastructures of Daily Life In Cohousing." *Environment and Planning A,* 43 (3) (2011): 560–577, 560.

18. Quoted in Julian Agyeman, Duncan McLaren, and Adrianne Schaefer-Borrego, "Sharing Cities." Thinkpiece for Friends of the Earth Big Ideas Project, 2013, 12. Available at http://www.foe.co.uk/sites/default/files/downloads/agyeman_sharing_cities.pdf.

19. Ibid.

20. City of Copenhagen, "Copenhagen City of Cyclists: Bicycle Account 2012." Available at http://subsite.kk.dk/sitecore/content/Subsites/CityOfCopenhagen/SubsiteFrontpage/LivingInCopenhagen/CityAndTraffic/~/media/4ADB52810C484064B5085F2A900CB8FB.ashx.

21. Justin Gerdes, "Copenhagen's Ambitious Push to Be Carbon-Neutral by 2025." *The Guardian,* 12 April 2013. Available at http://www.theguardian.com/environment/2013/apr/12/copenhagen-push-carbon-neutral-2025.

22. Sally McGrane, "Commuters Pedal to Work on Their Very Own Superhighway." *The New York Times,* 17 July 2012. Available at http://www.nytimes.com/2012/07/18/world/europe/in-denmark-pedaling-to-work-on-a-superhighway.html.

23. Cited by McGrane, "Commuters Pedal Own Superhighway."

24. Eleanor Beardsley, "In Bike-Friendly Copenhagen, Highways for Cyclists." National Public Radio. 1 September 2012. Available at http://www.npr.org/2012/09/01/160386904/in-bike-friendly-copenhagen-highways-for-cyclists.

25. Elizabeth Floyd, "Top Ten Bike and Bus Observations from Copenhagen." Mobility Lab, 17 January 2013. Available at http://mobilitylab.org/2013/01/17/top-10-bike-and-bus-observations-from-copenhagen.

26. The City of Copenhagen, "Copenhagen City of Cyclists: Bicycle Account 2012."

27. Leslie Braunstein, "Parking Yields to the Sharing Economy." *Urban Land,* November 2013. Available at http://urbanland.uli.org/infrastructure-transit/parking-yields-to-the-sharing-economy.

28. Floyd, "Top Ten Bike and Bus Observations."

29. CW, "Bicycle Bridges to Be Finished by End of Year." *The Copenhagen Post,* 13 May 2014. Available at http://cphpost.dk/news/bicycle-bridges-to-be-finished-by-end-of-year.9547.html.

30. Lasses S. Hauschildt, "New Bikesharing System in Denmark's Two Largest Cities." Cycling Embassy of Denmark, 22 April 2014. Available at http://www.cycling-embassy.dk/2014/04/22/new-bike-sharing-system-in-denmarks-two-largest-cities.

31. J. C. Decaux, "Copenhagen to Launch a New Bike-Sharing System." *Urban Radar*—Mobility & Trends blog. 29 August 2013. Available at http://www.mobility-trends.com/index.php/2013/08/copenhagen-to-launch-a-new-bike-sharing-system.

32. Cited in Floyd, "Top Ten Bike and Bus Observations."

33. Cited in Gerdes, "Copenhagen's Ambitious Push to Be Carbon-Neutral."

34. Gerdes, "Copenhagen's Ambitious Push to Be Carbon-Neutral."

35. Ibid.

36. Julian Isherwood, "Copenhagen: Inviting the World to See How It's Done." The Official Website of Denmark (undated). Available at http://denmark.dk/en/green-living/copenhagen/green-capital.

37. City of Copenhagen, "*What Is Sharing Copenhagen?*" Sharing Copenhagen 2014. Available at http://www.sharingcopenhagen.dk/english/about.

Chapter 3, Sharing Politics: The City as Public Realm

1. Paul Wheatley, *The Pivot of the Four Quarters: A Preliminary Enquiry into the Origins and Character of the Ancient Chinese City* (Edinburgh: Edinburgh University Press, 1971).

2. Ash Amin, "Collective Culture and Urban Public Space." *City* 12(1) (2008): 5–24.

3. Michael Sandel, "A New Citizenship. Reith Lecture, and Transcript of Discussion." *BBC Radio Broadcast*, 9 June 2009. Transcript available at http://downloads.bbc.co.uk/rmhttp/radio4/transcripts/20090609_thereithlectures_marketsandmorals.rtf, 15.

4. Anton Rosenthal, "Spectacle, Fear, and Protest: A Guide to the History of Urban Public Space in Latin America." *Social Science History* 24(1) (2000): 33–73.

5. Stacy Passmore, "The Social Life of Public Space in West Africa." Planning Pool blog, 16 February 2011. Available at http://planningpool.com/2011/02/land-use/african-public-space.

6. Galen Cranz, *The Politics of Park Design: A History of Urban Parks in America* (Cambridge, MA: MIT Press 1989); Anna Minton, *The Privatisation of Public Space* (London: Royal Institution of Chartered Surveyors, 2006).

7. Julian Agyeman, *Introducing Just Sustainabilities: Policy, Planning and Practice* (London: Zed Books, 2013).

8. Henry Shaftoe, *Convivial Urban Spaces*. (London: Routledge/Earthscan, 2008).

9. Ibid., 16.

10. Happy Wall is an interactive public artwork in Kongens Nytorv, distinct from The Wall (described in the case study preceding this chapter) which is a mobile exhibit curated by Copenhagen Museum.

11. Doreen Massey, "Places and Their Pasts." *History Workshop Journal* 39(1) (1995): 182–192.

12. Harvey, *Rebel Cities*, 137.

13. Jane Jacobs, *The Death and Life of Great American Cities* (New York: Random House, 1961), 50.

14. Doreen Massey, *For Space*. (London: Sage, 2005).

15. Amin, "Collective Culture and Urban Public Space," 11.

16. Jane Jacobs, "Downtown Is for People." *Fortune*, April 1958. Republished online 18 September 2011, http://fortune.com/2011/09/18/downtown-is-for-people-fortune-classic-1958.

17. Caroline Chen, "Dancing in the Streets of Beijing: Improvised Uses within the Urban System," in *Insurgent Public Space: Guerilla Urbanism and the Remaking of Contemporary Cities,* ed. Jeffrey Hou (New York: Routledge, 2010), 21–35.

18. Blaine Merker, "Taking Place: Rebar's Absurd Tactics in Generous Urbanism," in *Insurgent Public Space: Guerilla Urbanism and the Remaking of Contemporary Cities,* ed. Jeffrey Hou (New York, NY: Routledge, 2010), 47–57.

19. Stokes et al., *Making Sense of the UK Collaborative Economy,* 40.

20. Susanna Rustin, "If Women Built Cities, What Would Our Urban Landscape Look Like?" *The Guardian,* 5 December 2014. Available at http://www.theguardian.com/cities/2014/dec/05/if-women-built-cities-what-would-our-urban-landscape-look-like?

21. Mohamed S. El-Khatib, "Tahrir Square as Spectacle: Some Exploratory Remarks on Place, Body and Power." *Theater Research International,* 38 (2013): 104–115; Fady El Sadek, *The Streets of Cairo and the Battle for Public Space* (Masters thesis, University of Cambridge, 2011). Available at http://www.scribd.com/doc/113348910/El-Sadek-Fady-Streets-of-Cairo#scribd; Castells, *Networks of Outrage and Hope.*

22. Castells, *Networks of Outrage and Hope.*

23. Ibid., 59.

24. Ibid., 60.

25. David Graeber, "Occupy Democracy is Not Considered Newsworthy. It Should Be." *The Guardian,* 27 October 2014. Available at http://www.theguardian.com/commentisfree/2014/oct/27/occupy-democracy-london-parliament-square.

26. Castells, *Networks of Outrage and Hope,* 220.

27. Graeber, *The Democracy Project.*

28. Castells, *Networks of Outrage and Hope,* 245.

29. Debora MacKenzie, "Brazil Uprising Points to Rise of Leaderless Networks." *New Scientist,* 26 June 2013. Available at http://www.newscientist.com/article/mg21829234.300-brazil-uprising-points-to-rise-of-leaderless-networks.html.

30. Grodach et al. "The Location Patterns of Artistic Clusters," 3.

31. Ibid.

32. Gabriel, *Why I Buy,* 119.

33. Sherwin et al. "Cultural Creative Clusters."

34. Ibid.

35. In the UK the role of squatting in music culture has been highlighted in recent resistance to the criminalisation of squatting. See Ben Myers "Criminalising squatters will hurt British pop music" *The Guardian,* 3 September 2012. Available at: http://www.theguardian.com/music/musicblog/2012/sep/03/criminalising-squatters-hurts-british-music.

36. Yates McKee, "The Arts of Occupation: Occupy Wall Street Isn't Just Producing Art Work, It's Challenging the Boundaries of Art and Activism." *The Nation,* 11 December 2011. Available at http://www.thenation.com/article/165094/arts-occupation#.

37. Harvey, *Rebel Cities,* 111.

38. Gabriel *Why I Buy.*

39. Georgică Mitrache, "Architecture, Art, Public Space." *Procedia-Social and Behavioral Sciences* 51 (2012): 562–566.

40. Taylor Parkes, "A British Disaster: Blur's Parklife, Britpop, Princess Di & The 1990s." *The Quietus*, 28 April 2014. Available at http://thequietus.com/articles/15092-blur-parklife-anniversary-review.

41. Benkler, "Sharing Nicely," 351.

42. Ibid., 351.

43. Ibid., 352.

44. John McDuling, "'The Problem With Music' Has Been Solved by the Internet." *Quartz,* 29 April 2014. Available at http://qz.com/202194/steve-albini-the-problem-with-music-has-been-solved-by-the-internet.

45. Sofar Sounds website, http://www.sofarsounds.com.

46. Sofar Creative website, http://www.sofarcreative.com.

47. Thompson, *Smarter Than You Think,* 223–224.

48. Kahne, cited in Thompson, *Smarter Than You Think.*

49. Thompson, *Smarter Than You Think,* 263.

50. Castells, *Networks of Outrage and Hope,* 232.

51. Ibid., 233.

52. Cited in Thompson, *Smarter Than You Think,* 261.

53. Cristina Maza, "Why the Internet Makes the Personal Even More Political." *Open Democracy,* 3 September 2014. Available at https://www.opendemocracy.net/transformation/cristina-maza/why-internet-makes-personal-even-more-political.

54. Castells, *Networks of Outrage and Hope,* 106.

55. Ibid., 229.

56. Ibid., 107.

57. Tapscott and Williams, *Macrowikinomics*.

58. Jonathan Zittrain, "The Scary Future of Digital Gerrymandering—And How to Prevent It." *New Republic*, 1 June 2014. Available at http://www.newrepublic.com/article/117878/information-fiduciary-solution-facebook-digital-gerrymandering.

59. Thompson, *Smarter Than You Think*, 272.

60. Marianne Franklin with Robert Bodle and Dixie Hawtin, *The Charter of Human Rights and Principles for the Internet*. IRPC, 4th ed, 2014. Available at http://internetrightsandprinciples.org/site/wp-content/uploads/2014/08/IRPC_Booklet-English_4thedition.pdf.

61. Robert J. Deibert and Masashi Crete-Nishihata, "Global Governance and the Spread of Cyberspace Controls." *Global Governance* 18 (2012): 339–361.

62. Castells, *Networks of Outrage and Hope*, 21.

63. Tapscott and Williams, *Macrowikinomics*.

64. Castells, *Networks of Outrage and Hope*.

65. Tapscott and Williams, *Macrowikinomics*, 327.

66. Ibid., 329–330.

67. Castells, *Networks of Outrage and Hope*.

68. Uffe Elbæk and Neal Lawson, *The Bridge: How the Politics of the Future Will Link the Vertical to the Horizontal* (London: Compass and Alternativet, 2014, p. 3). Available at: http://www.compassonline.org.uk/wp-content/uploads/2014/03/Compass-The-Bridge2.pdf.

69. Ibid., 4.

70. Eurig Scandrett, *Citizen Participation and Popular Education in the City*. Big Ideas Thinkpiece for Friends of the Earth, London, 2013, p. 12. Available at http://www.foe.co.uk/sites/default/files/downloads/citizen_participation_and.pdf,

71. Ibid., 9.

72. Ibid., 19.

73. Belk, "Sharing."

74. Robert D. Putnam, "Bowling Alone: America's Declining Social Capital. An interview with Robert Putnam." *Journal of Democracy* 6(1) (1995): 65–78, 66.

75. Michael Woolcock and Deepa Narayan, "Social Capital: Implications for Development Theory, Research, and Policy." *The World Bank Research Observer*, 15(2) (2000): 225–249, 227.

76. Ibid., 230.

77. Ibid., 234.

78. Andreas Vårheim, "Gracious Space: Library Programming Strategies towards Immigrants as Tools in the Creation of Social Capital." *Library & Information Science Research, 33*(1) (2011): 12–18.

79. Kimberly J. Shinew, Troy D. Glover, and Diana C. Parry, "Leisure Spaces as Potential Sites for Interracial Interaction: Community Gardens in Urban Areas." *Journal of Leisure Research* 36(3) (2004): 336–55.

80. Jan Semenza and Tanya March, "An Urban Community-Based Intervention to Advance Social Interactions." *Environment and Behavior* 41(1) (2009): 22–42.

81. Daniel Sauter and Marco Huettenmoser, "Livable Streets and Social Inclusion." *Urban Design International* 13 (2008): 67–79.

82. Joshua Hart, "Driven to Excess: Impacts of Motor Vehicle Traffic on Residential Quality of Life in Bristol, UK." MSc dissertation in transport planning, University of the West of England, Bristol, 2008.

83. Shannon H. Rogers, John M. Halstead, Kevin H. Gardner, and Cynthia H. Carlson, "Examining Walkability and Social Capital as Indicators of Quality of Life at the Municipal and Neighborhood Scales." *Applied Research in Quality of Life* 6(2) (2011): 201–213.

84. Putnam, "Bowling Alone – Interview"; and *Bowling Alone.*

85. Botsman and Rogers, *What's Mine Is Yours*, 43.

86. Rosie Niven, "Case studies: Using Social Media to Increase Neighbourhood Co-Operation." *The Guardian,*Voluntary sector blog, 15 December 2011. Available at http://www.theguardian.com/voluntary-sector-network/community-action-blog/2011/dec/15/social-media-neighbourhood-cooperation.

87. Pumpipumpe website, http://www.pumpipumpe.ch/sticker-bestellen.

88. Streetlife website, https://www.streetlife.com.

89. Steve O'Hear, "Streetlife, the U.K. Local Social Network, Raises Further £600K from Archant Digital Ventures, Shohet & Cie." *Techcrunch*, 30 July 2013. Available at http://techcrunch.com/2013/07/30/streetlife/.

90. Joseph P. Schwieterman, "The Travel Habits of Gen Y." *Planning*, May/June (2011): 30–33.

91. Robb Willer, Francis J. Flynn, and Sonya Zak, "Structure, Identity, and Solidarity: A Comparative Field Study of Generalized and Direct Exchange." *Administrative Science Quarterly*, 57 (1) (2012): 119, doi: 10.1177/0001839212448626.

92. Botsman and Rogers, *What's Mine Is Yours,* 124.

93. Ibid., 130.

94. Stefano Bartolini, Ennio Bilancini, and Maurizio Pugno, "Did the Decline in Social Capital Depress Americans' Happiness." 2008. Available at SSRN: http://dx.doi.org/10.2139/ssrn.1210118, 7–8).

95. Ibid., 15.

96. Shaftoe, *Convivial Urban Spaces;* Office of the Deputy Prime Minister, *Safer Places: The Planning System and Crime Prevention* (London: Office of the Deputy Prime Minister and Home Office, 2004); John Pucher and John L. Renne, "Socioeconomics of Urban Travel: Evidence from the 2001 NHTS." *Transportation Quarterly* 57 (3) (2003): 49–77; Mariela Alfonzo, Marlon G. Boarnet, Kristen Day, Tracy McMillan, and Craig L. Anderson, "The Relationship of Neighbourhood Built Environment Features and Adult Parents' Walking." *Journal of Urban Design* 13 (1) (2008): 29–51.

97. Stephen Pinker, *The Better Angels of Our Nature: A History of Violence and Humanity* (London: Allen Lane, 2011).

98. Kevin Drum, "America's Real Criminal Element: Lead." *Mother Jones*, Jan/Feb 2013. Available at http://www.motherjones.com/environment/2013/01/lead-crime-link-gasoline.

99. Kahan, "The Logic of Reciprocity."

100. Wesley G. Skogan and Lynn Steiner, *CAPS at Ten: Community Policing in Chicago—An Evaluation of Chicago's Alternative Policing Strategy* (Prepared by the Chicago Community Policing Evaluation Consortium, 2004). Available at https://portal.chicagopolice.org/i/cpd/clearpath/Caps10.pdf.

101. Brian Resnick, "Chart: One Year of Prison Costs More Than One Year at Princeton." *The Atlantic*, 1 November 2011. Available at http://www.theatlantic.com/national/archive/2011/11/chart-one-year-of-prison-costs-more-than-one-year-at-princeton/247629.

102. Gabriel, *Why I Buy.*

103. Setha Low and Neil Smith, "Introduction: The imperative of Public Space," in *The Politics of Public Space*, eds. Setha Low and Neil Smith (Abingdon and New York: Routledge, 2006), 1–16, 1.

104. Gabriel, *Why I Buy.*

105. Anthony Giddens, *Modernity and Self-Identity: Self and Society in the Late Modern Age* (Stanford: Stanford University Press, 1991); Ulrich Beck, *Risk Society, Toward a New Modernity* (London: Sage Publications, 1992); Bauman, *Liquid Life.*

106. Manfred A. Max-Neef, "Development and Human Needs," in *Real-Life Economics: Understanding Wealth Creation,* 197–213, eds. Paul Ekins and Manfred Max-Neef (London: Routledge, 1992).

107. Rifkin, *The Empathic Civilization.*

108. Gabriel, *Why I Buy.*

109. Harvey, *Rebel Cities,* 14.

110. Botsman and Rogers, *What's Mine Is Yours.*

111. Capacity and Play Matters, cited in Lucie Ozanne and Paul Ballantine, "Sharing as a Form of Anti-consumption? An Examination of Toy Library Users." *Journal of Consumer Behavior* 9(6) (2010): 485–498, 488–489.

112. Ibid.

113. Ibid., 489.

114. Lucie Ozanne and Paul Ballantine, "Sharing as a Form of Anti-consumption?"

115. Ibid., 487.

116. Sandel, *What Money Can't Buy,* 172.

117. Ibid., 172–173.

118. Amy Curtis, "Five Years after Banning Outdoor Ads, Brazil's Largest City Is More Vibrant Than Ever." *New Dream Resources,* 2011. Available at http://www.newdream.org/resources/sao-paolo-ad-ban.

119. Cited in ibid.

120. Perkins, Anne, "Good Riddance Barclays. Boris Bike Sponsorship Was a Bad Deal for London." *The Guardian,* 11 December 2013. Available at http://www.theguardian.com/commentisfree/2013/dec/11/barclays-boris-bike-sponsorship-bad-deal-london.

121. Rifkin, *Zero Marginal Cost Society.*

122. Hurth et al. "Reforming Marketing for Sustainability."

123. Botsman and Rogers, *What's Mine Is Yours,* 98.

124. Sen, *Development as Freedom*; Amartya Sen, *The Idea of Justice* (London: Allen Lane, 2009).

125. Amin, "Collective Culture and Urban Public Space," 6.

126. Ibid., 6.

127. Tim Stonor, "Digital Urbanism." *Academy of Urbanism Journal* 2 (2013): 18–19 http://issuu.com/theaou/docs/129720640-aou-journal-issue-two, 19.

128. Amin, "Collective Culture and Urban Public Space," 7–8.

129. Ibid., 8.

130. Jan Gehl, "The Form and Use of Public Space." *Paper for European Transport Conference, Loughborough University,* 16 September 1998. Available at http://abstracts.aetransport.org/paper/download/id/719, 196.

131. Benkler, *Penguin and Leviathan,* 229.

132. Castells, *Networks of Outrage and Hope.*

133. Ben Hamilton-Baille, "Towards Shared Space." *Urban Design International* 13 (2008): 130–138.

134. Katie Williams and Carol Dair, "What is Stopping Sustainable Building in England? Barriers Experienced by Stakeholders in Delivering Sustainable Developments." *Sustainable Development* 15(3) (2006): 135–147.

135. Anastasia Loukaitou-Sideris and Renia Ehrenfeucht, *Sidewalks: Conflict and Negotiation over Public Space* (Cambridge, MA: MIT Press, 2009).

136. Ibid.

137. Paul Hess, "Avenues or Arterials: The Struggle to Change Street Building Practices in Toronto, Canada." *Journal of Urban Design* 14(1) (2009): 1–28; Anthony Flint, *This Land: The battle over Sprawl and the Future of America* (Baltimore, MD: Johns Hopkins University Press, 2006).

138. Stephen Zavestoski and Julian Agyeman, *Incomplete Streets: Processes, Practices, and Possibilities (*London: Routledge, 2014).

139. Christopher B. Leinberger, "Now Coveted: A Walkable, Convenient Place." *New York Times,* 25 May 2012. Available at www.nytimes.com/2012/05/27/opinion/sunday/now-coveted-a-walkable-convenient-place.html.

140. Anna Livia Brand, "The Most Complete Street in the World: A Dream Deferred and Co-opted," in *Incomplete Streets: Processes, Practices and possibilities,* ed. Stephen Zavestoski and Julian Agyeman (London: Routledge, 2014), 243–263.

141. Ibid., 259.

142. Mark Vallianatos, "Compl(eat)ing the Streets: Legalizing Sidewalk Food Vending in Los Angeles," in *Incomplete Streets: Processes, Practices and Possibilities,* ed. Stephen Zavestoski and Julian Agyeman (London. Routledge, 2014), 202–222, 204.

143. Collingwood Neighborhood House website, http://www.cnh.bc.ca/intercultural-community-learning.

144. Amin, "Collective Culture and Urban Public Space," 19.

145. James Tully, *Strange Multiplicity: Constitutionalism in an Age of Diversity* (Cambridge: Cambridge University Press, 1995).

146. Jude Bloomfield and Franco Bianchini, *Planning for the Cosmopolitan City: A Research Report for Birmingham City Council* (Leicester: Comedia, International Cultural Planning and Policy Unit, 2002), 6.

147. Sandercock, *Cosmopolis II*, 207–208.

148. Sennett, *Together*, 4.

149. S. Stouffer, E. Suchman, L. DeVinney, S. Star, R. Williams Jr. "The American Soldier: Adjustment During Army Life." 1949. Cited by T. F. Pettigrew, "Intergroup Contact Theory." *Annual Review of Psychology* 49 (1998): 65–85.

150. Allport, *The Nature of Prejudice*.

151. Thomas Pettigrew and Linda Tropp, "A Meta-analytic Test of Intergroup Contact Theory." *Journal of Personality and Social Psychology* 90(5) (2006): 751–783.

152. Christ et al. "Contextual Effect of Positive Intergroup Contact."

153. Jojanneke van der Toorn, Jaime L. Napier, and John F. Dovidio, "We the People: Intergroup Interdependence Breeds Liberalism." *Social Psychological and Personality Science* 5(5) (2014): 616–622, 616.

154. Ibid., 620.

155. Nancy Fraser, "How Feminism Became Capitalism's Handmaiden—And How to Reclaim It." *The Guardian*, 14 October 2013. Available at http://www.theguardian.com/commentisfree/2013/oct/14/feminism-capitalist-handmaiden-neoliberal.

156. Brittney Cooper, "On Bell, Beyoncé, and Bullshit." *Crunk Feminist Collective*, 20 May 2014. Available at http://www.crunkfeministcollective.com/2014/05/20/on-bell-beyonce-and-bullshit.

157. Paul Romer, "Technologies, Rules, and Progress: The Case for Charter Cities," Center for Global Development Working Paper, 2010. Available at www.cgdev.org/content/publications/detail/1423916.

158. Bulkeley et al., *Enhancing Urban Autonomy*.

159. Ibid., 2–4.

160. Tapscott and Williams, *Macrowikinomics*.

161. Ibid., 274.

162. Rifkin, *Zero Marginal Cost Society*.

163. Ibid., 22.

164. Romer, *The Case for Charter Cities*.

165. Benjamin Barber, *If Mayors Ruled the World: Dysfunctional Nations, Rising Cities* (New Haven, CT: Yale University Press, 2013).

166. Sassen, *The Global City*.

167. See for example https://bitcointalk.org/index.php?topic=344879.0.

168. Bulkeley et al., *Enhancing Urban Autonomy*.

169. Wright, *Envisioning Real Utopias*.

170. Yves Sintomer, Carsten Herzberg, and Anja Röcke, "Participatory Budgeting in Europe: Potentials and Challenges." *International Journal of Urban and Regional Research* 32 (2008): 164–178, 166–167.

171. Wright, *Envisioning Real Utopias*.

172. Rebecca Abers, *Inventing Local Democracy: Grassroots Politics in Brazil* (Boulder, CO: Lynne Rienner Publishers, 2000); Gianpaolo Baiocchi, "The Porto Alegre Experiment and Deliberative Democratic Theory." *Politics & Society* 29 (2001): 43–72.

173. Diether Beuermann and Maria Amelina, "Does Participatory Budgeting Improve Decentralized Public Service Delivery?" IDB Working Paper Series No 547, November 2014. Available at http://publications.iadb.org/bitstream/handle/11319/6699/Does-Participatory-Budgeting-Improve-Decentralized-Public-Service-Delivery.pdf.

174. Harvey, 2012, 140–150.

175. See, for example, CIC website, http://cooperativa.cat/en/whats-cic/general-principles.

176. Castells, *Networks of Outrage and Hope*, 129.

177. Ibid.

178. Graeber, *The Democracy Project*.

179. Castells, *Networks of Outrage and Hope*, 143.

180. Ibid., 144.

181. Alanna Krause, "When Business Met Occupy: Innovating for True Collaborative Decision-Making," 2014. Available at http://www.mixprize.org/story/when-business-met-occupy-innovating-true-collaborative-decision-making-and-true-empowerment.

182. Bulkeley et al., *Enhancing Urban Autonomy*.

183. Quilligan "People Sharing Resources."

184. Benkler, "Sharing Nicely."

185. Benkler, *Penguin and Leviathan*, 149.

186. Ostrom, "A General Framework for Analyzing Sustainability."

187. Benkler (in "Sharing Nicely"), however, criticizes Ostrom for framing commons governance merely as "stable" in certain conditions, rather than as more desirable than private ownership, and for placing commons on the periphery of the economy, and seeing them as a model of mutual ownership, rather than as an alternative to property.

188. Quilligan "People Sharing Resources," 37.

189. Ibid., 37.

190. Benkler, "Sharing Nicely."

191. David Bollier, "Bauwens Joins Ecuador in Planning a Commons-Based, Peer Production Economy." Blog entry, 20th September 2013. Available at http://bollier.org/blog/bauwens-joins-ecuador-planning-commons-based-peer-production-economy.

192. Alex von Tunzelmann, "An Uncertain Glory: India and Its Contradictions, by Jean Drèze and Amartya Sen, Review." *The Telegraph*, 1 August 2013. Available at http://www.telegraph.co.uk/culture/books/10211435/An-Uncertain-Glory-India-and-itsContradictions-by-Jean-Dreze-and-Amartya-Sen-review.html.

193. Bulkeley et al., *Enhancing Urban Autonomy*.

194. Barber, *If Mayors Ruled the World*.

195. Benjamin Barber's website, http://benjaminbarber.org/books/if-mayors-ruled-the-world.

196. *Collaborative Consumption*, "Shareable Cities Resolution: Passed." News post, 26 June 2013. Available at http://www.collaborativeconsumption.com/2013/06/26/shareable-cities-resolution-passed.

197. Tapscott and Williams, *Macrowikinomics*.

198. Gerry Hassan, "The Third Scotland: Self-Organising, Self-Determining, Suspicious of the SNP." *The Guardian*, 24 April 2014. Available at http://www.theguardian.com/commentisfree/2014/apr/24/third-scotland-self-determining-suspicious-of-snp.

199. Ibid.

200. Castells, *Networks of Outrage and Hope.*

201. MacKenzie, "Brazil Uprising—Leaderless Networks."

202. Tapscott and Williams, *Macrowikinomics,* 261.

203. Ibid.

204. Aarefa Johari, "One Way to Improve Women's Safety in India: Plot Each Reported Rape on a Map." *Quartz,* 25 March 2014. Available at http://qz.com/191724/one-way-to-improve-womens-safety-in-india-plot-each-reported-rape-on-a-map.

205. Tapscott and Williams, *Macrowikinomics,* 285.

206. Michelle Nijhuis, "How the Five-Gallon Plastic Bucket Came to the Aid of Grassroots Environmentalists." *Grist,* 23 July 2003. Available at http://grist.org/article/the19.

207. Tapscott and Williams, *Macrowikinomics.*

208. Tom Llewellyn, "#MapJam 2.0 to Put the New Economy on the Map!" *Shareable,* 22 September 2014. Available at http://www.shareable.net/blog/mapjam-20-to-put-the-new-economy-on-the-map.

209. Tapscott and Williams, *Macrowikinomics.*

210. Peter Newell, Philipp Pattberg, and Heike Schroeder, "Multiactor Governance and the Environment." *Annual Review of Environment and Resources* 37(1) (2012): 365–387, doi:10.1146/annurev-environ-020911-094659.

211. The Metropolitan Council, "Fiscal Disparities. Tax Base Sharing in the Twin Cities Metropolitan Area." 2012. Available at http://www.metrocouncil.org/Communities/Planning/Local-Planning-Assistance/Fiscal-Disparities-%282%29.aspx.

212. Sun Ki Kwon, "The Impact of the Shared Property Tax System on the Localities' Fiscal Capacity." Capstone project, 2012. Martin School of Public Policy and Administration University of Kentucky. Available at http://www.martin.uky.edu/centers_research/Capstones_2012/Kwon.pdf.

213. Orsi et al. *Policies for Shareable Cities.*

214. Ibid., 12.

215. Ibid., 18.

216. Ibid., 18.

217. Harvey, *Rebel Cities.*

218. Campaign to Take Back Vacant Land, website, http://takebackvacantland.org/?page_id=457.

219. Michel Bauwens, "The City as Commons," *Commons Transition*, 2 March 2015. Available at http://commonstransition.org/the-city-as-commons-with-professor-christian-iaione.

220. Orsi et al. *Policies for Shareable Cities*.

221. Peter Newman and Jeffrey R. Kenworthy, *Cities and Automobile Dependence: A Sourcebook* (Aldershot UK: Gower, 1989); Harley Sherlock, *Cities Are Good for Us* (London: Paladin, 1991); Tim Elkin and Duncan McLaren with Mayer Hillman, *Reviving the City: Toward Sustainable Urban Development* (London: Friends of the Earth Trust, 1991).

222. Seong-Hoon Cho, Seung Gyu Kim, and Roland K. Roberts, "Measuring the Effects of a Land Value Tax on Land Development." *Applied Spatial Analysis and Policy* 4(1) (2011): 45–64.

223. Joshua Vincent, "Neighborhood Revitalization and New Life: A Land Value Taxation Approach." *American Journal of Economics and Sociology* 71(4) (2012): 1073–1094.

224. Martinez, "Squatting for Justice."

225. Rob Sharp, "How to Build a House … with 20 Friends: Families across the Country Are Turning to Co-housing." *The Independent*, 26 April 2014. http://www.independent.co.uk/news/uk/home-news/how-to-build-a-housewith-20-friends-families-across-the-country-are-turning-to-cohousing-9279042.html.

226. Matthieu Lietaert, "The Growth of Cohousing in Europe." *The Cohousing Association of the United States*, 16 December 2007. Available at http://www.cohousing.org/node/1537.

227. Kollektivhus website, http://www.kollektivhus.nu/index.html.

228. Cohousing UK website, http://www.cohousing.org.uk/about.

229. Kelly Scott Hanson and Chris Scott Hanson. *The Cohousing Handbook: Building a Place for Community* (Gabriola Island, BC: New Society Publishers, 2nd revised edition, 2004).

230. LILAC (low impact living affordable community). Briefing Sheet: Low Impact Living (undated). Available at http://www.lilac.coop/documents/public/107-briefing-sheet-low-impact-living/file.html; Lancaster Cohousing, "Standards," (undated). Available at http://www.lancastercohousing.org.uk/Project/Standards.

231. Patsy Healey, *Collaborative Planning* (London: Macmillan, 1997): 3.

Chapter 4, Case Study: Medellín

1. Sibylla Brodzinsky, "From Murder Capital to Model City: Is Medellín's Miracle Show or Substance?" *The Guardian,* 17 April 2014. Available at http://www.theguardian.com/cities/2014/apr/17/medellin-murder-capital-to-model-city-miracle-un-world-urban-forum.

2. Greg Scruggs, "Latin America's New Superstar: How Gritty, Crime-Ridden Medellín Became a Model for 21st-Century Urbanism." *The Next City,* 31 March 2014. Available at http://nextcity.org/features/view/medellins-eternal-spring-social-urbanism-transforms-latin-america.

3. Brodzinsky, "From Murder Capital to Model City."

4. Scruggs, "Latin America's New Superstar."

5. Ed Vulliamy, "Medellín, Colombia: Reinventing the World's Most Dangerous City." *The Guardian,* 9 Jun 2013. Available at http://www.theguardian.com/world/2013/jun/09/medellin-colombia-worlds-most-dangerous-city.

6. Quoted in John Otis, "Medellín's Outdoor Escalator Part of Plan to Remake City." *Public Radio International,* 1 January 2013. Available at http://www.pri.org/stories/2013-01-01/medell-ns-outdoor-escalator-part-plan-remake-city.

7. Michael Kimmelman, "A City Rises, along with Its Hopes." *New York Times,* 20 May 2012. Available at http://www.nytimes.com/2012/05/20/arts/design/fighting-crime-with-architecture-in-medellin-colombia.html.

8. Scruggs, "Latin America's New Superstar."

9. Joseph Stiglitz, "Medellín's Metamorphosis Provides a Beacon for Cities across the Globe." *The Guardian,* 8 May 2014. http://www.theguardian.com/business/2014/may/08/medellin-livable-cities-colombia.

10. Quoted in Vulliamy, "Reinventing the World's Most Dangerous City."

11. Agyeman, *Introducing Just Sustainabilities.*

12. Scruggs, "Latin America's New Superstar."

13. Jorge Madrid, "Medellín's Amazing Metro System: Colombia Uses Public Transport to Drive Societal Change." *Climate Progress,* 13 March 2012. Available at http://thinkprogress.org/climate/2012/03/13/443330/Medell%C3%ADn-metro-system-colombia-public-transport.

14. John Henley, "Medellín: The Fast Track from the Slums." *The Guardian,* 31 July 2013. Available at http://www.theguardian.com/world/2013/jul/31/medellin-colombia-fast-track-slums-escalators; Otis, "Medellín's Outdoor Escalator."

15. Henley, "The Fast Track from the Slums."

16. Otis, "Medellín's Outdoor Escalator."

17. Scruggs, "Latin America's New Superstar."

18. Agyeman, *Introducing Just Sustainabilities.*

19. Maria Sendin, "The Multipurpose Cable Car of Rio's Largest Favela." *Not Only about Architecture,* 30 January 2014. Available at http://www.patriciasendin.com/2014/01/the-multipurpose-cable-car-of-rios.html.

20. Madrid, "Medellín's Amazing Metro System."

21. Ibid.

22. Elizabeth A. Ferruelo, "The Transformation of Medellín, and the Surprising Company Behind It." *Forbes,* 27 January 2014. Available at http://www.forbes.com/sites/ashoka/2014/01/27/the-transformation-of-medellin-and-the-surprising-company-behind-it.

23. Kimmelman, "A City Rises."

24. Quoted in Ferruelo, "The Transformation of Medellín."

25. Brodzinsky, "From Murder Capital to Model City."

26. Jota Samper, "Granting of Land Tenure in Medellin, Colombia's Informal Settlements: Is Legalization the Best Alternative in a Landscape of Violence?" *Informal Settlements Research,* 30 January 2014. Available at http://informalsettlements.blogspot.se/2011/01/v-behaviorurldefaultvmlo.html.

27. Paula Alvarado, "Medellín to Build Massive 46-Mile Urban Park Surrounding the City." *Urban Design* [blog], Tree Hugger, 2012. Available at http://www.treehugger.com/urban-design/medellin-build-massive-46-miles-urban-park-surrounding-city.html.

28. Charles Parkinson, "Monorail Plans Have Some Asking, Is Medellín Too Obsessed with 'Innovative' Transit?" The Next City—Resilient Cities, 29 April 2014. Available at http://nextcity.org/daily/entry/monorail-plans-have-some-asking-is-medellin-too-obsessed-with-innovative-tr.

29. Ferruelo, "The Transformation of Medellín."

30. Milford Bateman, "Medellín Emerges as a Latin American Trailblazer for Local Economic Growth." *The Guardian,* 3 April 2012. Available at http://www.theguardian.com/global-development/poverty-matters/2012/apr/03/medellin-trailblazer-local-economic-growth.

31. Stories Coop, Cooperativa Medica Social (Coomsocial IPS CTA), webpage, http://stories.coop/cooperatives/cooperativa-medica-social-%C2%93coomsocial-ips-cta%C2%94.

32. Bateman, "Medellín Emerges as a Latin American Trailblazer."

33. Quoted in Nick Parker, "Medellín Reborn: Colombian City Moves Out of Escobar Shadow." CNN, International Edition—Latin America, 22 October 2013. Available at http://www.cnn.com/2013/10/18/world/americas/colombia-medellin-regeneration.

34. Alex Schmidt, "Participatory Budgeting Is Music to Medellín's Poor." National Public Radio, 20 April 2011. Available at http://www.npr.org/2011/04/20/135152789/participatory-budgeting-is-music-to-medellin-s-poor.

35. Scruggs, "Latin America's New Superstar."

36. Shannon Young, "Art & Music as Alternatives to Violence in Medellín, Colombia." Public Radio International's The World, 11 December 2012. Available at http://www.pri.org/stories/2012-12-11/art-music-alternatives-violence-medell-n-colombia.

37. Quoted in ibid.

38. Kimmelman, "A City Rises."

39. Melanie Starkey, "Medellín Voted City of the Year." Urban Land Institute, 1 March 2013. Available at http://uli.org/urban-land-magazine/Medellin-named-most-innovative-city.

40. David Weiss, "Innovating in an Urbanized World: Leading from the Middle." Huffington Post blog, 23 May 2014. Available at http://www.huffingtonpost.com/david-weiss/innovating-in-an-urbanizi_b_5375486.html.

41. United Nations (UN) Habitat, "Medellín Declaration: Seventh World Urban Forum." Habitat, 2014. Available at http://unhabitat.org/7th-world-urban-forum-medellin-declaration.

42. IAI (International Association of Inhabitants), "The Alternative People's Urban Social Forum Concludes Successfully in Medellín." 14 April, 2014. Available at http://www.habitants.org/news/inhabitants_of_americas/the_alternative_people_s_urban_social_forum_concludes_successfully_in_medellin.

43. IAI (International Association of Inhabitants), "Political Declaration—People's Alternative Urban Social Forum. Let's Build Cities for a Life of Dignity!" 10 April, 2014. Available at http://www.habitants.org/the_urban_way/people_s_alternative_urban_social_forum_2014/political_declaration_-people_s_alternative_urban_social_forum._let_s_build_cities_for_a_life_of_dignity.

44. Luisa Sotomayor, "Medellín: The New Celebrity?" Spatial Planning in Latin America, 26 August 2013. Available at http://planninglatinamerica.wordpress.com/2013/08/26/medellin-the-new-celebrity.

45. Parkinson, "Monorail Plans Have Some Asking."

Chapter 4, Sharing Society: Reclaiming the City

1. Samuel Fraiberger and Arun Sundararajan, "Peer-to-Peer Rental Markets in the Sharing Economy." *SSRN*, 2015 http://papers.ssrn.com/sol3/papers.cfm?abstract_id=2574337.

2. Benita Matofska, "The Sharing Economy: Disrupting the Disruptors ... Can the Sharers Share?" Disruptive Innovation Festival session, webcast, 7 November 2014.

3. World Commission on Environment and Development, *Our Common Future* (Oxford: Oxford University Press, 1987).

4. Julian Agyeman, Robert D. Bullard, and Bob Evans, eds., *Just Sustainabilities: Development in an Unequal World* (Cambridge, MA: MIT Press, 2003), 5.

5. Agyeman, *Introducing Just Sustainabilities*.

6. Rifkin, *Zero Marginal Cost Society*.

7. John Rawls, *A Theory of Justice* (Cambridge MA: Harvard University Press, Revised edition, 2009), 87.

8. Richard Wilkinson and Kate Pickett, *The Spirit Level: Why Equality Is Better for Everyone* (London: Allen Lane, 2009).

9. Thomas Piketty, *Capital in the 21st Century* (Cambridge MA: Harvard University Press, 2014); Jonathan D. Ostry, Andrew Berg, and Charalambos G. Tsangarides, "Redistribution, Inequality, and Growth." IMF Staff Discussion Note. SDN/14/02, 2014. Available at http://www.imf.org/external/pubs/ft/sdn/2014/sdn1402.pdf; OECD, Directorate for Employment, Labour, and Social Affairs, "Does Income Inequality Hurt Economic Growth?" Focus on Inequality and Growth, 2014. Available at http://www.oecd.org/els/soc/Focus-Inequality-and-Growth-2014.pdf.

10. Jonathan Haidt, *The Righteous Mind: Why Good People are Divided by Politics and Religion*. (London: Allen Lane, 2012).

11. Etzioni, "Communitarianism."

12. Ibid., 228.

13. Arturo Escobar, "Latin America at a Crossroads." *Cultural Studies*, 24 (1) (2010): 1–65.

14. Sandel, *Justice*, 261.

15. Ibid., 263–264.

16. Sen, *Development as Freedom*, 155, citing former Philippines president Fidel Ramos.

17. Sandel, *Justice*, 265.

18. Max Holleran, "Let's Share! Please Provide Your Credit Card Information to Get Started." Open Democracy, 21 July, 2014. Available at https://www.opendemocracy.net/transformation/max-holleran/lets-share-please-provide-your-credit-card-information-to-get-started.

19. Sandel, *Justice*, 266–267.

20. Sen, *Development as Freedom; The Idea of Justice*.

21. Martha C. Nussbaum, *Women and Human Development: The Capabilities Approach* (Cambridge: Cambridge University Press, 2000) and *Creating Capabilities: The Human Development Approach* (Cambridge MA: Harvard University Press, 2011).

22. Chang, *23 Things*.

23. Sen, *Development as freedom*, 63.

24. Sennett, *Together*, 29.

25. Sen, *Development as Freedom*, 49–50.

26. Young, Iris Marion, *Justice and the Politics of Difference* (Princeton, NJ: Princeton University Press, 1990).

27. David Schlosberg, "Reconceiving Environmental Justice: Global Movements and Political Theories." *Environmental Politics* 13(3) (2004): 517–540.

28. David Schlosberg, *Defining Environmental Justice: Theories, Movements, and Nature* (Oxford: Oxford University Press, 2007).

29. Nancy Fraser, *The Fortunes of Feminism: From Women's Liberation to Identity Politics to Anti-Capitalism* (New York: Verso, 2013).

30. Zander Navarro, "In Search of a Cultural Interpretation of Power: The Contribution of Pierre Bourdieu." *IDS Bulletin* 37 (6) (2006): 11–22.

31. Andy Wightman, *The Poor Had No Lawyers. Who Owns Scotland and How They Got It* (Edinburgh: Birlinn, 2010).

32. Elizabeth Dunn, Lara Aknin, and Michael Norton, "Spending Money on Others Promotes Happiness." *Science* 319 (5870) (2008): 1687–1688; Andreas Mogensen, "Giving Without Sacrifice? The Relationship between Income, Happiness, and Giving." Giving What We Can, 2012. Available at http://www.givingwhatwecan.org/sites/givingwhatwecan.org/files/attachments/giving-without-sacrifice.pdf.

33. Clifford Cobb, Ted Halstead, and Jonathan Rowe, "If the GDP Is Up, Why Is America Down?" *The Atlantic Monthly*, October 1995: 59–78; Wilkinson and Pickett, *The Spirit Level*; Jackson, *Prosperity without Growth;* Richard Layard, *Happiness: Lessons from a New Science* London: Penguin, 2011.

34. Agyeman et al., *Sharing Cities*.

35. Schor, "Debating the Sharing Economy."

36. Schor, *True Wealth*; Holt, "Why the Sustainable Economy Movement Hasn't Scaled."

37. Sacks, "The Sharing Economy," citing Frost and Sullivan.

38. PWC (Price Waterhouse Coopers), "How Did We Develop Our Sharing Economy Revenue Projections: A Detailed Methodology," 2014. Available at http://www.pwc.co.uk/issues/megatrends/assets/how-did-we-develop-our-sharing-economy-revenue-projections-a-detailed-methodology.pdf.

39. Elizabeth Heideman, "Uber and Airbnb Leave Disabled People Behind." *The Daily Beast*, 4 October 2014. Available at http://www.thedailybeast.com/articles/2014/10/04/uber-and-airbnb-leave-disabled-people-behind.html.

40. Leo Mirani and Herman Wong, "Uber's Usage Maps Are a Handy Tool for Finding the World's Rich, Young People." *Quartz*, 23 April 2014. Available at http://qz.com/202187/ubers-usage-maps-are-a-handy-tool-for-finding-the-worlds-rich-young-people.

41. Glen Fleishman, "Uber and the Appropriation of Public Space." BoingBoing, 30 June 2014. Available at http://boingboing.net/2014/06/30/ubervalued-the-appropriation.html.

42. Noam Scheiber, "Uber Just Hired Obama's Most Ruthless Political Genius." *New Republic*, 19 August 2014. Available at http://www.newrepublic.com/article/119149/david-plouffe-will-serve-uber-he-served-obama.

43. Arielle Duhaime-Ross, "Driven: How Zipcar's Founders Built and Lost a Car-Sharing Empire." *The Verge*, 1 April 2014. Available at http://www.theverge.com/2014/4/1/5553910/driven-how-zipcars-founders-built-and-lost-a-car-sharing-empire.

44. Martin and Shaheen, "Impact of Carsharing on Vehicle Ownership."

45. Tapscott and Williams, *Macrowikinomics*.

46. Martin and Shaheen, "Impact of Carsharing on Vehicle Ownership."

47. Fraiberger and Sundararajan, "Peer-to-Peer Rental Markets."

48. C. V. Harquail, "Why Zipcar Is Not the 'Sharing Economy.'" Authentic Organisations, 9 January 2013. http://authenticorganizations.com/harquail/2013/01/09/zipcar-is-not-the-sharing-economy.

49. Bardhi and Eckhardt, "Access-Based Consumption."

50. Paul Keegan, "Zipcar—The Best New Idea in Business." *CNN Money*, 27 August 2009. Available at http://money.cnn.com/2009/08/26/news/companies/zipcar_car_rentals.fortune.

51. Nick Gibbs, "Daimler, BMW Bullish on Car Sharing." *Automotive News Europe*, 13 August, 2013. Available at http://europe.autonews.com/article/20130813/ANE/308139999/daimler-bmw-bullish-on-car-sharing.

52. Keegan, "Zipcar—The Best New Idea."

53. Wernau, "Enterprise Buying Chicago's I-Go,"

54. Botsman, "Collaborative Economy Services Changing the Way We Travel."

55. *The Economist*, "Boom and Backlash." 26 April 2014. Available at http://www.economist.com/news/business/21601254-consumers-and-investors-are-delighted-startups-offering-spare-rooms-or-rides-across-town.

56. Shane Hickey, "The Innovators: BlaBlaCar Is to Car Hire What Airbnb Is to the Hotel Industry." *The Guardian*, 13 April 2014. Available at http://www.theguardian.com/business/2014/apr/13/blablacar-hire-airbnb-hotel-car-share-service.

57. MitfahrGelegenheit. "Über Uns." Available at http://www.mitfahrgelegenheit.de/pages/about.

58. Albert Cañigueral, "7 Claves para la regulación del consumo colaborativo." Consumo Colaborativo Blog, 19 March 2014. Available at http://www.consumocolaborativo.com/2014/03/19/7-Claves-En-Materia-De-Regulacion-Del-Consumo-Colaborativo.

59. Gunjan Parik, "Spotlight on the C40 Transportation Initiative." C40 Cities blog, 14 August 2013. Available at http://www.c40.org/blog_posts/spotlight-on-the-c40-transportation-initiative.

60. Egged, "Who Are We?" Available at http://www.egged.co.il/Article-830-Who-Are-We.aspx.

61. Adam Greenfield, "Helsinki's Ambitious Plan to Make Car Ownership Pointless in 10 Years." *The Guardian*, 10 July 2014. Available at http://www.theguardian.com/cities/2014/jul/10/helsinki-shared-public-transport-plan-car-ownership-pointless.

62. Cat Johnson, "Are Uber and Lyft Now Offering Real Ridesharing?" Shareable, 7 August 2014. Available at http://www.shareable.net/blog/are-uber-and-lyft-now-offering-real-ridesharing.

63. Casey Newton, "This Is Uber's Playbook for Sabotaging Lyft." *The Verge*, 26 August 2014. Available at http://www.theverge.com/2014/8/26/6067663/this-is-ubers-playbook-for-sabotaging-lyft.

64. Danny Vinik, "Why We Should Celebrate Uber's Strategy to Poach Lyft Drivers: It's Ethical, Savvy, and Good for Workers." *New Republic*, 27 August 2014. Available at http://www.newrepublic.com/article/119235/ubers-strategy-poach-lyft-drivers-good-workers.

65. Grace Wyler, "Uber Drivers Are Revolting Against Their Bosses." *Vice*, 11 September 2014. Available at http://www.vice.com/en_uk/read/uber-drivers-are-revolting-against-their-bosses.

66. Evgeny Morozov, "The 'Sharing Economy' Undermines Workers' Rights—My FT Oped." Notes EM, 14 October 2013. Available at http://evgenymorozov.tumblr.com/post/64038831400/the-sharing-economy-undermines-workers-rights-my.

67. Ibid.

68. Quoted in Tapscott and Williams, *Macrowikinomics*, 367.

69. Schor, "Debating the Sharing Economy," 9–10.

70. Benkler, "Sharing Nicely," 342.

71. April Rinne, "Share Economy: Band-Aid Solution or Promoting Sustainability?" Interview on CBC Podcast, *The Current*, 3 March 2014. Available at http://www.cbc.ca/thecurrent/episode/2014/03/03/share-economy-band-aid-solution-to-real-economic-problems-or-promoting-sustainability.

72. Schor, "Debating the Sharing Economy," i.

73. Orsi, "The Sharing Economy Just Got Real."

74. Matofska, "The Sharing Economy: Disrupting the Disruptors."

75. Horowitz, "What Is New Mutualism?"

76. Michael Skapinker, "Unions Suffer for Lack of Killer App Amid Rise of Sharing Economy." *Financial Times*, 26 January 2015. Available at: http://www.ft.com/cms/s/0/b108e026-a24d-11e4-bbb8-00144feab7de.html.

77. Jacob Gershman, "Uber, Lyft Cases Could Set Far-Reaching Precedent." *Wall Street Journal Law Blog*, 16 March 2015. Available at: http://blogs.wsj.com/law/2015/03/16/uber-lyft-cases-could-set-far-reaching-precedent; Jordan Crook, "Uber Driver Deemed Employee by California Labor Commission." *TechCrunch*, 17 June 2015. Available at: http://techcrunch.com/2015/06/17/uber-drivers-deemed-employees-by-california-labor-commission/; Carmel DeAmicis, "Homejoy shuts down after battling worker classification lawsuits." *re/code*, 17 July 2015. Available at http://recode.net/2015/07/17/cleaning-services-startup-homejoy-shuts-down-after-battling-worker-classification-lawsuits/.

78. SolidarityNYC, quoted in Burns, "The 'Sharing' Hype."

79. Orsi, "The Sharing Economy Just Got Real."

80. Ibid.

81. Ibid.

82. Wright, *Envisioning Real Utopias*, citing Neamtan.

83. Orsi, "The Sharing Economy Just Got Real."

84. Ibid.

85. Wright, *Envisioning Real Utopias.*

86. Gorenflo quoted in Burns, "The 'Sharing' Hype."

87. Tapscott and Williams, *Macrowikinomics*, 348.

88. Schor "Debating the Sharing Economy," 11.

89. Katy Watson, "Waiting for a Shanty Home in Brazil's 'Tent Cities.'" *BBC News*: Business, 7 May 2014. Available at http://www.bbc.com/news/business-27223012.

90. Catlina Gomez, "Landslide prevention, environmental risk management and reforestation in Rio." *Urb.im Blog*, 18th May 2012. Available at http://www.urb.im/rj/120518la.

91. Julia Elyachar, *Markets of Dispossession: NGOs, Economic Development, and the State in Cairo* (Durham, NC: Duke University Press, 2005); and "Next Practices."

92. Elyachar, "Next Practices," 122.

93. Ananya Roy, "Ethical Subjects: Market Rule in an Age of Poverty." *Public Culture* 24 (1 66) (2012): 105–108, 106.

94. Ramachandra Guha and Joan Martínez-Alier, *Varieties of Environmentalism: Essays North and South* (London: Earthscan, 1997).

95. Sen, *Development as Freedom.*

96. Ibid.

97. Elyachar, "Next Practices."

98. Anon, "Occupy Reality!" Revolução, 28 October 2012. Available at http://spanishrevolution11.wordpress.com/2012/10/28/occupy-reality.

99. Graeber, "On the Invention of Money."

100. Benkler," Sharing Nicely"; Schor, "Debating the Sharing Economy."

101. Nembhard, "The Deep Roots of African American Cooperative Economics."

102. Sennett, *Together.*

103. Nembhard, "The Deep Roots of African American Cooperative Economics."

104. Ibid.

105. Ewald Engelen, Sukhdev Johal, Angelo Salento, and Karel Williams, "How to Build a Fairer City." *The Guardian*, 24 September 2014. Available at: http://www.theguardian.com/cities/2014/sep/24/manifesto-fairer-grounded-city-sustainable-transport-broadband-housing.

106. Peter Utting, "What Is Social and Solidarity Economy and Why Does It Matter?" World Bank Blogs, 30 April 2013. Available at http://blogs.worldbank.org/taxonomy/term/10911/taxonomy/term/10911/all/feed.

107. Golam Sarwar, "Paradoxes of Social Entrepreneurship." Paper prepared for the UNRISD Conference Potential and Limits of Social and Solidarity Economy, 6–8 May 2013, Geneva, Switzerland. Available at http://www.unrisd.org/80256B3C005BCCF9/search/5B21D317C8E40255C1257B5F0050347B?OpenDocument.

108. Wright, *Envisioning Real Utopias.*

109. Ibid.

110. Ibid.

111. Satterthwaite, *The Ten and a Half Myths.*

112. David Satterthwaite, "If We Don't Count the Poor, the Poor Don't Count." IIED blog, 12 June 2014. Available at http://www.iied.org/if-we-don-t-count-poor-poor-dont-count.

113. Satterthwaite, *The Ten and a Half Myths,* 20.

114. Ibid.

115. Ibid.

116. Ibid., 25.

117. Ibid., 25.

118. Mike Davis, "Planet of Slums: Interviewed by NPQ editor Nathan Gardels." *New Perspectives Quarterly* 23(2) (2006): 6–11, 7–8.

119. Satterthwaite, "If We Don't Count The Poor."

120. Davis, "Planet of Slums: Interviewed," 10.

121. Ibid., 11.

122. Stephens et al. *Co-Production: A Manifesto,* 12.

123. Salim Alimuddin, Arif Hasan, and Asiya Sadiq, "The Work of the Anjuman Samaji Behbood and the Larger Faisalabad Context, Pakistan." IIED Working Paper 7 on Poverty Reduction in Urban Areas. 2001. Available at http://pubs.iied.org/9073IIED.html?a=M.

124. Ibid., xii.

125. Ibid., xii.

126. Ibid., xiii.

127. Tapscott and Williams, *Macrowikinomics*; Thompson, *Smarter Than You Think.*

128. Chris Michael, "Missing Maps: Nothing Less than a Human Genome Project for Cities." *The Guardian*, 6 October 2014. Available at http://www.theguardian.com/cities/2014/oct/06/missing-maps-human-genome-project-unmapped-cities.

129. Gabriel Harp, "Civic Labs: Bangalore." Institute for the Future, 31 July 2012. Available at http://www.iftf.org/future-now/article-detail/civic-labs-bangalore.

130. Wilkinson and Pickett, *The Spirit Level*; Jackson, *Prosperity without Growth*.

131. Kathryn Zickuhr and Aaron Smith, *Digital Differences*. Pew Internet and American Life Project, 2012. Available at http://www.pewinternet.org/files/old-media/Files/Reports/2012/PIP_Digital_differences_041312.pdf.

132. Streetbank website, http://www.streetbank.com.

133. Sitrin, "The Commons and Defending the Public."

134. Katharine Ainger, "In Spain They Are All Indignados Nowadays." *The Guardian*, 28 April 2013. Available at http://www.guardian.co.uk/commentisfree/2013/apr/28/spain-indignados-protests-state-of-mind.

135. Sitrin, "The Commons and Defending the Public."

136. Ibid.

137. Michael Silverberg, "The World's Tallest Slum—A 'Pirate Utopia'—Is Being Cleared by the Venezuelan Government." *Quartz*, 23 July 2014. Available at http://qz.com/239103/the-worlds-tallest-slum-a-pirate-utopia-is-being-cleared-by-the-venezuelan-government.

138. Barbara Speed, "Radical Architects, Skyscraper Slums and Informal Cities: An Interview with Justin McGuirk." *CityMetric*, 21 October 2014. Available at http://www.citymetric.com/horizons/radical-architects-skyscraper-slums-and-informal-cities-interview-justin-mcguirk-402.

139. Ibid.

140. Fernanda Canofre, "How a Group of Squatters Convinced Brazilian Authorities to Seize a Vacant Building for Public Housing." *Global Voices*, 9 September 2014. Available at http://globalvoicesonline.org/2014/09/09/how-a-group-of-squatters-convinced-brazilian-authorities-to-seize-a-vacant-building-for-public-housing.

141. Sebastian Galiani and Ernesto Schargrodsky, "Property Rights for the Poor: Effects of Land Titling." *Journal of Public Economics* 94 (2010): 700–729.

142. Ibid.

143. Sen, *Development as Freedom*.

144. Harvey, *Rebel Cities*.

145. Harvey, *Rebel Cities*, 20.

146. Elyachar, *Markets of Dispossession.*

147. Samper, "Granting of Land Tenure in Medellin," citing Mukhija 2002.

148. Ibid.

149. Freire, *Pedagogy of the Oppressed.*

150. Scandrett, *Citizen Participation and Popular Education in the City,* 26.

151. Pret-a-Manger website, http://www.pret.com/sustainability/waste.htm.

152. Katharine Hibbert, "I Eat Out Of Bins Too. So What?" *The Guardian,* 15 February 2011. Available at http://www.theguardian.com/commentisfree/2011/feb/15/bins-freegans-leftover-food.

153. Estelle, "Guest Blog from the Real Junk Food Project." *Incredible Edible,* 19 February 2014. Available at http://www.incredible-edible-todmorden.co.uk/blogs/guest-blog-from-the-real-junk-food-project.

154. Garcia-Bardidia et al. "Consumer Resistance and Anti-consumption."

155. Ibid., citing IPSOS survey, 2009.

156. Harvey, *Rebel Cities,* 15.

157. Brian Covert, "Luxury Apartment Building Will Have Separate Door for Poor Residents." Think Progress, 21 July 2014. Available at http://thinkprogress.org/economy/2014/07/21/3462120/new-york-city-poor-door.

158. Loretta Lees, "Gentrification and Social Mixing: Towards an Inclusive Urban Renaissance?" *Urban Studies,* 45(12) (2008): 2449–2470, 2249.

159. Anne Power, "Does Demolition or Refurbishment of Old and Inefficient Homes Help to Increase Our Environmental, Social and Economic Viability?" *Energy Policy* 36 (2008): 4487–4501.

160. Leo Hollis, "Why Startup Urbanism Will Fail Us." *Shareable,* 4 August 2014. Available at http://www.shareable.net/blog/why-startup-urbanism-will-fail-us.

161. Ibid.

162. Newman, Andrew, "Contested Ecologies: Environmental Activism and Urban Space in Immigrant Paris." *City & Society* 23 (2) (2011): 192–209, 193.

163. Ibid., 200.

164. Ibid., 192.

165. Ibid., 192.

166. Ibid., 202.

167. Winifred Curran and Trina Hamilton, "Just Green Enough: Contesting Environmental Gentrification in Greenpoint, Brooklyn." *Local Environment: The International Journal of Justice and Sustainability* 17 (9) (2012): 1027–1042, 1028 (our emphasis).

168. Sarah Dooling, "Ecological Gentrification: A Research Agenda Exploring Justice in the City." *International Journal of Urban and Regional Research* 33 (3) (2009): 621–39, 621.

169. Ibid., 631.

170. Omidi, "Anti-homeless Spikes 'Defensive Urban Architecture.'"

171. Martinez, "Squatting for Justice."

172. Dooling, "Ecological Gentrification," 634–635.

173. Ibid., 634.

174. Escobar, "Latin America at a Crossroads."

175. Olivia Laing, "Reviewed: *The Gentrification of the Mind* by Sarah Schulman: Paradise lost." *New Statesman*, 7 March 2013. Available at http://www.newstatesman.com/culture/2013/03/reviewed-gentrification-mind-sarah-schulman.

176. Ibid.

177. Harvey, *Rebel Cities*, 17.

178. Heideman, "Uber and Airbnb Leave Disabled Behind."

179. Empathy and recognition appear to have generated separate literatures in different disciplinary traditions. And neither, as yet, has explicitly addressed sharing, as far as we found, so our suggestions here are based on our own interpretations of these concepts.

180. Paul Piff, Daniel Stancato, Stéphane Côté, Rodolfo Mendoza-Denton, and Dacher Keltner, "Higher Social Class Predicts Increased Unethical Behaviour." *PNAS* 109 (11) (2010): 4086–4091. Available at http://www.pnas.org/content/109/11/4086.

181. Castells, *Networks of Outrage and Hope*, 230; Giddens, *Modernity and Self-Identity*.

182. Castells, *Networks of Outrage and Hope*, 230.

183. Ibid., 230–231.

184. Bauman, *Liquid Life*.

185. Charles Blow, "The Self Sort." *New York Times*,11 April 2014. Available at http://www.nytimes.com/2014/04/12/opinion/blow-the-self-sort.html.

186. Human Library website, http://humanlibrary.org/what-is-the-living-library.html.

187. Krznaric, *Empathy: A Handbook for Revolution.*

188. Rifkin, *The Empathic Civilization.*

189. William Yardley, "Racial Shift in a Progressive City Spurs Talks," *New York Times,* 29 May 2008. Available at http://www.nytimes.com/2008/05/29/us/29portland.html.

190. Adam Baird, "Negotiating Pathways to Manhood: Rejecting Gangs and Violence in Medellín's Periphery." *Journal of Conflictology* 3(1) (2012): 30–41.

191. James Siddle, "I Know Where You Were Last Summer: London's Public Bike Data Is Telling Everyone Where You've Been." The Variable Tree, 10 April 2014. Available at http://vartree.blogspot.se/2014/04/i-know-where-you-were-last-summer.html.

192. Mayer-Schönberger and Cukier, *Big Data.*

193. Ibid.

194. Leo Kelion, "London Police Trial Gang Violence 'Predicting' Software." *BBC News,* 29 October 2014. Available at http://www.bbc.com/news/technology-29824854.

195. Mayer-Schönberger and Cukier, *Big Data.*

196. John Podesta, "Findings of the Big Data and Privacy Working Group Review." The White House Blog, 1 May 2014. Available at http://www.whitehouse.gov/blog/2014/05/01/findings-big-data-and-privacy-working-group-review.

197. Manoush Zomorodi, "Uber Is Upending That Old Motto about the Customer Always Being Right." *Quartz,* 9th May 2014. Available at http://qz.com/207764/uber-is-upending-that-old-motto-about-the-customer-always-being-right/#/h/67642,1.

198. Quoted in Siraj Datoo, "Smart cities: Are You Willing to Trade Privacy for Efficiency?" *The Guardian,* 4 April 2014. Available at http://www.theguardian.com/news/2014/apr/04/if-smart-cities-dont-think-about-privacy-citizens-will-refuse-to-accept-change-says-cisco-chief.

199. Mayer-Schönberger and Cukier, *Big Data.*

200. Zittrain, "The Scary Future of Digital Gerrymandering."

201. Katie Collins, "Brazil Passes 'Internet Constitution.'" *Wired,* 23 April 2014. Available at http://www.wired.co.uk/news/archive/2014-04/23/brazil-internet-bill-netmundial.

202. Over 270 examples of complementary currencies are data-based at http://www.complementarycurrency.org/ccDatabase/les_public.html.

203. Marco Sachy, "The New Frontier in Payment Systems: Virtual Currency Schemes, the C3 Uruguay case and the Potential Impact on SSE." Working Paper: Special session on Alternative Finance and Complementary Currencies. International Conference on Potential and Limits of Social and Solidarity Economy, UNRISD and ILO (May 2013) Available at http://www.unrisd.org/80256B3C005BCCF9/search/70F675806EAEFFF0C1257B600055B9D5.

204. Sennett, *Together*, 252.

205. Jem Bendall, "How Collaborative Credit Can Heal—Rather Than Just Disrupt—Capitalism." *The Guardian*, 3 July 2014. Available at http://www.theguardian.com/media-network/media-network-blog/2014/jul/03/collaborative-credit-heal-capitalism.

206. Ibid.

207. Botsman and Rogers, *What's Mine Is Yours*, 160.

208. Schifferes, "Sharing Our Way to Prosperity."

209. Lucie K. Ozanne, "Learning to Exchange Time: Benefits and Obstacles to Time Banking." *International Journal of Community Currency Research* 14 (A) (2010): 1–16.

210. Schor, "Debating the Sharing Economy."

211. Stokes et al., *Making Sense of the UK Collaborative Economy*, 14.

212. Gary Alexander, "Complementary Currencies, LETS and Timebanks." REconomy Project, undated. Available at http://www.reconomy.org/economic-enablers/alternative-means-of-exchange/complementary-currencies.

213. Schor, "Debating the Sharing Economy."

214. Charles Eisenstein, "Development in the Ecological Age." *Kosmos*, Spring/Summer 2014. Available at http://www.kosmosjournal.org/article/development-in-the-ecological-age.

Chapter 5, Case Study: Amsterdam

1. Lee and Hancock, *Toward a Framework for Smart Cities*.

2. Neal Gorenflo, "The Tale of Two Sharing Cities, Part One." *Shareable*, 9 April 2014. Available at http://www.shareable.net/blog/the-tale-of-two-sharing-cities-part-one.

3. John Gilderbloom, Matthew Hanka, and Carrie Beth Lasley, "Amsterdam: Planning and Policy for the Ideal City? *Local Environment: The International Journal of Justice and Sustainability* 14 (6) (2009): 473–493, 473.

4. Ibid., 481, citing Lofland 2000.

5. Ibid., 482.

6. I Amsterdam, "Diversity in the City," undated. Available at http://www.iamsterdam.com/en-GB/living/about-amsterdam/people-culture/diversity-in-the-city.

7. Ibid.

8. Andrew Chung, "Amsterdam's Newcomers Thrive Even as Immigration Gets Tougher." *The Star*, 23 June 2012. Available at http://www.thestar.com/news/world/2012/06/23/amsterdams_newcomers_thrive_even_as_immigration_gets_tougher.html.

9. Ibid., citing Eurislam.

10. Laure Michon and Floris Vermeulen, "Explaining Different Trajectories in Immigrant Political Integration: Moroccans and Turks in Amsterdam." *West European Politics* 36 (3) (2013): 597–614.

11. Ibid., 599.

12. Quoted in Chung, "Amsterdam's Newcomers Thrive."

13. Philip Lawton, "Understanding Urban Practitioners' Perspectives on Social-Mix Policies in Amsterdam: The Importance of Design and Social Space." *Journal of Urban Design*. 18 (1) (2012): 98–118.

14. Chung, "Amsterdam's Newcomers Thrive."

15. Lex Veldboer and Machteld Bergstra, "Does Income Diversity Increase Trust in the Neighborhood? The Social Impact of Gentrification in Amsterdam." Paper presented at RC21 Conference: The Struggle to Belong: Dealing with Diversity in Twenty-first Century Urban Settings, University of Amsterdam, 2011.

16. Lawton, "Understanding Urban Practitioners' Perspectives on Social-Mix Policies."

17. Veldboer and Bergstra, "Does Income Diversity Increase Trust in the Neighborhood?"

18. Ibid.

19. Gilderbloom et al. "Amsterdam: Planning and Policy for the Ideal City?"

20. Lawton, "Understanding Urban Practitioners' Perspectives on Social-Mix Policies."

21. Elizabeth Austerberry, "Netherlands Follows Britain's Lead on Social Housing." *The Guardian*, 21 June 2013. Available at http://www.theguardian.com/housing-network/2013/jun/21/netherlands-britain-social-housing-provision.

22. Manuel Aalbers and Rinus Deurloo, "Concentrated and Condemned? Residential Patterns of Immigrants from Industrial and Non-Industrial Countries in Amsterdam." *Housing, Theory and Society* 20(4) (2010): 197–208.

23. Lawton, "Understanding Urban Practitioners' Perspectives on Social-Mix Policies."

24. Austerberry, "Netherlands Follows Britain's Lead on Social Housing."

25. Dorit Fromm, "Central Living in the Netherlands: The Influence of Architecture on Social Structure." The Fellowship for Intentional Community, undated. Available at http://www.ic.org/wiki/central-living-netherlands-influence-architecture-social-structure.

26. Lietaert, "The Growth of Cohousing in Europe."

27. Peter Bakker, "Cohousing in the Netherlands." Presentation at International Cohousing Summit, Seattle, Washington, 2009. Available at http://www.lvcw.nl/teksten/Cohousing%20in%20the%20Netherlands%20-%20as%20presentated%20at%20the%20Summit.pdf.

28. Pieter van de Glind, "Amsterdam Embraces Sharing Economy." Collaborative Consumption, 14 February 2014. Available at http://www.collaborativeconsumption.com/2014/02/14/amsterdam-embraces-sharing-economy; Harrison Weber, "After a Rough Few Months, Airbnb Receives Amsterdam's Blessing. Will Other Cities Follow?" The Next Web blog, 7 June 2013. Available at http://thenextweb.com/eu/2013/06/07/four-months-after-its-hunt-for-illegal-hotels-amsterdam-lightens-restrictions-on-airbnb-rentals.

29. Quoted in van de Glind, "Amsterdam Embraces Sharing Economy."

30. Josh Ong, "Amsterdam Adopts New Private Rental Policy That Benefits Airbnb Hosts and the Sharing Economy." The Next Web blog, 13 February 2014. Available at http://thenextweb.com/insider/2014/02/13/amsterdam-adopts-new-private-rental-policy-benefits-airbnb-hosts.

31. Airbnb, press release, 18 December 2014. https://www.airbnb.dk/press/news/amsterdam-and-airbnb-sign-agreement-on-home-sharing-and-tourist-tax.

32. Karla Zabludovsky, "NYC is Battling Airbnb, but the Home-Sharing Firm Got a Green Light in Amsterdam." *Newsweek,* 1 May 2014. http://www.newsweek.com/nyc-battling-airbnb-home-sharing-firm-got-green-light-amsterdam-249256.

33. Ibid.

34. van de Glind, "Amsterdam Embraces Sharing Economy;" Anna Bergren Miller, "Amsterdam is now Europe's first named 'Sharing City'." *Shareable,* 24 February 2015. Available at http://www.shareable.net/blog/amsterdam-is-now-europes-first-named-sharing-city.

35. Carlien Roodink, "To share or not to share? Ouishare." Amsterdam Economic Board, 8 May 2014. Available at http://www.amsterdameconomicboard.com/nieuws/6309/to-share-or-not-to-share-ouishare.

36. Repair Café website, http://repaircafe.org.

37. Colin Webster, "Amsterdam: Exploring the Sharing City." Disruptive Innovation Festival video, 30 October 2014. Available at http://thinkdif.co/pages/top-10-sessions.

38. Ibid.

39. Cat Johnson, "How a New Dutch Library Smashed Attendance Records." *Shareable*, 21 July 2014. Available at http://www.shareable.net/blog/how-a-new-dutch-library-smashed-attendance-records.

40. Ibid.

41. Pieter van de Glind, "The Consumer Potential of Collaborative Consumption." Research MSc in Sustainable Development—Environmental Governance, Utrecht University, August, 2014. Available at http://www.collaborativeconsumption.com/2013/09/24/study-the-consumer-potential-of-collaborative-consumption, 3.

42. Lee and Hancock, *Toward a Framework for Smart Cities.*

43. Amsterdam Smart City, "Open Data," undated. Available at http://amsterdamsmartcity.com/projects/theme/label/open-data.

44. Lee and Hancock, *Toward a Framework for Smart Cities.*

45. Amsterdam Smart City, "Amsterdam Free Wifi," undated. Available at http://amsterdamsmartcity.com/projects/detail/id/63/slug/amsterdam-free-wifi.

46. Amsterdam Smart City, "IJburg—Smart Work@IJburg," undated. Available at http://amsterdamsmartcity.com/projects/detail/id/21/slug/ijburg-smart-workijburg.

47. Alexandra Gowling, "Amsterdam: Europe's Second 'Smartest' City." *I Am Expat*, 5 February, 2014. Available at http://www.iamexpat.nl/read-and-discuss/expat-page/news/amsterdam-europe-second-smartest-city.

48. Lee and Hancock, *Toward a Framework for Smart Cities*; Webster, "Amsterdam: Exploring the Sharing City."

Chapter 5, The Sharing City: Understanding and Acting on the Sharing Paradigm

1. Harvey, *The Enigma of Capital.*

2. Griffiths, *The Great Sharing Economy*; Diplock, *The Sharing Project*; Heinrichs and Grunenberg, *Sharing Economy;* Latitude, *The New Sharing Economy;* The People Who

Share, *State of the Sharing Economy;* Sunrun, "Sunrun Survey—Disownership"; Owyang, *Sharing is the New Buying;* and Stokes et al., *Making Sense of the UK Collaborative Economy.*

3. Latitude, *The New Sharing Economy.*

4. Stokes et al., *Making Sense of the UK Collaborative Economy.*

5. Griffiths, *The Great Sharing Economy.*

6. Diplock, *The Sharing Project.*

7. Latitude, *The New Sharing Economy,* 9.

8. Ibid., 9.

9. Stokes et al., *Making Sense of the UK Collaborative Economy,* 27.

10. Benkler, "Sharing Nicely."

11. Ethan Ligon and Laura Schechter, "Motives for Sharing in Social Networks." *Journal of Development Economics* 1 (2011): 1–14.

12. Harvey, *The Enigma of Capital.*

13. Joseph Stiglitz, *The Roaring Nineties: A New History of the World's Most Prosperous Decade* (London and New York: WW Norton, 2004).

14. Wright, *Envisioning Real Utopias.*

15. Duncan McLaren, "Delivering Transformation: a Challenge to Friends of the Earth Europe." Internal Paper for FOE Europe, Brussels, 2011.

16. Wright, *Envisioning Real Utopias.*

17. Tom Slee, "Share Economy: Band-Aid Solution or Promoting Sustainability?" Interviewed on CBC Podcast, *The Current,* 3 March 2014. http://www.cbc.ca/thecurrent/episode/2014/03/03/share-economy-band-aid-solution-to-real-economic-problems-or-promoting-sustainability.

18. Powercube. "Bourdieu and 'Habitus.'" Website created by Participation, Power and Social Change team, at the Institute of Development Studies, University of Sussex (undated). Available at http://www.powercube.net/other-forms-of-power/bourdieu-and-habitus.

19. Michèle Lamont and Virág Molnár, "How Blacks Use Consumption to Shape Their Collective Identity: Evidence from African-American Marketing Specialists." *Journal of Consumer Culture* 1(1) (2001): 31–45, 42.

20. Milo Yiannopoulos, "Don't Believe the Hype: Here's What's Wrong with the 'Sharing Economy.'" *TheNextWeb,* 6 June 2013. Available at http://thenextweb.com/insider/2013/06/06/dont-believe-the-hype-heres-whats-wrong-with-the-sharing-economy.

21. Ibid.

22. de Rugy, Veronique, "Airbnb versus San Francisco: The Fight Continues." *National Review*, 9 May 2014. Available at http://www.nationalreview.com/corner/377575/airbnb-versus-san-francisco-fight-continues-veronique-de-rugy.

23. Juliet Schor, "The New Politics of Consumption." *Boston Review*, Summer 1999.

24. Ibid.

25. Bardhi and Eckhardt, "Access-Based Consumption."

26. Belk, "Why Not Share," 131.

27. Belk, "Possessions and the Extended Self."

28. Yiannopoulos, "Don't Believe the Hype."

29. Martin Gilens and Benjamin I. Page, "Testing Theories of American Politics: Elites, Interest Groups, and Average Citizens." *Perspectives on Politics* 12 (03) September 2014: 564–581.

30. Wilkinson and Pickett, *The Spirit Level*; Layard, *Happiness*.

31. Rydin, *The Future of Planning*.

32. Ibid., 130–134.

33. Moira Herbst, "Let's Get Real: The 'Sharing Economy' Won't Solve Our Jobs Crisis." *The Guardian*, 7 January 2014. Available at http://www.theguardian.com/commentisfree/2014/jan/07/sharing-economy-not-solution-to-jobs-crisis.

34. Sen, *Development as Freedom*, 80.

35. Benkler, "Sharing Nicely."

36. Joseph Stiglitz, Amartya Sen, and Jean-Paul Fitoussi, Report by the Commission on the Measurement of Economic Performance and Social Progress, 2011. Available at: http://www.stiglitz-sen-fitoussi.fr/documents/rapport_anglais.pdf.

37. Project website, www.iiicitadel.com/index.html.

38. Séverine Deneulin and Roy Maconachie, "Gated Communities Lock Cities into Cycles of Inequality." *CityMetric*, 3 November 2014. Available at http://www.citymetric.com/horizons/gated-communities-lock-cities-cycles-inequality-439.

39. Michele Provoost, "From Welfare City to Neoliberal Utopia." Strelka talk, 2013. Available at http://vimeo.com/64392842.

40. Jathan Sadowski and Paul Manson, "3-D Print Your Way to Freedom and Prosperity: The Hidden Politics of the 'Maker' Movement." *Al Jazeera America*, 17 May 2014. Available at http://america.aljazeera.com/opinions/2014/5/3d-printing-politics.html.

41. Ibid.

42. Ibid.

43. Ibid.

44. David Runciman, "Politics or Technology—Which Will Save the World?" *The Guardian*, 23 May 2014. Available at http://www.theguardian.com/books/2014/may/23/politics-technology-save-world-david-runciman.

45. Ibid.

46. Acker, "Why San Francisco's Tech Community Is Creating Problems."

47. Mazzucato, *The Entrepreneurial State*; Runciman, "Politics or Technology."

48. Alastair McIntosh, *Hell and High Water: Climate Change, Hope and the Human Condition* (Edinburgh: Birlinn, 2008).

49. Benkler, "Sharing Nicely," 341, citing Kahan.

50. Ibid.

51. Ibid., 324.

52. Samuel Bowles, "Policies Designed for Self-Interested Citizens May Undermine "The Moral Sentiments": Evidence from Economic Experiments." *Science* 320 (5883) (2008): 1605–1609.

53. Ibid., 1605.

54. Kahan, "The Logic of Reciprocity."

55. Ibid., 11–12.

56. Ibid., 15.

57. Benkler, *Penguin and Leviathan*, 240.

58. Sandel, *What Money Can't Buy*, 124, citing Richard Titmuss.

59. Cited in Sandel, *What Money Can't Buy*, 130.

60. Ibid., 128.

61. Cited in ibid., 129.

62. The Circle Economy website, http://www.circle-economy.com.

63. Wilkinson and Pickett, *The Spirit Level*.

64. Piketty, *Capital in the 21st Century*.

65. Hurth et al. "Reforming Marketing For Sustainability."

66. Lindsey B. Carfagna, Emilie A. Dubois, Connor Fitzmaurice, Monique Y. Ouimette, Juliet B. Schor, Margaret Willis, and Thomas Laidley, "An Emerging Eco-habitus: The Reconfiguration of High Cultural Capital Practices Among Ethical Consumers." *Journal of Consumer Culture* 14(2) (2014): 158–178.

67. Holt, "Why the Sustainable Economy Movement Hasn't Scaled."

68. Twenge et al. "Declines in Trust in Others and Confidence in Institutions."

69. Botsman and Rogers, *What's Mine Is Yours*.

70. Ibid., 203.

71. Onora O'Neill, "What We Don't Understand about Trust." TED transcript, September 2013. Available at http://www.ted.com/talks/onora_o_neill_what_we_don_t_understand_about_trust/transcript.

72. Tapscott and Williams, *Macrowikinomics*, 33.

73. O'Neill, "What We Don't Understand about Trust" (at 7 minutes 29 seconds).

74. Jason Tanz, "How Airbnb and Lyft Finally Got Americans to Trust Each Other." *Wired*, 23 April 2014. Available at http://www.wired.com/2014/04/trust-in-the-share-economy.

75. Ostrom, "A General Framework for Analyzing Sustainability."

76. Botsman and Rogers, *What's Mine Is Yours*, 136.

77. Ibid., 143.

78. Paolo Parigi and Bogdan State, "Disenchanting the World: The Impact of Technology on Relationships." *SocInfo* 2014, LNCS 8851: 166–182.

79. Tanz, "How Airbnb and Lyft Finally Got Americans to Trust."

80. Ibid.

81. Ibid.

82. Ibid.

83. Quoted in Mary Hoff, "Premal Shah: Loans That Change Lives." *Ensia*, 4 November 2013. Available at http://ensia.com/interviews/premal-shah-loans-that-change-lives.

84. Benkler, "Sharing Nicely"; Tanz, "How Airbnb and Lyft Finally Got Americans to Trust."

85. Tanz, "How Airbnb and Lyft Finally Got Americans to Trust."

86. Tonkinwise, "Sharing You Can Believe In."

87. Botsman and Rogers, *What's Mine Is Yours*, 175, citing Tim O'Reilly.

88. Ibid., 176.

89. Ibid.

90. Botsman and Rogers, *What's Mine Is Yours*, 201.

91. Quoted in Taylor Soper, "Lyft Co-founder: Ride-Sharing Startups Will Make Taxi Companies More Money." *GeekWire*, 29 August 2013. Available at http://www.geekwire.com/2013/lyft.

92. Olson and Connor, *The Disruption of Sharing*, 13.

93. Bulajewski, "The Cult of Sharing."

94. Tonkinwise, "Sharing You Can Believe In."

95. Schor "Debating the Sharing Economy," 10.

96. Parigi and State, "Disenchanting the World."

97. O'Neill, "What We Don't Understand about Trust."

98. Adam Grant, "Raising a Moral Child." *New York Times*, 11 April 2014. Available at http://www.nytimes.com/2014/04/12/opinion/sunday/raising-a-moral-child.html.

99. Brendan Nyhan and Jason Reifler, *Blank Slates or Closed Minds? The Role of Information Deficits and Identity Threat in the Prevalence of Misperceptions*. 2013. Working paper. Available at http://www.dartmouth.edu/~nyhan/opening-political-mind.pdf.

100. Stokes et al., *Making Sense of the UK Collaborative Economy;* van de Glind, "The Consumer Potential of Collaborative Consumption."

101. Gabriel, *Why I Buy*.

102. Quoted in Geron, "Airbnb and the Unstoppable Rise of the Share Economy."

103. Jenny Morris, "Do as You Would Be Done By: Poverty and Disability," in *Poverty in the UK: Can it be eradicated?* ed. Jonathan Derbyshire (York: JRF and Prospect, 2013), 60–64. Available at: http://www.prospectmagazine.co.uk/wp-content/uploads/2013/11/jrf_web.pdf.

104. Martinez, "Squatting for Justice."

105. Ibid.

106. Ibid.

107. Don Peppers and Martha Rogers, *Extreme Trust: Honesty as a Competitive Advantage* (London: Penguin, 2012).

108. Ibid.

109. Trust Cloud website, https://trustcloud.com.

110. E-rated website, http://www.erated.co.

111. Matofska, "The Sharing Economy: Disrupting the Disruptors."

112. Cited by Thompson, *Smarter Than You Think*.

113. John Naughton, "Google Privacy Ruling Is Just the Thin End of a Censorship Wedge." *The Guardian*, 17 May 2014. Available at http://www.theguardian.com/technology/2014/may/17/google-privacy-ruling-thin-end-censorship-wedge.

114. Viktor Mayer-Schönberger, "Omission of Search Results Is Not a 'Right to Be Forgotten' or The End of Google." *The Guardian*, 13 May 2014. Available at http://www.theguardian.com/commentisfree/2014/may/13/omission-of-search-results-no-right-to-be-forgotten.

115. Castells, *Networks of Outrage and Hope*.

116. Paul Bernal, *Internet Privacy Rights: Rights to Protect Autonomy* (Cambridge: Cambridge University Press, 2014).

117. As suggested by Tapscott and Williams, *Macrowikinomics*.

118. Cited by Thompson, *Smarter Than You Think*, 237.

119. Ger Baron, "'Smartness' from the bottom up: A few Insights into the Amsterdam Smart City Programme." *Metering International* 3 (2013): 98–101, 100.

120. Benjamin G. Edelman and Michael Luca, "Digital Discrimination: The Case of Airbnb.com" Harvard Business School NOM Unit Working Paper No. 14–054, 2014. Available at http://papers.ssrn.com/sol3/papers.cfm?abstract_id=2377353.

121. Ibid.

122. Latoya Peterson, "Cab Drivers, Uber, and the Costs of Racism." Racialicious, 28 November 2012. Available at http://www.racialicious.com/2012/11/28/cab-drivers-uber-and-the-costs-of-racism.

123. Ibid.

124. Christ et al. "Contextual effect of positive intergroup contact."

125. Paul Bernal, *Internet Privacy Rights*.

126. In this respect Airbnb's response to Edelman and Luca was disappointing—in that they refused to release data that would allow the claim to be verified or rebutted. (Adrianne Jeffries, "Study Says Black Airbnb Hosts Earn Less Than Their White Counterparts." *The Verge*, 21 Jan 2014. Available at http://www.theverge.com/2014/1/21/5331106/study-says-black-hosts-earn-12-percent-less-than-white-hosts-on-airbnb).

127. Botsman and Rogers, *What's Mine Is Yours*.

128. Anja Kollmuss and Julian Agyeman, "Mind the Gap: Why Do People Act Environmentally and What Are the Barriers to Pro-environmental Behavior?" *Environmental Education Research* 8 (2002): 239–260.

129. Chris Rose and Pat Dade, "Using Values Modes," undated. Available at http://www.campaignstrategy.org/articles/usingvaluemodes.pdf.

130. Solitaire Townsend, *The Naked Environmentalist* (London: Futerra, 2013).

131. Tom Crompton, *Common Cause* (Godalming: WWF, 2010). Available at http://assets.wwf.org.uk/downloads/common_cause_report.pdf.

132. Schwartz, "Les valeurs de base de la personne."

133. Ibid.

134. Rinne, "Share Economy: Band-Aid Solution or Promoting Sustainability?"

135. Grant, "Raising a Moral Child."

136. Weick, Karl. *Sensemaking in Organisations* (London: Sage, 1995).

137. Benkler, *Penguin and Leviathan*, 161.

138. Ibid., 61.

139. Schifferes, "Sharing Our Way to Prosperity." We note that Benita Matofska's Compare and Share marketplace aims to occupy this niche. See http://www.compareandshare.com.

140. Paul Ormerod, *Positive Linking: How Networks Can Revolutionise the World* (London: Faber and Faber, 2012).

141. Benkler, "Sharing Nicely."

142. Wright, *Envisioning Real Utopias*.

143. Botsman and Rogers, *What's Mine Is Yours*, 216.

144. Benkler, "Sharing Nicely," 331.

145. Alina M. Chircu and Robert J. Kauffman, "Strategies for Internet Middlemen in the Intermediation / Disintermediation / Reintermediation Cycle." *Electronic Markets* 9(1–2) (1999): 109–117.

146. Botsman and Rogers, *What's Mine Is Yours*, 166).

147. Orsi et al., *Policies for Shareable Cities*.

148. Jones, "How Capitalism and Regulation Will Reshape Sharing."

149. Quoted in Tess Riley, "Sharing Economies Are Here to Stay." *The Guardian*, 7 May 2014. Available at http://www.theguardian.com/sustainable-business/behavioural-insights/sharing-economy-sustainable-alternative-economics.

150. Ibid.

151. Latitude, *The New Sharing Economy*.

152. Orsi et al., *Policies for Shareable Cities*; April Rinne, "Top 10 Things a City Can Do to Become a Shareable City." Collaborative Consumption, 5 February 2014. Available at http://www.collaborativeconsumption.com/2014/02/05/top-10-things-a-city-can-do-to-become-a-shareable-city; Botsman and Rogers, *What's Mine Is Yours*.

153. Daniel Rauch and David Schleicher, "Like Uber, But for Local Governmental Policy: The Future of Local Regulation of the Sharing Economy," George Mason Law & Economics Research Paper No. 15-01, 14 January 2015. Available at http://papers.ssrn.com/sol3/papers.cfm?abstract_id=2549919.

154. Uber's estimated value leapt from $17bn in Spring 2014, to $40bn in its next funding call in Fall 2014.

155. Orsi et al., *Policies for Shareable Cities*.

156. *The Economist*, "All Eyes on the Sharing Economy."

157. Orsi et al., *Policies for Shareable Cities*, 27.

158. Ibid.

159. Ibid., 27.

160. Ibid., 27.

161. *The Economist*, "Boom and Backlash."

162. Ibid.

163. Orsi et al. *Policies for Shareable Cities*.

164. Walker, "These Five Ideas for Smarter Cities Just Won Millions."

165. Sundararajan, "Trusting the 'Sharing Economy' to Regulate Itself."

166. Eli Lehrer, "A Big Conservative City Is about to Strangle Its Sharing Economy." *National Review Online*, 3 November 2014. Available at http://www.nationalreview.com/corner/391748/big-conservative-city-about-strangle-its-sharing-economy-eli-lehrer.

167. Sundararajan, "Trusting the 'Sharing Economy' to Regulate Itself."

168. Stokes et al., *Making Sense of the UK Collaborative Economy*, 31.

169. Arun Sundararajan, "The New 'New Deal'? Sharing Responsibility in the Sharing Economy." Policy Network, 30 October 2014. Available at http://www.policy-network.net/pno_detail.aspx?ID=4762&title=The-new-New-Deal-Sharing-responsibility-in-the-sharing-economy.

170. Debbie Wosskow, *Unlocking the Sharing Economy: An Independent Review.* London: Department For Business, Innovation and Skills BIS/14/1227, 2014. Available at https://www.gov.uk/government/uploads/system/uploads/attachment_data/file/378291/bis-14-1227-unlocking-the-sharing-economy-an-independent-review.pdf.

171. Ibid., 10.

172. Cañigueral, "7 Claves Para La Regulación Del Consumo Colaborativo."

173. Tarun Wadhwa, "Who's Looking Out for Consumers in the Sharing Economy?" Huffington Post Blog, 5 May 2014. Available at http://www.huffingtonpost.com/tarun-wadhwa/whos-looking-out-for-cons_b_5269138.html.

174. Richard Eskow, "Let's Nationalize Amazon and Google: Publicly Funded Technology Built Big Tech" *Salon*, 8 July, 2014. Available at http://www.salon.com/2014/07/08/lets_nationalize_amazon_and_google_publicly_funded_technology_built_big_tech.

175. Cited by Mike Gurstein, "Ooh-la-la, the French Get (Inter)Net Neutrality Right: It's All About the Platform Monopolies—Google, Amazon, Facebook, Twitter etc." Gurstein's Community Informatics, 27 August 2014. Available at https://gurstein.wordpress.com/2014/08/27/ooh-la-la-the-french-get-internet-neutrality-right-its-all-about-the-monopolies-google-amazon-facebook-twitter-etc.

176. Tarleton Gillespie, "Facebook's Algorithm: Why Our Assumptions Are Wrong and Our Concerns Are Right." Culture Digitally, 4 July 2014. Available at http://culturedigitally.org/2014/07/facebooks-algorithm-why-our-assumptions-are-wrong-and-our-concerns-are-right.

177. Zittrain, "The Scary Future of Digital Gerrymandering."

178. Ostrom, "A General Framework for Analyzing Sustainability."

179. Ibid.

180. Bulkeley et al., *Enhancing Urban Autonomy.*

181. Our thinking here is influenced by the multilevel perspective of Frank Geels (see "Processes and Patterns in Transitions and System Innovations: Refining the Co-Evolutionary Multi-Level Perspective," *Technology Forecasting and Social Change* 72 (2007): 681–696), and its application to spatial planning (see Rob Roggema, Tim Vermeend, and Andy van den Dobbelsteen, "Incremental Change, Transition or Transformation? Optimising Change Pathways for Climate Adaptation in Spatial Planning." *Sustainability* 4 (10) (2012), 2525–2549), as well as by Wright's (2010) approach to social transformation.

182. Phillip Späth and Harald Rohracher, "'Energy Regions': The Transformative Power of Regional Discourses on Socio-technical Futures." *Research Policy* 39 (2010): 449–458.

183. Harvey, *The Enigma of Capital*; and *Rebel Cities.*

Chapter 6, Case Study: Bengaluru

1. Srinath Perur, "The IT Crowd Gets Political: Bangalore's Techies Seek Indian Election Sweep." *The Guardian,* 9 April 2014. Available at http://www.theguardian.com/cities/2014/apr/09/the-it-crowd-gets-political-bangalores-techies-seek-indian-election-sweep.

2. Ibid.

3. Bharati Mukherjee, "Bangalore: A Bridge Culture to a New India." *The Wall Street Journal,* 16 May 2011. Available at http://blogs.wsj.com/speakeasy/2011/05/16/bangalore-a-bridge-culture-to-a-new-india.

4. Shilpa Kannan, "Bangalore: India's IT Hub Readies for the Digital Future." *BBC News,* 3 September 2013. Available at http://www.bbc.com/news/technology-23931499.

5. N. V. Krishnakumar, "Can Bangalore Become India's First Smart City?" *Deccan Herald,* 21 February 2014. Available at http://www.deccanherald.com/content/387477/can-bangalore-become-india039s-first.html.

6. Sujit John and Shilpa Phadnis, "Bangalore among Top 8 Clusters." *The Times of India.* 2 August 2013. Available at http://timesofindia.indiatimes.com/tech/tech-news/Bangalore-among-top-8-tech-clusters/articleshow/21544103.cms.

7. Ibid.

8. Anirudha Dutta, "Mumbai vs. Bangalore." *Forbes* Beyond the Numbers blog, 2 July 2013. Available at http://forbesindia.com/blog/beyond-the-numbers/mumbai-vs-bangalore.

9. Maheswara Y. Reddy, "An Inconvenient Truth: Bangalore's Future Is in the Gutter." DNA, Diligent Media Corporation, 23 June 2013. Available at http://www.dnaindia.com/bangalore/report-an-inconvenient-truth-bangalores-future-is-in-the-gutter-1851841.

10. Cited in ibid.

11. Aravind Gowda, "Bangalore Police Guard Landfills after Residents Voice Outrage at 'Health Problems' Caused by Illegal Dumping." *The Daily Mail,* 5 June 2014. Available at http://www.dailymail.co.uk/indiahome/indianews/article-2649895/Bangalore-Police-guard-landfills-residents-streets-outrage-health-problems-caused-illegal-dumping.html.

12. Saritha Rai, "Garden City, Garbage City." *The Indian Express,* 9 June 2014. Available at http://indianexpress.com/article/opinion/columns/garden-city-garbage-city.

13. Gowda, "Bangalore Police Guard Landfills after Outrage."

14. Perur, "The IT Crowd Gets Political."

15. Cited in ibid.

16. Cited in ibid.

17. Harp, "Civic Labs: Bangalore." Available at http://www.iftf.org/future-now/article-detail/civic-labs-bangalore.

18. Casey Tolan, "Cities of the future? Indian PM pushes plan for 100 'smart cities,'" *CNN,* 18 July 2014. Available at: http://edition.cnn.com/2014/07/18/world/asia/india-modi-smart-cities.

19. Krishnakumar, "Can Bangalore Become India's First Smart City?"

20. IBN Live, "Second Phase of Bangalore Metro Project Cleared." 4 January 2012. Available at http://ibnlive.in.com/news/second-phase-of-bangalore-metro-project-cleared/217527-62-129.html.

21. Business Standard, "Bangalore Metro Ph-1 completion pushed to Q-1 2016." 22 September, 2014. Available at http://www.business-standard.com/article/companies/bangalore-metro-ph-1-completion-pushed-to-q1-2016-114092200747_1.html.

22. Renuka Phadnis, "Namma Metro to accept BMTC smart cards." *The Hindu,* 27 June 2014. Available at http://www.thehindu.com/news/cities/bangalore/namma-metro-to-accept-bmtc-smart-cards/article6145961.ece.

23. *The Hindu,* "Free Wifi on M.G. Road and Brigade Road from Friday." 23 January 2014. Available at http://www.thehindu.com/news/cities/bangalore/free-wifi-on-mg-road-and-brigade-road-from-friday/article5606757.ece.

24. Krishnakumar, "Can Bangalore Become India's First Smart City?"

25. Ayona Datta, "India's Smart City Craze: Big, Green and Doomed from the Start?" *The Guardian.* 17 April 2014. Available at http://www.theguardian.com/cities/2014/apr/17/india-smart-city-dholera-flood-farmers-investors.

26. Richard Sennett, "No One Likes a City That's Too Smart." The Guardian, 4 December 2012. Available at http://www.theguardian.com/commentisfree/2012/dec/04/smart-city-rio-songdo-masdar.

27. Quoted in Daniel Nye Griffiths, "*City Cynic: 'Against The Smart City,'* by Adam Greenfield (review)." *Forbes,* 12 February 2013. Available at http://www.forbes.com/sites/danielnyegriffiths/2013/12/02/city-cynic-against-the-smart-city-by-adam-greenfield-review.

28. Adam Greenfield, "The Smartest Cities Rely on Citizen Cunning and Unglamourous Technology," *The Guardian*, 22 December 2014. Available at http://www.theguardian.com/cities/2014/dec/22/the-smartest-cities-rely-on-citizen-cunning-and-unglamorous-technology.

29. Mathieu Lefevre, "Moving Beyond the 'Smart City' Paradigm." New Cities Foundation blog, 9 July 2014. Available at http://www.newcitiesfoundation.org/moving-beyond-smart-city-paradigm.

30. Hug March and Ribera-Fumaz, Smart Contradictions: The Politics of Making Barcelona a Self-Sufficient City. *European Urban and Regional Studies*, 20 November 2014, 1–15. doi: 10.1177/0969776414554488.

31. Provoost, "From Welfare City to Neoliberal Utopia."

32. Sennett, "No One Likes a City That's Too Smart."

33. Ibid.

34. This list of sharing infrastructures is based on predictions for the future of collaborative consumption in Botsman and Rogers, *What's Mine Is Yours*.

Bibliography

Acker, Lizzy. "Why San Francisco's Tech Community Is Creating Problems, Not Solving Them." *Policy Mic,* October 22, 2013. Available at http://www.mic.com/articles/68969/why-san-francisco-s-tech-community-is-creating-problems-not-solving-them.

Agyeman, Julian. *Introducing Just Sustainabilities: Policy, Planning and Practice.* London: Zed Books, 2013.

Agyeman, Julian, Duncan McLaren, and Adrianne Schaefer-Borrego. *Sharing Cities.* Thinkpiece for Friends of the Earth Big Ideas Project, 2013. Available at http://www.foe.co.uk/sites/default/files/downloads/agyeman_sharing_cities.pdf.

Alkon, Alison H., and Julian Agyeman, eds. *Cultivating Food Justice: Race, Class and Sustainability.* Cambridge, MA: MIT Press, 2011.

Allport, Gordon W. *The Nature of Prejudice.* Cambridge, MA: Addison-Wesley, 1954.

Amin, Ash. "Collective Culture and Urban Public Space." *City* 12 (1) (2008): 5–24.

Barber, Benjamin. *If Mayors Ruled the World: Dysfunctional Nations, Rising Cities.* New Haven, CT: Yale University Press, 2013.

Bardhi, Fleura, and Giana M.Eckhart. "Access-Based Consumption: The Case of Car Sharing." *Journal of Consumer Research* 39 (2012): 881–898.

Bauman, Zygmunt. *Liquid Life.* Cambridge: Polity Press, 2005.

Belk, Russell W. "Why Not Share Rather Than Own." *Annals of the American Academy of Political and Social Science* 611 (2007): 126–140.

Belk, Russell W. "Sharing." *Journal of Consumer Research* 5 (2010): 715–734.

Belk, Russell W. "Possessions and the Extended Self." *Journal of Consumer Research* 15 (2) (1988): 139–168.

Benkler, Yochai. "Sharing Nicely: On Shareable Goods and the Emergence of Sharing as a Modality of Economic Production." *Yale Law Journal* 114 (2) (2004): 273–358.

Benkler, Yochai. *The Penguin and the Leviathan. The Triumph of Cooperation Over Self-Interest*. New York, NY: Crown Business, 2011.

Bernal, Paul. *Internet Privacy Rights: Rights to Protect Autonomy*. Cambridge: Cambridge University Press, 2014.

Botsman, Rachel. "The Case for Collaborative Consumption." TEDx Sydney, May 2010. Available at http://www.ted.com/talks/rachel_botsman_the_case_for_collaborative_consumption.html.

Botsman, Rachel. "Collaborative Economy Services Changing the Way We Travel." *Collaborative Consumption*, June 25, 2014. Available at http://www.collaborativeconsumption.com/2014/06/25/collaborative-economy-services-changing-the-way-we-travel.

Botsman, Rachel, and Roo Rogers. *What's Mine Is Yours: The Rise of Collaborative Consumption*. London: Harper Business, 2010.

Brewin, Kester. *Mutiny! Why We Love Pirates and How They Can Save Us*. London: Vaux / Kester Brewin, 2012.

Brodzinsky, Sibylla. "From Murder Capital to Model City: Is Medellín's Miracle Show or Substance?" *The Guardian*, April 17, 2014. Available at http://www.theguardian.com/cities/2014/apr/17/medellin-murder-capital-to-model-city-miracle-un-world-urban-forum.

Bulkeley, Harriet, Andres Luque, Colin McFarlane, and Gordon MacLeod. *Enhancing Urban Autonomy: Towards a New Political Project for Cities*. Friends of the Earth and Durham University, 2013. Available at http://www.foe.co.uk/sites/default/files/downloads/autonomy_briefing.pdf.

Burns, Rebecca. "The 'Sharing' Hype: Do Companies like Lyft and Airbnb Help Democratize the Economy?" *In These Times*, January 27, 2014. Available at http://inthesetimes.com/article/16111/the_sharing_economy_hype.

Cañigueral, Albert. "7 Claves Para La Regulación Del Consumo Colaborativo." *Consumo Colaborativo* [blog], March 19, 2014. Available at http://www.consumocolaborativo.com/2014/03/19/7-Claves-En-Materia-De-Regulacion-Del-Consumo-Colaborativo.

Castells, Manuel. *Networks of Outrage and Hope*. Cambridge: Polity Press, 2012.

Chang, Ha-Joon. *23 Things They Don't Tell You about Capitalism*. London: Penguin, 2010.

Christ, Oliver, Katharina Schmid, Simon Lolliot, Hermann Swart, Dietlind Stolle, Nicole Tausch, Ananthi Al Ramiah, Ulrich Wagner, Steven Vertovec, and Miles Hewstone. "Contextual Effect of Positive Intergroup Contact on Outgroup Prejudice." *Proceedings of the National Academy of Sciences of the United States of America* 111 (11) (March 18, 2014): 3996–4000.

Dervojeda, Kristina, Diederik Verzijl, Fabian Nagtegaal, Mark Lengton, Elco Rouwmaat, Erica Monfardini, and Laurent Frideres. *The Sharing Economy: Accessibility Based Business Models for Peer-to-Peer Markets*. Price Waterhouse Coopers for Business Innovation Observatory, 2013. Available at http://ec.europa.eu/enterprise/policies/innovation/policy/business-innovation-observatory/files/case-studies/12-she-accessibility-based-business-models-for-peer-to-peer-markets_en.pdf.

Diplock, Chris. *The Sharing Project: A Report on Sharing in Vancouver*. Vancouver: The Sharing Project, 2012. Available at http://thesharingproject.ca/TheSharingProject_Report.pdf.

Dooling, Sarah. "Ecological Gentrification: A Research Agenda Exploring Justice in the City." *International Journal of Urban and Regional Research* 33 (3) (2009): 621–639.

Economist, The. "All Eyes on the Sharing Economy." March 9, 2013. Available at http://www.economist.com/news/technology-quarterly/21572914-collaborative-consumption-technology-makes-it-easier-people-rent-items.

Economist, The. "Boom and Backlash," April 26, 2014. Available at http://www.economist.com/news/business/21601254-consumers-and-investors-are-delighted-startups-offering-spare-rooms-or-rides-across-town.

Elyachar, Julia. *Markets of Dispossession: NGOs, Economic Development, and the State in Cairo*. Durham, NC: Duke University Press, 2005.

Elyachar, Julia."Next Practices: Knowledge, Infrastructure, and Public Goods at the Bottom of the Pyramid." *Public Culture* 24 (1) (2012): 109–129.

Escobar, Arturo. "Latin America at a Crossroads." *Cultural Studies* 24 (1) (2010): 1–65.

Etzioni, Amitai. "Communitarianism." In *Encyclopedia of Community: From the Village to the Virtual World*. vol. 1. ed. Karen Christensen and David Levinson, 224–228. Thousand Oaks, CA: Sage Publications, 2003.

Freire, Paulo. *Pedagogy of the Oppressed*. New York: Continuum, 2007. First published in Portuguese in 1968 as *Pedagogia do Oprimido*.

Gabriel, Rami. *Why I Buy: Self, Taste, and Consumer Society in America*. Chicago: University of Chicago Press, 2013.

Gerdes, Justin. "Copenhagen's Ambitious Push to Be Carbon-Neutral by 2025." *The Guardian*, April 12, 2013. Available at http://www.theguardian.com/environment/2013/apr/12/copenhagen-push-carbon-neutral-2025.

Geron, Tomio. "Airbnb and the Unstoppable Rise of the Share Economy." *Forbes*, January 23, 2013. Available at http://www.forbes.com/sites/tomiogeron/2013/01/23/airbnb-and-the-unstoppable-rise-of-the-share-economy.

Giddens, Anthony. *Modernity and Self-Identity: Self and Society in the Late Modern Age*. Stanford: Stanford University Press, 1991.

Gilderbloom, John I., Matthew J. Hanka, and Carrie Beth Lasley. "Amsterdam: Planning and Policy for the Ideal City?" *Local Environment: The International Journal of Justice and Sustainability* 14 (6) (2009): 473–493.

Graeber, David. "On the Invention of Money—Notes on Sex, Adventure, Monomaniacal Sociopathy and the True Function of Economics." *Naked Capitalism*, September 13, 2011. http://www.nakedcapitalism.com/2011/09/david-graeber-on-the-invention-of-money-%E2%80%93-notes-on-sex-adventure-monomaniacal-sociopathy-and-the-true-function-of-economics.html.

Grant, Adam. "Raising a Moral Child." *New York Times*, April 11, 2014. Available at http://www.nytimes.com/2014/04/12/opinion/sunday/raising-a-moral-child.html?_r=0.

Griffiths, Rachel. *The Great Sharing Economy: A Report into Sharing across the UK*. Manchester: Cooperatives UK, 2011. Available at http://www.uk.coop/sites/storage/public/downloads/sharing_0.pdf.

Grodach, Carl, Elizabeth Currid-Halkett, Nicole Foster, and James Murdoch, III. "The Location Patterns of Artistic Clusters: A Metro- and Neighborhood-Level Analysis." *Urban Studies* (Edinburgh, Scotland) (2014): 1–22. doi:.10.1177/0042098013516523

Harp, Gabriel. "Civic Labs: Bangalore." *Institute for the Future*, July 31, 2012. Available at http://www.iftf.org/future-now/article-detail/civic-labs-bangalore.

Harvey, David. *The Enigma of Capital and the Crises of Capitalism*. London: Profile Books, 2011.

Harvey, David. *Rebel Cities: From the Right to the City to the Urban Revolution*. London: Verso, 2012.

Heideman, Elizabeth. "Uber and Airbnb Leave Disabled People Behind." *The Daily Beast*, October 4, 2014. Available at http://www.thedailybeast.com/articles/2014/10/04/uber-and-airbnb-leave-disabled-people-behind.html.

Heinrichs, Harald, and Heiko Grunenberg. *Sharing Economy: Towards a New Culture of Consumption?* Lüneburg: Centre for Sustainability Management, 2013.

Heinrichs, Harald. "Sharing Economy: A Potential New Pathway to Sustainability." *Gaia* 22 (4) (2013): 228–231.

Holt, Douglas, "Why the Sustainable Economy Movement Hasn't Scaled and How to Fix It: Toward a Campaign Strategy That Empowers Main Street." In *Sustainable Lifestyles and the Quest for Plenitude: Case Studies of the New Economy*, ed. Juliet B. Schor and Craig J. Thompson, 202–232. New Haven, CT: Yale University Press, 2014.

Horowitz, Sara. "What Is New Mutualism?" *Freelancers Broadcasting Network, November* 5, 2013. Available at https://www.freelancersunion.org/blog/2013/11/05/what-new-mutualism.

Hurth, Victoria, David Jackman, Jules Peck, and Enrico Wensing. *Reforming Marketing for Sustainability*. Big Ideas Thinkpiece. London: Friends of the Earth, 2015. Available at http://www.foe.co.uk/sites/default/files/downloads/reforming-marketing-sustainability-full-report-76676.pdf.

Jackson, Tim. *Prosperity Without Growth: Economics for a Finite Planet*. Oxford, New York: Routledge, 2009.

John, Nicholas A. "The Social Logics of Sharing." *Critical Studies in Media Communication* 16 (3) (2013): 113–131.

Johnson, Cat. "Is Seoul the Next Great Sharing City?" *Shareable,* July 16, 2013. Available at http://www.shareable.net/blog/is-seoul-the-next-great-sharing-city.

Johnson, Cat. "Sharing City Seoul: a Model for the World." *Shareable*, June 3, 2014. Available at http://www.shareable.net/blog/sharing-city-seoul-a-model-for-the-world.

Jones, Mike, "How Capitalism and Regulation Will Reshape the Sharing Economy." *Forbes*, October 9, 2013. Available at http://www.forbes.com/sites/ciocentral/2013/10/09/how-capitalism-and-regulation-will-reshape-the-sharing-economy.

Kahan, Dan M. "The Logic of Reciprocity: Trust, Collective Action, and Law," John M. Olin Center for Studies in Law, Economics, and Public Policy Working Paper 281 (2002). Available at http://digitalcommons.law.yale.edu/lepp_papers/281.

Krishnakumar, N. V. "Can Bangalore Become India's First Smart City?" *Deccan Herald,* February 21, 2014. Available at http://www.deccanherald.com/content/387477/can-bangalore-become-india039s-first.html.

Krznaric, Roman. *Empathy: A Handbook for Revolution*. London: Random House, 2013.

Latitude. "The New Sharing Economy: A Study by Latitude in Collaboration with Shareable Magazine" (2013). Available at http://latdsurvey.net/pdf/Sharing.pdf.

Layard, Richard. *Happiness: Lessons from a New Science*. 2nd ed.London: Penguin, 2011.

Lee, Jung-Hoon, and Marguerite Gong Hancock. *Toward a Framework for Smart Cities: A Comparison of Seoul, San Francisco & Amsterdam*, Yonsei University. Seoul: Korea and Stanford Program on Regions of Innovation and Entrepreneurship, 2012. Available at http://iis-db.stanford.edu/evnts/7239/Jung_Hoon_Lee_final.pdf.

Lietaert, Matthieu. "The Growth of Cohousing in Europe." The Cohousing Association of the United States, December 16, 2007. Available at http://www.cohousing.org/node/1537.

MacKenzie, Debora. "Brazil Uprising Points to Rise of Leaderless Networks." *New Scientist*, June 26, 2013. http://www.newscientist.com/article/mg21829234.300-brazil-uprising-points-to-rise-of-leaderless-networks.html.

Martin, Elliot W., and Susan A. Shaheen. "The Impact of Carsharing on Household Vehicle Ownership." *Access* 38 (2011): 22–27.

Martinez, Miguel. "Squatting for Justice, Bringing Life to the City." *ROARMAG*, May 13, 2014. Available at http://roarmag.org/2014/05/squatting-urban-justice-commons/.

Matofska, Benita. "The Sharing Economy: Disrupting the Disruptors ... Can the Sharers Share?" Disruptive Innovation Festival session, webcast, November 7, 2014.

Mayer-Schönberger, Viktor, and Kenneth Cukier. *Big Data: A Revolution That Will Transform How We Live, Work and Think*. London: John Murray, 2013.

Mazzucato, Mariana. *The Entrepreneurial State: Debunking Public vs. Private Sector Myths*. London: Anthem, 2013.

McLaren, Duncan. "Environmental Space, Equity and the Ecological Debt." In *Just Sustainabilities: Development in an Unequal World,* ed. Julian Agyeman, Robert D Bullard, and Bob Evans, 19–37. London: Earthscan, 2003.

Morozov, Evgeny. "Don't Believe the Hype, the 'Sharing Economy' Masks a Failing Economy." *The Guardian,* September 28, 2014. Available at http://www.theguardian.com/commentisfree/2014/sep/28/sharing-economy-internet-hype-benefits-overstated-evgeny-morozov.

Nembhard, Jessica Gordon. "The Deep Roots of African American Cooperative Economics." Interviewed by Mira Lund. *Shareable*, April 28, 2014. http://www.shareable.net/blog/interview-the-deep-roots-of-african-american-cooperative-economics.

Olson, Michael J., and Andrew D. Connor. "The Disruption of Sharing: An Overview of the New Peer-to-Peer 'Sharing Economy' and the Impact on Established Internet Companies." Piper Jaffray Investment Research, 2013. Available at https://piper2.bluematrix.com/docs/pdf/35ef1fcc-a07b-48cf-ab80-04d80e5665c4.pdf.

Omidi, Maryam. "Anti-homeless Spikes Are Just the Latest in 'Defensive Urban Architecture.'" *The Guardian*, June 12, 2014. Available at http://www.theguardian.com/cities/2014/jun/12/anti-homeless-spikes-latest-defensive-urban-architecture.

Orsi, Janelle. "The Sharing Economy Just Got Real." *Shareable*, September 16, 2013. Available at http://www.shareable.net/blog/the-sharing-economy-just-got-real.

Orsi, Janelle, Yassi Eskandari-Qajar, Eve Weissman, Molly Hall, Ali Mann, and Mira Luna. *Policies for Shareable Cities: A Sharing Economy Policy Primer for Urban Leaders*. Oakland, CA: Shareable and the Sustainable Economies Law Center, 2013. Available at http://www.shareable.net/blog/policies-for-a-shareable-city.

Ostrom, Elinor. "A General Framework for Analyzing Sustainability of Social-Ecological Systems." *Science* 325 (5939) (2009): 419–422. doi:.10.1126/science.1172133

Owyang, Jeremiah. *Sharing Is the New Buying*. (March 3, 2014) Available at http://www.web-strategist.com/blog/2014/03/03/report-sharing-is-the-new-buying-winning-in-the-collaborative-economy.

Pagel, Mark. *Wired for Culture: The Natural History of Human Cooperation*. London: Allen Lane, 2012.

Parkinson, Charles. "Monorail Plans Have Some Asking, Is Medellín Too Obsessed With 'Innovative' Transit?" *The Next City—Resilient Cities*, April 29, 2014. http://nextcity.org/daily/entry/monorail-plans-have-some-asking-is-Medellín-too-obsessed-with-innovative-tr.

People Who Share, The,*The State of the Sharing Economy: Food Sharing in the UK* (2013) Available at http://www.thepeoplewhoshare.com/tpws/assets/File/TheStateoftheSharingEconomy_May2013_FoodSharingintheUK.pdf.

Piketty, Thomas. *Capital in the 21st Century*. Cambridge, MA: Harvard University Press, 2014.

Provoost, Michele. "From Welfare City to Neoliberal Utopia." Strelka Talk, 2013. Available at http://vimeo.com/64392842.

Putnam, Robert D. *Bowling Alone: The Collapse and Revival of American Community*. New York: Simon & Schuster, 2000.

Quilligan, James B. "People Sharing Resources: Towards a New Multilateralism of the Global Commons." *Kosmos* (Stockholm) (Fall/Winter) (2009): 36–43.

Ridley, Matt. *The Origins of Virtue*. London: Viking, 1996.

Rifkin, Jeremy. *The Empathic Civilization: The Race to Global Consciousness in a World in Crisis*. Cambridge: Polity Press, 2010.

Rifkin, Jeremy. *The Zero Marginal Cost Society: The Internet of Things, the Collaborative Commons, and the Eclipse of Capitalism*. New York: Palgrave Macmillan, 2014.

Rinne, April. "Share Economy: Band-Aid Solution or Promoting Sustainability?" Interview on CBC Podcast. *The Current*, March 3, 2014. Available at http://www.cbc.ca/thecurrent/episode/2014/03/03/share-economy-band-aid-solution-to-real-economic-problems-or-promoting-sustainability.

Runciman, David. "Politics or Technology—Which Will Save the World?" *The Guardian*, May 23, 2014. Available at http://www.theguardian.com/books/2014/may/23/politics-technology-save-world-david-runciman.

Rydin, Yvonne. *The Future of Planning: Beyond Growth Dependence*. Bristol: Policy Press, 2013.

Sacks, Danielle. "The Sharing Economy." *Fast Company*, April 18, 2011. Available at http://www.fastcompany.com/1747551/sharing-economy.

Samper, Jota. "Granting of Land Tenure in Medellin, Colombia's Informal Settlements: Is Legalization the Best Alternative in a Landscape of Violence?" *Informal Settlements Research*, January 30, 2014. Available at http://informalsettlements.blogspot.se/2011/01/v-behaviorurldefaultvmlo.html.

Sandel, Michael. *Justice: What's the Right Thing to Do?* London: Penguin, 2010.

Sandel, Michael. *What Money Can't Buy: The Moral Limits of Markets*. London: Penguin, 2012.

Sandercock, Leonie. *Cosmopolis II: Mongrel Cities of the 21st Century*. London: Continuum, 2003.

Sassen, Saskia. *The Global City: New York / London / Tokyo*. Princeton, NJ: Princeton University Press, 2001.

Satterthwaite, David. *The Ten and a Half Myths That May Distort the Urban Policies of Governments and International Agencies*. UCL, 2003. Available at http://www.ucl.ac.uk/dpu-projects/drivers_urb_change/urb_infrastructure/pdf_city_planning/IIED_Satterthwaite_Myths_complete.pdf.

Satterthwaite, David. "If We Don't Count the Poor, the Poor Don't Count." IIED blog, June 12, 2014. Available at http://www.iied.org/if-we-dont-count-poor-poor-dont-count.

Scandrett, Eurig. *Citizen Participation and Popular Education in the City*. Big Ideas Thinkpiece. London: Friends of the Earth, 2013. Available at http://www.foe.co.uk/sites/default/files/downloads/citizen_participation_and.pdf.

Schifferes, Jonathan, "Sharing Our Way to Prosperity" (Part 1) and "Profiting from Sharing" (Part 2). RSA blog entries. August 6, 2013. Available at: http://www.rsablogs.org.uk/2013/social-economy/sharing-prosperity.

Schor, Juliet. *True Wealth*. London: Penguin, 2011.

Schor, Juliet. "Debating the Sharing Economy." *Great Transition Initiative*. October 2014. Available at http://www.greattransition.org/publication/debating-the-sharing-economy.

Schwartz, Shalom H. "Les valeurs de base de la personne: Théorie, mesures et applications." [Basic Human Values: Theory, Measurement, and Applications] *Revue Francaise de Sociologie* 42 (2006): 249–288.

Scruggs, Greg. "Latin America's New Superstar: How Gritty, Crime-Ridden Medellín Became a Model for 21st-Century Urbanism." *The Next City*, March 31, 2014. Available at http://nextcity.org/forefront/view/Medellíns-eternal-spring-social-urbanism-transforms-latin-america.

Sen, Amartya. *Development as Freedom*. Oxford: Oxford University Press, 2001. (first published 1999).

Sen, Amartya. *The Idea of Justice*. London: Allen Lane, 2009.

Sennett, Richard. *Together: The Rituals, Pleasures and Politics of Cooperation*. London: Penguin, 2013. (First published 2012).

Shaheen, Susan, Stacey Guzman, and Hua Zhang. "Bikesharing in Europe, the Americas, and Asia: Past, Present, and Future." *Institute of Transportation Studies Report*, 2010. Available at http://escholarship.org/uc/item/79v822k5.

Sherwin, Wilson, Maria-Valerie Schegk, and Olivia Sandri. "Cultural Creative Clusters for Dummies." *MCR Digipaper*, March 7, 2011. Available at http://mcrdigipaper.wordpress.com/2011/03/07/cultural-creative-clusters-for-dummies.

Sitrin, Marina. "The Commons and Defending the Public." *Telesur*, November 17, 2014. Available at http://www.telesurtv.net/english/opinion/The-Commons-and-Defending-the-Public-20141117-0020.html.

Stephens, Lucie, Josh Ryan-Collins, and David Boyle. *Co-Production: A Manifesto for Growing the Core Economy*. London: New Economics Foundation, 2008.

Stokes, Kathleen, Emma Clarence, Lauren Anderson, and April Rinne. *Making Sense of the UK Collaborative Economy*. London: Collaborative Lab and NESTA, 2014. http://www.nesta.org.uk/sites/default/files/making_sense_of_the_uk_collaborative_economy_14.pdf.

Sundararajan, Arun. "Trusting the 'Sharing Economy' to Regulate Itself." *Economix blog, New York Times*, 3 March 2014. Available at http://economix.blogs.nytimes.com/2014/03/03/trusting-the-sharing-economy-to-regulate-itself/?_r=0.

Sunrun. "Sunrun Survey Finds Nearly 92 Million Americans Plan to Participate in "Disownership" This Summer." Press release, June 15, 2013. Available at http://www.sunrun.com/why-sunrun/about/news/press-releases/sunrun-survey-finds-nearly-92-million-americans-plan-to-participate-in-disownership-this-summer.

Tapscott, Don, and Anthony D. Williams, *Wikinomics: How Mass Collaboration Changes Everything*. 2006. London: Atlantic Books.

Tapscott, Don, and Anthony D. Williams, *Macrowikinomics: Rebooting Business and the World*. London: Atlantic Books, 2010 (citations from 2011 Atlantic Paperback edition).

Thompson, Clive. *Smarter Than You Think: How Technology Is Changing Our Minds for the Better*. London: William Collins Books, 2014.

Thompson, Derek, and Jordan Weissmann. "The Cheapest Generation: Why the Millennials Aren't Buying Cars or Houses and What that Means for the Economy." *The Atlantic,* September 2012. Available at http://www.theatlantic.com/magazine/archive/2012/09/the-cheapest-generation/309060.

Tonkinwise, Cameron. "Sharing You Can Believe In: The Awkward Potential within Sharing Economy Encounters." *Medium*, July 1, 2014. Available at https://medium.com/@camerontw/sharing-you-can-believe-in-9b68718c4b33.

Twenge, Jean M., W. Keith Campbell, and Nathan T. Carter. "Declines in Trust in Others and Confidence in Institutions Among American Adults and Late Adolescents, 1972–2012." *Psychological Science* 25 (10) (2014): 1914–1923.

US Chamber of Commerce Foundation. *Millennials Report,* 2012. Available at http://www.uschamberfoundation.org/MillennialsReport.

Volker, Beate, and Henk Flap. "Sixteen Million Neighbors: A Multilevel Study of the Role of Neighbors in the Personal Networks of the Dutch." *Urban Affairs Review* 43 (2007): 256. Available at http://uar.sagepub.com/content/43/2/256.

Walker, Alissa. "These Five Ideas for Smarter Cities Just Won Millions in Funding." *Gizmodo,* September 19, 2014. http://gizmodo.com/these-five-ideas-for-smarter-cities-just-won-millions-i-1636858366.

Webster, Colin. "Amsterdam: Exploring the Sharing City." Disruptive Innovation Festival video. October 30, 2014. Available at http://thinkdif.co/emf-stages/sharing-economy-and-the-mobile-internet-the-amsterdam-experience (last accessed November 21, 2014).

Wernau, Julie. "Enterprise Buying Chicago's I-Go Car Service." *Chicago Tribune,* May 28, 2013. Available at http://articles.chicagotribune.com/2013-05-28/business/chi-igo-enterprise-enterprise-buys-chicagos-igo-car-service-20130528_1_feigon-phillycarshare-enterprise-holdings.

Wilkinson, Richard, and Kate Pickett. *The Spirit Level: Why Equality Is Better for Everyone*. London: Allen Lane, 2009.

Wright, Erik Olin. *Envisioning Real Utopias*. London: Verso, 2010.

Yglesias, Matthew. "There Is No 'Sharing Economy,'" *Slate*, December 26, 2013. Available at http://www.slate.com/blogs/moneybox/2013/12/26/myth_of_the_sharing_economy_there_s_no_such_thing.html.

Yiannopoulos, Milo. "Don't Believe the Hype: Here's What's Wrong with the 'Sharing Economy,'" *TheNextWeb*, June 6, 2013. Available at http://thenextweb.com/insider/2013/06/06/dont-believe-the-hype-heres-whats-wrong-with-the-sharing-economy.

Zavestoski, Stephen, and Julian Agyeman. *Incomplete Streets: Processes, Practices, and Possibilities*. London: Routledge, 2014.

Zittrain, Jonathan. "The Scary Future of Digital Gerrymandering—and How to Prevent It." *New Republic*, June 1, 2014. Available at http://www.newrepublic.com/article/117878/information-fiduciary-solution-facebook-digital-gerrymandering.

Index

Note: page numbers followed by "f" and "t" refer to figures and tables, respectively.

Urban and Industrial Environments

Series editor: Robert Gottlieb, Henry R. Luce Professor of Urban and Environmental Policy, Occidental College

Maureen Smith, *The U.S. Paper Industry and Sustainable Production: An Argument for Restructuring*

Keith Pezzoli, *Human Settlements and Planning for Ecological Sustainability: The Case of Mexico City*

Sarah Hammond Creighton, *Greening the Ivory Tower: Improving the Environmental Track Record of Universities, Colleges, and Other Institutions*

Jan Mazurek, *Making Microchips: Policy, Globalization, and Economic Restructuring in the Semiconductor Industry*

William A. Shutkin, *The Land That Could Be: Environmentalism and Democracy in the Twenty-First Century*

Richard Hofrichter, ed., *Reclaiming the Environmental Debate: The Politics of Health in a Toxic Culture*

Robert Gottlieb, *Environmentalism Unbound: Exploring New Pathways for Change*

Kenneth Geiser, *Materials Matter: Toward a Sustainable Materials Policy*

Thomas D. Beamish, *Silent Spill: The Organization of an Industrial Crisis*

Matthew Gandy, *Concrete and Clay: Reworking Nature in New York City*

David Naguib Pellow, *Garbage Wars: The Struggle for Environmental Justice in Chicago*

Julian Agyeman, Robert D. Bullard, and Bob Evans, eds.,

Just Sustainabilities: Development in an Unequal World

Barbara L. Allen, *Uneasy Alchemy: Citizens and Experts in Louisiana's Chemical Corridor Disputes*

Dara O'Rourke, *Community-Driven Regulation: Balancing Development and the Environment in Vietnam*

Brian K. Obach, *Labor and the Environmental Movement: The Quest for Common Ground*

Peggy F. Barlett and Geoffrey W. Chase, eds., *Sustainability on Campus: Stories and Strategies for Change*

Steve Lerner, *Diamond: A Struggle for Environmental Justice in Louisiana's Chemical Corridor*

Jason Corburn, *Street Science: Community Knowledge and Environmental Health Justice*

Peggy F. Barlett, ed., *Urban Place: Reconnecting with the Natural World*

David Naguib Pellow and Robert J. Brulle, eds., *Power, Justice, and the Environment: A Critical Appraisal of the Environmental Justice Movement*

Eran Ben-Joseph, *The Code of the City: Standards and the Hidden Language of Place Making*

Nancy J. Myers and Carolyn Raffensperger, eds., *Precautionary Tools for Reshaping Environmental Policy*

Kelly Sims Gallagher, *China Shifts Gears: Automakers, Oil, Pollution, and Development*

Kerry H. Whiteside, *Precautionary Politics: Principle and Practice in Confronting Environmental Risk*

Ronald Sandler and Phaedra C. Pezzullo, eds., *Environmental Justice and Environmentalism: The Social Justice Challenge to the Environmental Movement*

Julie Sze, *Noxious New York: The Racial Politics of Urban Health and Environmental Justice*

Robert D. Bullard, ed., *Growing Smarter: Achieving Livable Communities, Environmental Justice, and Regional Equity*

Ann Rappaport and Sarah Hammond Creighton, *Degrees That Matter: Climate Change and the University*

Michael Egan, *Barry Commoner and the Science of Survival: The Remaking of American Environmentalism*

David J. Hess, *Alternative Pathways in Science and Industry: Activism, Innovation, and the Environment in an Era of Globalization*

Peter F. Cannavò, *The Working Landscape: Founding, Preservation, and the Politics of Place*

Paul Stanton Kibel, ed., *Rivertown: Rethinking Urban Rivers*

Urban and Industrial Environments

Kevin P. Gallagher and Lyuba Zarsky, *The Enclave Economy: Foreign Investment and Sustainable Development in Mexico's Silicon Valley*

David N. Pellow, *Resisting Global Toxics: Transnational Movements for Environmental Justice*

Robert Gottlieb, *Reinventing Los Angeles: Nature and Community in the Global City*

David V. Carruthers, ed., *Environmental Justice in Latin America: Problems, Promise, and Practice*

Tom Angotti, *New York for Sale: Community Planning Confronts Global Real Estate*

Paloma Pavel, ed., *Breakthrough Communities: Sustainability and Justice in the Next American Metropolis*

Anastasia Loukaitou-Sideris and Renia Ehrenfeucht, *Sidewalks: Conflict and Negotiation over Public Space*

David J. Hess, *Localist Movements in a Global Economy: Sustainability, Justice, and Urban Development in the United States*

Julian Agyeman and Yelena Ogneva-Himmelberger, eds., *Environmental Justice and Sustainability in the Former Soviet Union*

Jason Corburn, *Toward the Healthy City: People, Places, and the Politics of Urban Planning*

JoAnn Carmin and Julian Agyeman, eds., *Environmental Inequalities Beyond Borders: Local Perspectives on Global Injustices*

Louise Mozingo, *Pastoral Capitalism: A History of Suburban Corporate Landscapes*

Gwen Ottinger and Benjamin Cohen, eds., *Technoscience and Environmental Justice: Expert Cultures in a Grassroots Movement*

Samantha MacBride, *Recycling Reconsidered: The Present Failure and Future Promise of Environmental Action in the United States*

Andrew Karvonen, *Politics of Urban Runoff: Nature, Technology, and the Sustainable City*

Daniel Schneider, *Hybrid Nature: Sewage Treatment and the Contradictions of the Industrial Ecosystem*

Catherine Tumber, *Small, Gritty, and Green: The Promise of America's Smaller Industrial Cities in a Low-Carbon World*

Sam Bass Warner and Andrew H. Whittemore, *American Urban Form: A Representative History*

John Pucher and Ralph Buehler, eds., *City Cycling*

Stephanie Foote and Elizabeth Mazzolini, eds., *Histories of the Dustheap: Waste, Material Cultures, Social Justice*

David J. Hess, *Good Green Jobs in a Global Economy: Making and Keeping New Industries in the United States*

Joseph F. C. DiMento and Clifford Ellis, *Changing Lanes: Visions and Histories of Urban Freeways*

Joanna Robinson, *Contested Water: The Struggle Against Water Privatization in the United States and Canada*

William B. Meyer, *The Environmental Advantages of Cities: Countering Commonsense Antiurbanism*

Rebecca L. Henn and Andrew J. Hoffman, eds., *Constructing Green: The Social Structures of Sustainability*

Peggy F. Barlett and Geoffrey W. Chase, eds., *Sustainability in Higher Education: Stories and Strategies for Transformation*

Isabelle Anguelovski, *Neighborhood as Refuge: Community Reconstruction, Place-Remaking, and Environmental Justice in the City*

Kelly Sims Gallagher, *The Global Diffusion of Clean Energy Technology: Lessons from China*

Vinit Mukhija and Anastasia Loukaitou-Sideris, eds., *The Informal City: Settings, Strategies, Responses*

Roxanne Warren, *Rail and the City: Shrinking Our Carbon Footprint and Reimagining Urban Space*

Marianne Krasny and Keith Tidball, *Civic Ecology: Adaptation and Transformation from the Ground Up*

Duncan McLaren and Julian Agyeman, *Sharing Cities: A Case for Truly Smart and Sustainable Cities*